Gwasanaethau Llyfrgell a Gwybodaeth
Library and Information Services

Prif Swyddfa
Headquarters

Hafod Elfyn
Penrhos Road
Bangor
Gwynedd
LL57 2BQ

Tel: (01248) 385522

Fax: (01248) 385510

Cyngor Cefn Gwlad Cymru
Countryside Council for Wales

Ecological Studies, Vol. 173

Analysis and Synthesis

Ecological Studies

Volumes published since 1998 are listed at the end of this book.

W.W. Weisser E. Siemann (Eds.)

Insects and Ecosystem Function

With 50 Figures and 12 Tables

Springer

Professor Dr. Wolfgang W. Weisser
Institute of Ecology
Friedrich-Schiller-University
Dornburger Str. 159
07743 Jena
Germany

Professor Dr. Evan Siemann
Department of Ecology and Evolutionary Biology – MS 170
Rice University
6100 Main Street
Houston, TX 77005-1892
USA

ISSN 0070-8356
ISBN 3-540-21672-3 Springer-Verlag Berlin Heidelberg New York

Library of Congress Control Number: 2004105871

Springer-Verlag Berlin Heidelberg New York
Springer-Verlag is a part of Springer Science+Business Media

springeronline.com

Printed in Germany

Editor: Dr. Dieter Czeschlik, Heidelberg, Germany
Desk editor: Dr. Andrea Schlitzberger, Heidelberg, Germany
Typesetting and Production: Friedmut Kröner, Heidelberg, Germany
Cover design: *design & production* GmbH, Heidelberg

31/3150 YK – 5 4 3 2 1 0 – Printed on acid free paper

Preface

In the last decade there has been a dramatic increase in our understanding of the role of plant diversity and plant community composition in determining local ecosystem processes. In contrast, our understanding of the role of insects in ecosystem processes is relatively primitive. Because insects are a dominant component of biodiversity in terrestrial ecosystems, this state of knowledge is unsatisfactory. Insects perform a number of activities that influence ecosystem processes such as nutrient cycling, and insects are likely to play a key role in mediating the relationship between plants and ecosystem processes by influencing the physiology, population dynamics and competitive relationships of plants. For herbivorous insects, the scarcity of manipulative studies on the role of insects in ecosystem processes contrasts sharply with the existing knowledge of insect effects on individual plants and populations. A primary reason for the under-representation of insects in biodiversity-ecosystem function studies is the practical problems associated with controlled experiments. Insects do not lend themselves easily to diversity or composition manipulations. Because insects are small, mobile, and often capable of flight, maintaining a gradient of insect diversity is more difficult than maintaining a gradient in plant diversity. However, researchers in the field of insect herbivory have developed a number of methods to reduce, enhance or simulate insect herbivory that could be used to explore the relationship between ecosystem function and local insect diversity and composition. All of these methods have their advantages and disadvantages which have to be kept in mind when an experiment is designed. It is our firm belief that functional biodiversity research can benefit greatly, both conceptually and methodically, from the insights gained in the many years of research on insect—plant interactions.

The need for a such a synthetic vision of the future of research on the relationship between insect biodiversity and ecosystem function became clear in an ESF LINKECOL-funded workshop on "Manipulating insect herbivory in biodiversity-ecosystem function experiments" held in Jena, Germany, in September 2001. The purpose of the workshop was to highlight current research on insect herbivores in community and ecosystem ecology and to explore the strengths and weaknesses of different methods of manipulating insect herbivory. The talks and interactions that took place at this workshop were the inspiration for this book. In particular, it became clear to us that there was a wealth of information that could be brought together to provide an overview of the current status of the field and help guide future research in this area. Thus, the aim of the book is to both review the known effects of insects on ecosystem functioning and to provide a detailed discussion of the strengths and weaknesses of various techniques of manipulating insect herbivory. We hope that this book is a useful tool for both students and researchers.

This volume is divided into five sections. The first section gives an introduction to the roles that insects play in ecosystem processes and discusses the progress that has been made to date in understanding the importance of insects in a variety ecosystem processes. The second section reviews some of the recent experimental work on the interactions between above-ground insect herbivory, the below-ground community and ecosystem processes. The third section provides a number of examples of the complex interactions between plants and insects that have both direct and indirect effects on ecosystem functioning. The fourth section reviews established and novel methods to manipulate the interactions between plants and insect herbivores. Particular emphasis is given to the advantages and disadvantages of the various methods to reduce, enhance or simulate the effects herbivory on plant communities. The fifth and final section is a synthesis of the other areas of the book and includes our thoughts on future directions of this area of research.

A number of people have been instrumental in the creation of this book. We would like to thank Andrea Schlitzberger at Springer for her editorial assistance, Maria Hartley for contributing illustrations for the cover, Lutz Merbold for help with the cover, Rachel Tardif for help in the assembly of the manuscripts, Ian Baldwin for his guidance and support as coordinating editor, and Detlef Schulze for his encouragement, especially in the early stages of development. We would like to extend our thanks to all who provided reviews of chapters. Our great appreciation goes out to the chapter authors for their hard work and contributions without which the book could never have happened. Finally, we would like to thank friends and family for their support, in particular Anne, Bea, Dylan, Megan and Sarah for their patience and understanding during many trans-Atlantic meetings.

June 2004

Jena, Germany Wolfgang Weisser
Houston, Texas, USA Evan Siemann

Contents

Contributors

ALPHEI, J.

Abteilung Ökologie, Institut für Zoologie, Universität Göttingen,
Berliner Strasse 28, 37073 Göttingen, Germany, e-mail: jalphei@gwdg.de

BALDWIN, I.T.

Max-Planck-Institute for Chemical Ecology, Winzerlaer Straße 10,
Beutenberg Campus, Jena, Germany, e-mail: baldwin@ice.mpg.de

BOALT, E.

Department of Natural Sciences, Södertörn University College,
14189 Huddinge, Sweden, e-mail: elin.boalt@sh.se

BARDGETT, R.D.

Institute of Environmental and Natural Sciences, Lancaster University,
Lancaster LA1 4YQ, UK, e-mail: r.bardgett@lancaster.ac.uk

BONKOWSKI, M.

Institut für Zoologie, Technische Universität Darmstadt, Schnittspahnstr. 3,
64287 Darmstadt, Germany, e-mail: bonkow@bio.tu-darmstadt.de

CARSON, W.P.

Biological Sciences, University of Pittsburgh, A234 Langley Hall,
Pittsburgh, Pennsylvania 15260, USA, e-mail: walt+@pitt.edu

CRONIN, J.P.

Biological Sciences, University of Pittsburgh, A234 Langley Hall,
Pittsburgh, Pennsylvania 15260, USA, e-mail: jpcronin@pitt.edu

Eber, S.

Agroecology, Georg-August-University, Waldweg 26, 37073 Göttingen, Germany, e-mail: s.eber@uaoe.gwdg.de

Hartley, S.E.

School of Life Sciences, University of Sussex, Falmer, Brighton BN1 9QG, UK, e-mail: s.hartley@sussex.ac.uk

Hector, A.

Institute of Environmental Sciences, University of Zurich, Winterthurerstrasse 190, 8057 Zurich, Switzerland, e-mail: ahector@uwinst.unizh.ch

Hjältén, J.

Department of Animal Ecology, Swedish University of Agricultural Sciences, 901 83 Umeå, Sweden, e-mail: joakim.hjalten@szooek.slu.se

Janssen, A.

Institute for Biodiversity and Ecosystem Dynamics, Section Population Biology, University of Amsterdam, P.O. Box 94084, 1090 GB, Amsterdam, The Netherlands, e-mail: janssen@science.uva.nl

Joshi, J.

Institute of Environmental Sciences, University of Zurich, Winter thurerstrasse 190, 8057 Zurich, Switzerland, e-mail: joshi@uwinst.unizh.ch

Jones, T.H.

Cardiff School of Biosciences, Cardiff University, P.O. Box 915, Cardiff CF10 3TL, UK, e-mail: jonesth@cardiff.ac.uk

Kluth, S.

Agroecology, Georg-August-University, Waldweg 26, 37073 Göttingen, Germany, e-mail: skluth@gmx.de

Koricheva, J.

Section of Ecology, Department of Biology, University of Turku, 20014 Turku, Finland, e-mail: julkoric@utu.fi

KRUESS, A.

Agroecology, Georg-August-University, Waldweg 26, 37073 Göttingen, Germany, e-mail: akruess@gwdg.de

LEHTILÄ, K.

Department of Natural Sciences, Södertörn University College, 14189 Huddinge, Sweden, e-mail: kari.lehtila@sh.se

LONG, Z.T.

Department of Ecology, Evolution, and Natural Resources, Rutgers University, New Brunswick, New Jersey 08901, USA, e-mail: zlong@rci.rutgers.edu

MASTERS, G.J.

Ecology, Systematics and Biodiversity, CABI Bioscience, UK Centre, Bakeham Lane, Egham, Surrey, TW20 9TY, UK, e-mail: g.masters@cabi.org

MICHALZIK, B.

Geographic Institute, University of Göttingen, Section of Landscape Ecology, Goldschmidtstr. 5, 37077 Göttingen, Germany, e-mail: bmichal@gwdg.de

MÜHLENBERG, E.

Bayreuth Institute for Terrestrial Ecosystem Research, University of Bayreuth, 95440 Bayreuth, Germany, e-mail: eva.muehlenberg@bitoek.uni-bayreuth.de

PFISTERER, A.B.

Institute of Environmental Sciences, University of Zurich, Winterthurerstrasse 190, 8057 Zurich, Switzerland, e-mail: pfisterer@uwinst.unizh.ch

ROGERS, W.E.

Department of Ecology and Evolutionary Biology, MS 170, Rice University, 6100 Main Street, Houston, Texas 77005-1892, USA, e-mail: wer@rice.edu

ROY, B.A.

Center for Ecology and Evolutionary Biology, 5289 University of Oregon, Eugene, Oregon 97405-5289, USA, e-mail: bit@darkwing.uoregon.edu

Otway, S.J.

Natural Environmental Research Council (NERC) Centre for Population Biology, Imperial College London, Silwood Park Campus, Ascot, Berkshire, SL5 7PY, UK, e-mail: Stuart.Otway@de.mod.uk

Sabelis, M.W.

Institute for Biodiversity and Ecosystem Dynamics, Section Population Biology, University of Amsterdam, P.O. Box 94084, 1090GB, Amsterdam, The Netherlands, e-mail: sabelis@bio.uva.nl

Scherer-Lorenzen, M.

Institute of Plant Sciences, Swiss Federal Institute of Technology Zurich, Universitätsstrasse 2, ETH Zentrum LFW C55.2, 8092 Zurich, Switzerland, e-mail: michael.scherer@ipw.agrl.ethz.ch

Scheu, S.

Institut für Zoologie, Technische Universität Darmstadt, Schnittspahnstr 3, 64287 Darmstadt, Germany, e-mail: scheu@bio.tu-darmstadt.de

Schmid, B.

Institute of Environmental Sciences, University of Zurich, Winterthurerstrasse 190, 8057 Zurich, Switzerland, e-mail: bschmid@uwinst.unizh.ch

Schmitz, O.J.

Yale University, School of Forestry and Environmental Studies, 370 Prospect Street, New Haven, Connecticut 06515, USA, e-mail: oswald.schmitz@yale.edu

Siemann, E.

Department of Ecology and Evolutionary Biology, MS 170, Rice University, 6100 Main Street, Houston, Texas 77005-1892, USA, e-mail: siemann@rice.edu

Spehn, E.

Institute of Botany, University of Basel, Schoenbeinstrasse 6, 4056 Basel, Switzerland, e-mail: eva.spehn@unibas.ch

STADLER, B.

Bayreuth Institute for Terrestrial Ecosystem Research,
University of Bayreuth, 95440 Bayreuth, Germany,
e-mail: bernhard.stadler@bitoek.uni-bayreuth.de

TSCHARNTKE, T.

Agroecology, Georg-August-University, Waldweg 26,
37073 Göttingen, Germany, e-mail: ttschar@gwdg.de

VOELCKEL, C.

Max-Planck-Institute for Chemical Ecology, Winzerlaer Straße 10,
Beutenberg Campus, Jena, Germany, e-mail: voelckel@ice.mpg.de

WARDLE, D.A.

Department of Forest Vegetation Ecology, Swedish University of Agricultural Sciences, S901-83 Umeå, Sweden, and Landcare Research, P.O. Box 69, Lincoln, New Zealand, e-mail: david.wardle@svek.slu.se

WEISSER, W.W.

Institute of Ecology, Friedrich-Schiller-University, Dornburger Str. 159,
07743 Jena, Germany, e-mail: wolfgang.weisser@uni-jena.de

ZIMMERMANN, H.

Plant Protection Research Institute, Private Bag X 134, Pretoria 0001,
South Africa, e-mail: riethgz@plant2.agric.za

ZWÖLFER, H.

Department of Animal Ecology I, University of Bayreuth, NW1,
95440 Bayreuth, Germany, e-mail: h.zwoelfer@freenet.de

Section I
Introduction

1 The Various Effects of Insects on Ecosystem Functioning

W.W. Weisser and E. Siemann

1.1 Summary

Insects represent a dominant component of biodiversity in most terrestrial ecosystems yet they have largely been neglected in studies on the role of biodiversity in nutrient cycling, or, more generally, the functioning of ecosystems. The scarcity of manipulative studies on the role of insects in ecosystem processes contrasts with the expert knowledge and large body of research already available, in particular in the field of insect herbivory. Insects are likely to play a key role in mediating the relationship between plants and ecosystem processes by influencing the physiology, activity and population dynamics of plants. The aim of this book is two-fold: (1) to summarize the effects that insects have on ecosystem functioning, focusing mainly, but not exclusively, on herbivorous insects. Authors with extensive experience in the field of plant–insect interactions will discuss the importance of insects in ecosystem functioning; and (2) to provide a detailed discussion of the advantages and disadvantages of various techniques of manipulating insect herbivory. Thus, the book aims to provide both a theoretical basis and practical advice for future manipulative studies on biodiversity–ecosystem functioning. This introductory chapter briefly summarizes the various effects of insects on ecosystem functioning and introduces the chapters in the various sections of this book.

1.2 Introduction

It appears to be obvious that the way matter flows through an ecosystem is influenced by the organisms within the system. Without autotrophic, chemotrophic or heterotrophic activity, matter fluxes through ecosystems would be very different. Nevertheless, research on the role of organisms in nutrient cycling or, more generally, the 'functioning' of ecosystems has had a

Ecological Studies, Vol. 173
W.W. Weisser and E. Siemann (Eds.)
Insects and Ecosystem Function

mixed history. The reason is the traditional division of ecology into community ecology and ecosystem ecology (Odum 1953; Likens 1992; Jones and Lawton 1995). While community ecologists generally consider organisms to be the main drivers of ecosystem processes, they rarely measure nutrient cycling as a function of the constituent community. Conversely, systems ecologists quantify energy and element flux rates through ecosystems, but typically do so over large spatial scales (e.g. a watershed), and often with little reference to the role of organisms within the ecosystem. Classic ecosystem studies such as those performed within the framework of the International Biological Programme (IBP) in the 1960s and 1970s were very important in determining the contribution of various groups of organisms (plants, animals, etc.) to ecosystem productivity or energy flux (e.g. Bormann and Likens 1967; Golley et al. 1975; Likens et al. 1977). However, because of the different focus and because experimental community manipulations are difficult at the scale of hundreds of hectares, these ecosystem studies have generally not addressed the question of whether a particular change in the biotic community would lead to measurable changes in matter fluxes through the ecosystem. As a consequence of the division into systems and community ecology, we know a considerable amount about the structure and dynamics of natural communities, and about input/output relations of mineral nutrients for a variety of ecosystems, but we still know relatively little about the interaction between community dynamics and nutrient cycling (Mooney 1991; Loreau et al. 2001).

In the last decade, however, interest in the question of how important a diverse biotic community is for processes at the ecosystem level has greatly increased. Much of this work has been inspired by the realization that whilst global biodiversity is undergoing dramatic changes (Ehrlich and Ehrlich 1981; Wilson 1988), science has been unable to predict the effects of these changes on the ecosystems concerned (Schulze and Mooney 1993). Recent years have seen a dramatic increase in the number of manipulative experimental studies that have investigated the effects of particular components of diversity or of diversity per se for various aspects of ecosystem functioning (see reviews in Kinzig et al. 1991; Loreau et al. 2001, 2002; Wardle 2002). These studies have shown convincingly that changes in the diversity and composition of a community can have consequences at the ecosystem level that are sometimes drastic. A common result is that a decrease in diversity causes a loss in ecosystem function such as productivity or nutrient retention in the soil (e.g. Loreau et al. 2001, 2002). While in the first studies the term 'ecosystem functioning' was restricted to processes at the ecosystem level, some confusion has recently arisen from a rather loose use of this term. In the more recent literature, ecosystem functioning has been used to describe a variety of ecological processes not only at the level of the ecosystem, but also at the level of the community, populations or even individuals. As the focus of biodiversity–ecosystem functioning research is on ecosystem-level consequences of a loss of biodiversity, it is useful to define an ecosystem function or ecosystem process as any ecosystem-level attribute that can be measured in and compared between ecosystems. Thus, the state of a particular species or population cannot be an ecosystem function as this particular species or population will only occur in certain ecosystems and can hence only be measured in these ecosystems. In contrast, community-level attributes such as the stability of the community present in the ecosystem or the presence of a functional group of organisms such as pollinators can be measured in any ecosystem and therefore fall under this definition of ecosystem function. In a more narrow sense, the term ecosystem function is used for processes related to nutrient cycling at the ecosystem level (Schulze and Mooney 1993). This restricted definition is the one we would like to adopt for this book.

While the pioneering biodiversity–ecosystem functioning experiment was conducted using model communities composed of a variety of organisms (Naeem et al. 1994), most subsequent studies have focused on the manipulation of plant communities. This is true in particular for some of the most influential studies, which were carried out in grasslands (Tilman and Downing 1994; Tilman et al. 1996, 1997; Hooper and Vitousek 1997; Hector et al. 1999). In these studies as well as in later ones, plant species richness and/or plant functional group richness and composition were the main variables manipulated. Because producers are ultimately determining the amounts of carbon that enter an ecosystem in each food web, the manipulation at the plant level provided an obvious starting point in the analysis of the relationship between biodiversity and ecosystem functioning. However, an additional reason why plant communities have been the main object of a manipulative approach is that plant communities can be easily manipulated in climate chambers, greenhouses and field experiments. In other groups of organisms, manipulating organismic diversity is much more difficult, for example when the organisms are mobile, as is the case for most animals. Practical advantages may also have been responsible, at least in part, for the bias in more recent biodiversity–ecosystem function studies towards microcosm experiments with microbial communities (e.g. McGrady-Steed et al. 1997; Naeem and Li 1997). Only recently have organisms of other trophic levels been incorporated into experiments at spatial scales larger than a Petri dish. Pioneering mesocosm studies involving mycorrhizal fungi (van der Heijden et al. 1998), fresh-water insects (Wallace and Webster 1996; Cardinale et al. 2002), terrestrial insects (Cardinale et al. 2003; Schmitz 2003) and soil fauna (Wardle 2002) have brought a new impetus to biodiversity–ecosystem functioning research. However, because the diversity of invertebrates, vertebrates and microorganisms exceeds that of plant diversity by far, the existing manipulative studies of the heterotrophic component of ecosystems represent only a first step towards a more comprehensive understanding of the relationship between organismic diversity and ecosystem functioning.

With about 1 million described species, insect diversity is higher than that of any other animal or plant taxon (Stork 1988). In terrestrial ecosystems insects function as herbivores, pollinators, seed dispersers, predators, parasites, detritivores or ecosystem engineers. In the past decades, there have been several reviews of how insects, in particular herbivores, can affect ecosystem function (Mattson and Addy 1975; Gosz et al. 1978; Lee 1979; Hutchinson and King 1982; Seastedt and Crossley 1984; Lamb 1985; Detling 1988; Urbanek 1988; Hutson 1989; Whelan 1989; Huntly 1991; Curry 1994; Lerdau 1996; Lewinsohn and Price 1996; Wallace and Webster 1996; Price 1997; Coleman and Hendrix 2000; Schowalter 2000a; Feller 2002). Nevertheless, despite the many roles that insects fulfil in terrestrial ecosystems, their importance in nutrient cycling is not universally recognized. One reason for the skepticism is that the total biomass of insects (the standing crop) appears to be small compared to plant biomass or the biomass of other animals. For example, in an IBP study of a meadow-steppe in the V.V. Alkhin Central Chernozem Reserve in the Central Russian Upland, above- and belowground invertebrate biomass was equivalent to about 10 % of the yearly plant productivity of 11–14 tons dry weight ha^{-1} (Zlotin and Khodashova 1980). More than 90 % of the animal biomass was found below ground, and earthworms accounted for 94 % of soil animal biomass or 80–90 % of total animal biomass. The biomass of above- and belowground insects constituted less than 2 % of animal biomass, or 0.2 % of plant productivity. For some ecosystems these percentages may be higher, but in terms of contribution to total standing crop, biomass does not generally suggest a prominent role for insects in nutrient cycling (cf. Petrusewicz 1967; Schowalter 2000a). A second reason why insects are often not considered to be very important for nutrient cycling is that the average proportion of net primary productivity (NPP) consumed by herbivorous insects is 10 % or less, except in outbreak situations (e.g. Wiegert and Evans 1967; Detling 1988; Curry 1994; but see Coupe and Cahill 2003). In contrast, large mammalian herbivores such as the North American bison or African mammals are known to consume up to 90 % of prairie or savannah NPP (Detling 1988). The observation that insects generally affect an insignificant fraction of NPP was one of the reasons why Hairston et al. (1960) proposed their hypothesis that herbivores are regulated top-down rather than bottom-up.

The only instances where insects are unequivocally recognized to have a large effect on ecosystem functions are outbreaks of particular species such as the gypsy moth, *Lymantria dispar* (L.) (Lepidoptera: Lymantriidae) or *Epirrata autumnata* (Bkh.) (Lep., Geometridae). During outbreaks, the proportion of leaf area removed by the extremely high population densities of the outbreaking herbivore species can reach 100 %, which has immediate and large effects on nutrient fluxes (e.g. Lovett and Ruesink 1995; Kosola et al. 2001; Christenson et al. 2002). Because of their regular outbreaks, Schowalter proposed that herbivorous insects act as cybernetic regulators for ecosystem processes (Schowalter 2000a, b). In his interpretation, insect outbreaks are 'feedbacks that maintain ecosystem production within sustainable ranges' (Schowalter 2000b). While this hypothesis is based on the somewhat controversial view that there is a tendency for homeostasis at supraorganismal levels, it is true that outbreaks not only cause an immediate release of nutrients previously fixed in plants, but also have longer-lasting effects by changing a number of parameters that affect matter fluxes such as soil nutrient availability, the physiology of long-living plants such as trees, or plant species composition (e.g. Carson and Root 2000; Kosola et al. 2001; Christenson et al. 2002). Outbreaks are therefore one example of large insect effects on ecosystem processes.

The view that insects have only minute effects at the ecosystem level also contrasts with the results of studies on individual plant–insect interactions, which document large effects of insects on plant traits such as investment into secondary metabolites, plant architecture and seed number (e.g. Crawley 1983, 1986; Karban and Baldwin 1997). Why do insects, in particular herbivorous insects, have large effects on individual plants but apparently small effects at the ecosystem level? In our view, one important reason for this apparent discrepancy is the way in which insect effects on nutrient cycling have been measured in most ecosystem studies. The main variable quantified in studies at the ecosystem level has been the reduction of plant standing crop, but this is only one of the many ways in which herbivorous insects impact ecosystem functioning. First, insects modify the way in which nutrients are distributed within the ecosystem, but these effects have rarely been quantified (Stadler et al. 2001). There is a shortage of studies that quantify not only herbivory but also rates of mineralization, respiration, or the forms and availability of nitrogen or phosphorus. Second, many of the insect effects on plants such as shifts in phenology or changes in plant resource allocation do not result in immediate and obvious changes in matter fluxes, and their ecosystem-level effects can only be quantified in longer-term studies. For example, studies that employed insecticides over a longer time span have shown that above- and belowground insect herbivores affect plant community composition over a time-scale of several years (Brown 1990). Such a change in plant community composition will have consequences for nutrient cycling in the ecosystem, but this can only be assessed when nutrient fluxes are measured over an extended time period. Longer-term consequences for nutrient cycling can also be expected from the role of insects as pollinators, seed dispersers, predators, parasites, detritivores or ecosystem engineers, and these consequences need to be explored in more detail. Finally, the full extent to which insects influence nutrient cycling can only be unravelled through manipulative studies where the number of insects in the ecosystem is either augmented or reduced, yet such studies are still very rare (e.g. Mulder et al. 1999; Belovsky and Slade 2000; Cardinale et al. 2002; 2003; Mitchell 2003; Montoya et al. 2003; Schmitz 2003). This lamentable lack of studies is one of the main reasons why we started this book project. Before we outline the structure of this book, we briefly summarize the various effects of insects on ecosystem functioning. These effects are discussed in greater depths in the different chapters of the book.

1.3 A Brief Overview of Insect Effects on Ecosystem Function

Insects affect nutrient cycling both directly and indirectly. Direct effects include the reduction of NPP by herbivores and the breakdown of litter by detritivores, but there are other direct effects of insects on nutrient cycling, for example through their role as ecosystem engineers (Jones and Lawton 1995). Indirect effects of insects on ecosystem functioning include, for example, changes in matter fluxes due to changes in plant species composition that are mediated by insect herbivory. Insects affect ecosystem functions mainly through their interaction with plants. However, interactions of insects with other organisms also have bearings on matter flux in ecosystems. Table 1.1 provides a list of insect effects on ecosystem function.

1.3.1 Insect Effects on Ecosystem Function Via Interactions with Plants

Insects interact with plants in a number of ways. The main interactions that have consequences for ecosystem functioning are herbivory (i.e. antagonistic phytophagy, in contrast to mutualistic phytophagy, in the terminology of Lewinsohn and Price 1996) and mutualism, in particular pollination, seed dispersal and plant protection. Both of these interactions between plants and insects have direct and indirect effects on ecosystem function. Direct effects on carbon storage and element cycling, in particular NPP, are well documented for herbivory, but only to a lesser extent for mutualistic interactions. Indirect effects on ecosystem function involve effects on plant species composition including succession, on plant resource allocation and on food web interactions. While direct effects have received most attention, indirect effects may also greatly influence nutrient cycling, but quantification of these effects is still rudimentary.

1.3.1.1 Herbivory

Insect herbivores consume living material and therefore have a direct effect on NPP. This direct effect has been measured in a large number of studies (e.g. Petrusewicz 1967; Wiegert and Evans 1967; Golley et al. 1975; Mattson and

Table 1.1. Insect effects on nutrient cycling

(1) Plant–insect interactions that affect nutrient cycling

Interaction	Main mechanisms through which nutrient cycling is affected	Examples
(a) Herbivory		
Herbivory on roots, shoots, flowers, seeds	Direct change of carbon storage, element cycles	Reduction of standing crop
		Partial or complete defoliation during outbreak
		Induction of leaf fall
		Increasing nutrient leaching from foliage
		Production of honeydew
		Production of methane in guts (termites)
		Defaecation, respiration (CO_2)
	Effect on plant species composition	Change in direction of succession
		Recreation of early successional state by outbreak
		Effects on plant species diversity
		Recruitment limitation and species loss in case of seed predation
		Rank–abundance relationships altered
	Effect on plant resource allocation	Increase in foliage growth
		Increase in root exudation
		Root/shoot allocation modified
		Allocation of resources to chemical, physical defence with consequences for litter quality
		Release of VOC
		Other changes in allocation patterns
	Food web interactions altered	Stimulation/change of activity of decomposer organisms
		Effects on soil microbial community due to (root) herbivory
		Modification of plant–mycorrhiza interactions
		Modification of interaction between plants and pollinators, etc.

Table 1.1. (*Continued*)

(1) Plant–insect interactions that affect nutrient cycling

Interaction	Main mechanisms through which nutrient cycling is affected	Examples
(b) Mutualism		
Pollination, seed dispersal	Direct change of carbon storage, element cycles	Effects on standing crop and productivity Redistribution of nutrients
	Effect on plant species composition	Recruitment limitation and species loss
Protective mutualisms	Direct change of carbon storage, element cycles	Reduction of biomass loss
	Effect on plant species composition	Competitive interactions altered

(2) Other direct and indirect effects of insects on nutrient cycling

Functional role of insects	Main mechanisms through which nutrient cycling is affected	Examples
Detritivores	Direct change of carbon storage, element cycles	Reduction of standing crops of recalcitrant materials Increase in decomposition rate Shift from net immobilization to net mineralization Removal of dung and carrion

Breakdown of litter and wood

Table 1.1. (*Continued*)

(2) Other direct and indirect effects of insects on nutrient cycling

Functional role of insects	Main mechanisms through which nutrient cycling is affected	Examples
Ecosystem engineers	Direct change of carbon storage, element cycles	Moving of nutrients (e.g. termites)
	Modification of habitat conditions for other organisms	Substrate and microorganism mixing
		Reduction and abrasion of particles
		Modification of abiotic conditions for soil microfauna and other fauna
		Increase of plant growth by changing soil aeration, nutrient availability and water infiltration
		Creation of fungal gardens
Predators/parasitoids	Food web interactions altered	Effects on soil food web
		Trophic cascades
Microbial feeders	Direct change of carbon storage, element cycles	Change in decomposition rates
	Food web interactions altered	Grazing on fungal hyphae
		Effects on plant–mycorhiza interactions
		Effects on interactions between other soil organisms
Dispersal agents	Dispersal of other organisms	Distribution of arbuscular mycorrhizal spores

Addy 1975; Gosz et al. 1978; Lee 1979; Hutchinson and King 1982; Seastedt and Crossley 1984; Detling 1988; Curry 1994; Price 1997; Coleman and Hendrix 2000; Schowalter 2000a). Direct effects of flower and nectar feeders, and of seed predators, on NPP are relatively small, because reproductive organs make up only a small proportion of total plant biomass. Herbivory also increases nutrient leaching from foliage and the rate of fall of leaves (Mattson and Addy 1975), and insect faeces make nutrients available for mineralization that were previously fixed in plants (Seastedt and Crossley 1984). As with the consumption of biomass, these direct effects of insects on nutrient cycling are small for nominal herbivory but may be large in outbreak situations (Mattson and Addy 1975; Schowalter et al. 1986; Christenson et al. 2002). Outbreaks may result in leaching of nutrients out of the ecosystem, although some studies show that nutrients are redistributed within the ecosystem rather than lost (Lerdau 1996; Christenson et al. 2002; Lovett et al. 2002). As discussed above, it is in outbreak situations that direct effects of herbivorous insects on nutrient cycling become most visible. While outbreak situations clearly show the potential of herbivorous insects to directly affect nutrient cycling, the effects of continuous low-level herbivory do need to be investigated in more detail (Stadler et al. 2001).

An important indirect effect of herbivory on nutrient cycling is due to the role of herbivores in plant–plant competition. A large number of studies have shown that insect herbivores influence competitive interactions within the plant community and therefore affect plant species composition (Crawley 1983; Whelan 1989; Brown 1990; Louda et al. 1990; Brown and Gange 1992; Davidson 1993; Bach 2001; Blatt et al. 2001; Dyer and Shugart 2002). This holds not only for foliovores but also for seed predators and other root, shoot, flower or nectar feeders (Sallabanks and Courtney 1992; Davidson 1993; Brown 1997; Irwin et al. 2001). A number of studies employing insecticides have shown that the direction of succession depends on the presence of insect herbivores, both above- and below ground (Brown et al. 1987, 1988; Brown 1990). In the extreme case, herbivory causes plant death, for example in seedlings, with clear consequences for plant species composition (Crawley 1986; Khan and Tripathi 1991). Changes in plant species composition caused by insect herbivores may only be visible in long-term studies (Brown 1990; Cain et al. 1991; McCullough and Werner 1998; Bach 2001). Outbreaks of herbivores occur only infrequently, but they may play an important role in structuring plant communities (Danell and Ericson 1990; Carson and Root 2000), as they usually reduce the growth and resource acquisition of dominant plants and therefore delay or redirect succession (Davidson 1993; Carson and Root 2000; Blatt et al. 2001). One fascinating example in this respect are the interactions between insects and fire in northern and boreal forests in North America (McCullough and Werner 1998). Accumulation of fuels following insect outbreaks may determine extent and intensity of subsequent fires, or fire may predispose trees to subsequent attack by insects.

Both processes greatly affect forest species. Changes in plant species composition will result in changes in community productivity and nutrient cycling, but these indirect effects of insect herbivores in forests have not been quantified.

Another important effect of insects on plants which indirectly affects ecosystem function is a change in plant resource allocation. Allocation to root, shoot or to flowers and seeds changes after herbivore attack and these alterations may be long-lasting (e.g. Abrahamson and McCrea 1986; Dyer et al. 1991; Trumble et al. 1993; Marquis 1996; Honkanen and Haukioja 1998; Strauss et al. 1999; Stowe et al. 2000; Kosola et al. 2001). Plants often increase root exudation in response to herbivory and this has effects on soil fauna (Wardle 2002). Inducible plant defences also require changes in plant resource allocation (Karban and Baldwin 1997; Baldwin and Preston 1999). Inducible defences and other herbivore-induced changes in plant resource allocation are costly and affect plant competitive ability (Karban and Baldwin 1997; Baldwin 1998). Thus, an adjustment of plant resource allocation affects plant productivity and, in the longer term, plant species composition. In addition, it may affect nutrient cycling by changing the uptake and release of chemical compounds, and by changing litter quality. An example is the release of volatile organic compounds (VOCs) which play a role in attracting natural enemies of herbivores (Kessler and Baldwin 2001). The consequences of herbivore-induced changes in plant resource allocation for nutrient cycling have yet to be quantified for the ecosystem level. Most present studies focus on individual plants or, more rarely, plant populations.

Finally, insect herbivory can affect nutrient cycling by affecting food web interactions (Wardle 2002). For example, root herbivory impacts soil microbial communities with consequences for decomposition and nutrient release (Grayston et al. 2001). Other examples include a modification of plant–mycorrhiza interactions (Gehring and Whitham 2002) and the modification of interactions between plants and pollinators (Hambäck 2001). As in the case of herbivore impact on plant resource allocation, the consequences of herbivore-induced modifications of species interactions for nutrient cycling at the ecosystem level need to be investigated in more detail.

Two well-known and related hypotheses about the effects of herbivorous insects on nutrient cycling are the nutrient acceleration hypothesis and the grazing optimization hypothesis. The nutrient acceleration hypothesis states that herbivorous insects increase nutrient cycling by returning nutrients fixed in plants back to producers at a rate faster than the nutrients would otherwise flow through the processes of senescence, litter fall and decomposition (e.g. Chew 1974; Mattson and Addy 1975). Such increases in nutrient cycling have been documented, mostly in short-term studies (e.g. Lightfoot and Whitford 1990; Belovsky and Slade 2000). The controversial grazing optimization hypothesis states that low levels of herbivory actually increase plant productivity such that the relationship between productivity and herbivory intensity is hump-shaped (McNaughton 1993). Mathematical modelling shows that several mechanisms could account for a positive effect of herbivores on plant production, such as a stimulation of plant growth, an increase in cycling of a limiting nutrient or changes in plant species composition (De Mazancourt and Loreau 2000).

1.3.1.2 Plant–Insect Mutualisms

Pollination and seed dispersal are two services that insects provide for plants within the context of mutualistic interactions. In addition, there are a number of protective mutualisms between insects and plants, such as those between ants and ant–plant mutualisms (Fonseca 1994). Pollination and seed dispersal probably have little direct effects on nutrient cycling, even though energy expenditure on nectar rewards may be high, and the quantities of seeds moved by insect dispersers, in particular ants, can be large (Harder and Barrett 2002). The lack of pollinators or seed dispersers, and their choosiness, may, however, affect plant species composition (Howe and Smallwood 1982; Willson 1992; Corbet 1997), which will have consequences for nutrient cycling (see previous section).

Plants also employ insects to defend themselves against herbivores, for example by providing them with domatia or by attracting them with volatile emissions (Turlings et al. 1990; Grostal and O'Dowd 1994; Agrawal and Karban 1997; Kessler and Baldwin 2001). In the absence of the mutualists, plants suffer higher damage and may be displaced by stronger competitors. Thus, the different types of mutualisms between insects and plants are likely to affect nutrient cycling mainly via their effect on plant species composition.

1.3.2 Other Direct and Indirect Effects of Insects on Ecosystem Function

Insects also affect ecosystem functions directly without the involvement of living plant tissues. Insect detritivores such as cockroaches or springtails are an important component of soil macro- and mesofauna and hence the soil food web, and their presence is essential for litter breakdown and a fast return of nutrients to the primary producers (Wardle 2002). Soil insects also acts as ecosystem engineers by affecting soil conditions and creating physical structures which provide a modified habitat for soil microflora and other soil fauna. This affects nutrient cycling both directly and indirectly. Because the interactions in the belowground food web are complex, disentangling the effects of insects from those other soil fauna requires detailed studies (Wardle 2002). The most important insect ecosystem engineers globally are probably termites which move large amounts of soil and affect the soil's structure and fertility (Lee and Wood 1971). All insects, whether herbivore, detritivores or predators, affect nutrient cycling through processing of food and making organic matter available for mineralization.

In addition to the effects listed above, insects have a number of indirect effects on nutrient cycling. Predators and parasitoids affect interactions between plants and herbivores, pollinators or seed dispersers and can therefore modulate the effects of their prey on nutrient cycling. In tritrophic interactions between plants, herbivores, and predators and parasitoids, changes in carnivore abundance often affect plant biomass through trophic cascades (Schmitz et al. 2000; Cardinale et al. 2003; Schmitz 2003), although it has been argued that such food web dynamics are more important in water than on land (Polis 1999; Halaj and Wise 2001). Insect predators are an important part of the soil food web and have been shown to affect rates of decomposition although the exact mechanisms are far from clear (Wardle 2002). Because insects are an integral part of any terrestrial food web, they affect a great number of food web interactions which, at least in theory, could have consequences for nutrient cycling (Montoya et al. 2003). Insect ecosystem engineers among insects also affect food web interactions with possible consequences for ecosystem functioning (Johnson et al. 2002). So far, these interactions have rarely been studied with respect to their role in nutrient cycling.

1.4 The Aim and Structure of this Book

As illustrated in the previous sections, insects have a large number of effects on ecosystem functioning, yet little is known about the magnitude and relative importance of the various ways insects impact the matter fluxes in an ecosystem. What is underrepresented in the literature are manipulative studies that establish causal relationships between insects and ecosystem functioning, and that unravel the mechanisms underlying these relationships. Thus, studies are needed in which appropriate techniques are employed through which the action of insects in ecosystems can be reduced, augmented or simulated. Such techniques have been developed in the field of plant–insect interactions, and have been successfully employed to study the effects of insects at the level of individual plants, plant populations and plant communities. It is our view that research on the functional aspects of biodiversity can benefit from employing some of these approaches used in the studies of plant–insect interactions. The aim of this book is therefore two-fold. First, it aims to provide a summary of what is known about insect effects on ecosystem function, in particular on nutrient cycling. Particular attention is given to the role of herbivorous insects in terrestrial ecosystems, but effects of predators, parasitoids and detritivores are also touched upon. The second aim of this book is to provide a methodological guide to manipulative ecosystem studies of insect herbivory in the field. The techniques used to manipulate or simulate herbivory are critically reviewed, and their advantages and disadvantages are discussed in detail. We hope that by reviewing the known and suspected ecosystem-level effects of insects, and by providing methodological advice, this book will stimulate further research into the relationship between insects and ecosystem function.

Sections 2 and 3 of this book review what we know about the effects of herbivorous insects on ecosystem function. As belowground organisms have finally been recognized as being the main drivers of ecosystem processes (Wardle 2002), Section 2 focuses on the interactions between insects, the belowground community and ecosystem functioning. The belowground community represents a complex food web that consists of a large number of taxa including bacteria, protozoa, invertebrates and vertebrates. Insects involved in the belowground food web act as herbivores (e.g. the larvae of tipulids and a number of beetle families), detritivores (e.g. collembola) and predators (e.g. predatory beetle larvae). Because of the complexity of this food web, it is not always possible to isolate the effects of insects from those of other organisms. While emphasizing the particular role of insects, the chapters in the first section therefore also discuss the involvement of other belowground organisms in ecosystem functioning, if this is necessary for the understanding of the mechanisms underlying the cycling of nutrients. In the first chapter of this section, Hartley and Jones (Chap. 2) provide quantifications of direct and indirect effects of herbivorous insects on nutrient cycling, focusing on decomposition, mineralization and plant productivity. The chapter describes how stable isotope techniques and inventive uses of controlled environment studies have led to new insights into the role of soil biota in both below- and aboveground processes. In Chapter 3, Wardle and Bardgett review the indirect effects of aboveground herbivory on the decomposer community and identify four types of mechanisms through which herbivory affects the quantity and quality of plant-derived resources entering the soil. The authors argue that because decomposers affect plant-available nutrient supply, and therefore the quality and quantity of foliage available for invertebrate herbivores, herbivores function as important participants in feedbacks between the above- and belowground subsystems. In Chapter 4, Bonkowski and Scheu look at above- and belowground interactions from the opposite side and review the effects of interactions in the rhizosphere for aboveground processes. Bonkowski and Scheu discuss how aboveground herbivore–plant interactions and plant productivity are affected by the processes in a number of distinct belowground subsystems such as the bacterial loop and the fungal food chain. In Chapter 5, the final chapter of Section 2, Masters focuses on the role of belowground insect herbivores for nutrient cycling. In his chapter, Masters reviews the different effects of low, medium and high level of belowground herbivory for plant growth and nutrient cycling. The chapter also shows how belowground insect herbivory affects higher trophic levels such as seed predators and parasitoids with subsequent effects on nutrient cycling. Together, the chapters of the first sections provide a comprehensive review of the importance of above- and belowground interactions for ecosystem functioning, and they point out the approaches needed for a deeper understanding of the role of insects in influencing these interactions.

The chapters in Section 3 discuss aboveground interactions between plants, herbivorous insects and higher trophic levels, and the consequences of these interactions for nutrient cycling. In Chapter 6, Joshi and coauthors review results from the European BIODEPTH project and investigate the relationships between plant diversity, community productivity, and the diversity and abundance of insects in artificially assembled grassland plant communities. Joshi et al. show that a reasonable starting point in unravelling the complex relationship between producer diversity and whole-ecosystem consequences is to measure the effects of changing plant-species diversity on primary productivity, the basis of each food web, and to explore the potential consequences of these effects on higher trophic levels. The next two chapters use examples from biological control to illustrate the sometimes dramatic effects of insects on ecological processes. In Chapter 7, Zwölfer and Zimmermann use three case studies to show how insect–plant interactions may strongly affect plant community composition and all its dependent ecosystem properties. From successful biological weed control examples Zwölfer and Zimmermann deduce insect and plant properties that make a drastic insect-mediated change in vegetation cover more likely. Kruess and coauthors, in Chapter 8, show how the interactions between insects and plants and between insects and pathogens may be modified by the spatial context of the landscape in which these interactions take place. Such modifications affect matter fluxes both directly and indirectly, but these effects are far from being explored. In Chapter 9, the theme of food web interactions is taken a step further. Janssen and Sabelis review the effects of higher trophic levels on plant biomass, plant diversity and ecosystem processes. Navigating this virtually unexplored territory, one of the authors' sobering conclusions is that while food web interactions such as apparent competition, omnivory, intraguild predation or plant–plant interactions are important for ecosystem processes, generalizing rules relating food web interactions to ecosystem processes probably do not exist. In contrast to this view that it is very difficult to deduce generalizing mechanisms from the patterns observed, Carson et al., in Chapter 10, propose a general rule that predicts when insect herbivores will have very strong top-down effects on plant communities with subsequent effects on ecosystem functioning. The authors argue that outbreaks of specialist insects, traditionally seen as being of little importance for the long-term regulation of plant communities, are in fact both widespread and frequent enough to exert strong top-down effects on ecosystem functioning across multiple spatial scales. In Chapter 11, the final chapter of Section 3, Stadler et al. discuss the role of insects in forests and examine in detail the way in which the ecology of the insect herbivore influences nutrient cycling under chronic herbivory. The authors argue that in order to understand nutrient cycling in forests, it is important to appreciate the role of small herbivores such as aphids or adelgids as these excrete enormous amounts of honeydew and can change the flow of water and nutrients in the canopy.

The fourth section of this book concerns methodology. The chapters in this section review established and novel methods to manipulate the interaction between plants and insect herbivores. Particular emphasis is given to the advantages and disadvantages of the various methods to reduce, enhance or simulate the effects of herbivory on plant communities. The first two chapters by Hjältén and by Lehtilä and Boalt discuss the chances and pitfalls of simulated herbivory. In Chapter 12, Hjältén identifies the most obvious advantages and disadvantages of using simulated insect herbivory and suggests ways in which some of the problems can be avoided. He argues that except for some limited circumstances, the drawbacks of using artificial herbivory outweigh the main benefit of this method, which is the relative ease with which the degree, timing and distribution of damage can be controlled. In Chapter 13, Lehtilä and Boalt present the results of a literature survey in which authors have documented differences in effects when the same plant trait was analysed both using artificial herbivory and employing real herbivores. This literature survey reveals which traits are most sensitive to the way in which damage is applied in herbivory manipulations. Taken together, Chapters 12 and 13 should present a guide for the use of artificial herbivory in future studies on the effects of herbivorous insects on ecosystem function. In Chapter 14, another technique commonly used to elucidate the role of insects for ecological processes, the use of cages to exclude insect herbivores, or to assemble particular communities of insect herbivores and natural enemies, is reviewed critically. Using examples from his own work on a food web in a New England meadow ecosystem, Schmitz shows how enclosure cage experiments can be an effective tool in an endeavour to predict effects of perturbations on whole ecosystem function. In addition to discussing the requirements of cage design, Schmitz provides guidelines to show how predictive insights can be gained into complex trophic interactions by performing a series of cage experiments that examine successively how different ecosystem components fit together to determine function. Another method to exclude insects commonly employed in insect–plant studies is the use of insecticides, which allows experiments to be carried out in the field on a larger scale than cage experiments. Insecticide use too has a number of side effects that may confound the results of the experiment. In Chapter 15, Siemann and coauthors investigate the general principles that apply to the interpretation of insecticide experiments, and review the artefacts of the method that may masquerade as release from herbivory.

While the use of insecticides, cages or simulated herbivory all have their benefits, for many studies on the effects of insects on ecosystem function it may not be sufficient to concentrate on one of the established techniques only.

Chapter 16, by Rogers and coauthors, illustrates how a combination of cage, insecticide and simulated herbivory experiments can be used to unravel the role insects play in a particular ecosystem. Finally, in the last chapter of this section, Chapter 17, Voelckel and Baldwin go beyond a discussion of techniques that are already well established in the plant–insect literature, and develop a perspective for future manipulative studies employing the rapidly developing knowledge of molecular ecology. The authors describe the state of the art in research on herbivore-specific transcriptional responses and discuss their research potential for future ecosystem studies.

During the compilation of the literature review for this book, it became apparent to the editors that in addition to the lack of experimental studies little theory has been developed to investigate theoretically the effects of insect herbivory on ecosystem functioning (Bachelet et al. 1989; De Mazancourt and Loreau 2000; Blatt et al. 2001; Dyer and Shugart 2002). As a consequence, the last chapter of this book develops some simple theory to predict how the interaction between plant and insect herbivore diversity impacts plant productivity. This theory is intended to act as a starting point for further theoretical research into the effects of insect biodiversity on ecosystem functioning. In addition to this theoretical work, the last chapter uses the conclusions derived by the authors of the book chapters to attempt an outline of the areas where further research into insect effects on ecosystem functioning is needed. The editors hope that this outline along with the insights provided by the chapters of this book will stimulate entomologists, biogeochemists and other researchers to more closely investigate the role of the most speciose component of terrestrial biodiversity for ecosystem nutrient cycling.

Acknowledgements. First of all, we thank all internal and external reviewers whose comments greatly helped to shape this book. We are indebted to the managing editor, Ian T. Baldwin, for his support throughout the genesis of this book, and to Andrea Schlitzberger of the Springer Editorial Office for the professional handling of the chapters and the correspondence. Steffen Hahn and Ingrid Jacobi helped with the literature review for this chapter. W.W.W. was supported by the Deutsche Forschungsgemeinschaft, through grants FOR 456 and WE 2618/3-2618/5. E.S. was supported by US NSF (NSF DEB-9981564, DEB-0315796), EPA (R82-8903) and USDA (2003-35320-13498).

References

Abrahamson WG, McCrea KD (1986) Nutrient and biomass allocation in *Solidago altissima*: effects of two stem gallmakers, fertilization, and ramet isolation. Oecologia 68:174–180

Agrawal AA, Karban R (1997) Domatia mediate plant–arthropod mutualism. Nature (Lond) 387:562–563

Bach CE (2001) Long-term effects of insect herbivory and sand accretion on plant succession on sand dunes. Ecology 82:1401–1416

Bachelet D, Hunt HW, Detling JK (1989) A simulation model of intraseasonal carbon and nitrogen dynamics of blue grama swards as influenced by above- and belowground grazing. Ecol Model 44:231–252
Baldwin IT (1998) Jasmonate-induced responses are costly but benefit plants under attack in native populations. Proc Natl Acad Sci USA 95:8113–8118
Baldwin IT, Preston CA (1999) The eco-physiological complexity of plant responses to insect herbivores. Planta 208:137–145
Belovsky GE, Slade JB (2000) Insect herbivory accelerates nutrient cycling and increases plant production. Proc Natl Acad Sci USA 97:14412–14417
Blatt SE, Janmaat JA, Harmsen R (2001) Modelling succession to include a herbivore effect. Ecol Model 139:123–136
Bormann FH, Likens GE (1967) Nutrient cycling. Science 155:424–429
Brown MJF (1997) Effects of harvester ants on plant species distribution and abundance in a serpentine grassland. Oecologia 112:237–243
Brown VK (1990) Insect herbivores, herbivory and plant succession. In: Gilbert F (ed) Insect life cycles. Springer, Berlin Heidelberg New York, pp 183–196
Brown VK, Gange AC (1992) Secondary plant succession: how is it modified by insect herbivory? Vegetatio 101:3–13
Brown VK, Leijn M, Stinson CSA (1987) The experimental manipulation of insect herbivore load by the use of an insecticide (Malathion): the effect of application on plant growth. Oecologia 72:377–381
Brown VK, Jepsen M, Gibson CWD (1988) Insect herbivory: effects on early old field succession demonstrated by chemical exclusion methods. Oikos 52:293–302
Cain ML, Carson WP, Root RB (1991) Long-term suppression of insect herbivores increases the production and growth of *Solidago-altissima* rhizomes. Oecologia 88:251–257
Cardinale BJ, Palmer MA, Collins SL (2002) Species diversity enhances ecosystem functioning through interspecific facilitation. Nature (Lond) 415:426–429
Cardinale BJ, Harvey CT, Gross K, Ives AR (2003) Biodiversity and biocontrol: emergent impacts of a multi-enemy assemblage on pest suppression and crop yield in a agroecosystem. Ecol Lett 6:857–865
Carson WP, Root RB (2000) Herbivory and plant species coexistence: community regulation by an outbreaking phytophagous insect. Ecol Monogr 70:73–99
Chew RM (1974) Consumers as regulators of ecosystems: an alternative to energetics. Ohio J Sci 74:359–370
Christenson L, Lovett GM, Mitchell MJ, Groffman PM (2002) The fate of nitrogen in gypsy moth frass deposited to an oak forest floor. Oecologia 131:444–452
Coleman PC, Hendrix PF (eds) (2000) Invertebrates as webmasters in ecosystems. CAB International, Wallingford, UK
Corbet SA (1997) Role of pollinators in species preservation, conservation, ecosystem stability and genetic diversity. In: Richards KW (ed) Proc Int Symp on Pollination, vol Acta Hort 437, ISHS, pp 219–229
Coupe MD, Cahill JFJ (2003) Effects of insects on primary production in temperate herbaceous communities: a meta-analysis. Ecol Entomol 28:511–521
Crawley MJ (1983) Herbivory. The dynamics of animal–plant interactions. Blackwell, Oxford
Crawley MJ (ed) (1986) Plant ecology. Blackwell, Oxford
Curry JP (1994) The grassland invertebrate community. In: Curry JP (ed) Grassland, invertebrates. Chapman and Hall, London
Danell K, Ericson L (1990) Dynamic relations between the antler moth and meadow vegetation in northern Sweden. Ecology 7:1068–1077

Davidson DW (1993) The effects of herbivory and granivory on terrestrial plant succession. Oikos 68:23–35
De Mazancourt C, Loreau M (2000) Effects of herbivory and plant species replacement on primary production. Am Nat 155:735–754
Detling JK (1988) Grasslands and savannas: regulation of energy flow and nutrient cycling by herbivores. In: Pomeroy LR, Alberts JJ (eds) Concepts of ecosystem ecology. Springer, Berlin Heidelberg New York, pp 131–148
Dyer MI, Shugart HH (2002) Multi-level interactions arising from herbivory: a simulation analysis of deciduous forests utilizing Foret. Ecol Appl 2:376–386
Dyer MI, Acra MA, Wang GM, Coleman DC, Freckman DW, McNaughton SJ, Strain BR (1991) Source-sink carbon relations in two *Panicum coloratum* ecotypes in response to herbivory. Ecology 72:1472–1483
Ehrlich P, Ehrlich A (1981) Extinction: the causes and consequences of the disappearance of species. Random House, New York
Feller IC (2002) The role of herbivory by wood-boring insects in mangrove ecosystems in Belize. Oikos 97:167–176
Fonseca CR (1994) Herbivory and the long-lived leaves of an Amazonian ant-tree. J Ecol 82:833–842
Gehring AC, Whitham TG (2002) Mycorrhizae–herbivore interactions: population and community consequences. In: Van der Heijden MGA, Sanders IR (eds) Mycorrhizal ecology. Springer, Berlin Heidelberg New York, pp 295–344
Golley FB, McGinnis JT, Clements RG, Child GI, Duever MJ (1975) Mineral cycling in a tropical forest ecosystem. University of Georgia Press, Athens
Gosz JR, Holmes RT, Likens GE, Bormann FH (1978) The flow of energy in a forest ecosystem. Sci Am 238:92–102
Grayston SJ, Dawson LA, Treonis AM, Murray PJ, Ross J, Reid EJ, McDougall R (2001) Impact of root herbivory by insect larvae on soil microbial communities. Eur J Soil Biol 37:277–280
Grostal P, O'Dowd DJ (1994) Plants, mites and mutualism: leaf domatia and the abundance and reproduction of mites on *Viburnum tinus* (Caprifoliaceae). Oecologia 97:308–315
Hairston NG, Smith FE, Slobodkin LB (1960) Community structure, population control, and competition. Am Nat 94:421–425
Halaj J, Wise DH (2001) Terrestrial trophic cascades: how much do they trickle? Am Nat 157:262–281
Hambäck PA (2001) Direct and indirect effects of herbivory: feeding by spittlebugs affects pollinator visitation rates and seedset of *Rudbeckia hirta*. Ecoscience 8:45–50
Harder LD, Barrett SCH (2002) The energy cost of bee pollination for *Pontederia cordata* (Pontederiaceae). Funct Ecol 6:226–233
Hector A, Schmid B, Beierkuhnlein C, Caldeira MC, Diemer M, Dimitrakopoulos PG, Finn JA, Freitas H, Giller PS, Good J, Harris R, Högberg P, Huss-Danell K, Joshi J, Jumpponen A, Körner C, Leadley PW, Loreau M, Minns A, Mulder CPH, O'Donovan G, Otway SJ, Pereira JS, Prinz A, Read DJ, Scherer-Lorenzen M, Schulze E-D, Siamantziouras A-SD, Spehn EM, Terry AC, Troumbis AY, Woodward FI, Yachi S, Lawton JH (1999) Plant diversity and productivity experiments in European grasslands. Science 286:1123–1127
Honkanen T, Haukioja E (1998) Intra-plant regulation of growth and plant–herbivore interactions. Ecoscience 5:470–479
Hooper DU, Vitousek PM (1997) The effects of plant composition and diversity on ecosystem processes. Science 277:1302–1305
Howe HF, Smallwood J (1982) Ecology of seed dispersal. Annu Rev Ecol Syst 13:201–228

Huntly N (1991) Herbivores and the dynamics of communities and ecosystems. Annu Rev Ecol Syst 22:477–503

Hutchinson KJ, King KL (1982) Invertebrates and nutrient cycling. In: Lee KE (ed) Proc 3rd Australasian Conf on Grassland Invertebrate Ecology, Adelaide, 30 Nov–4 Dec 1981. SA Government Printer, Adelaide, pp 331–338

Hutson BR (1989) The role of fauna in nutrient turnover. In: Majer JD (ed) Animals in primary succession. Cambridge University Press, Cambridge

Irwin RE, Brody AK, Waser NM (2001) The impact of floral larceny on individuals, populations, and communities. Oecologia 129:161–168

Johnson SN, Mayhew PJ, Douglas AE, Hartley SE (2002) Insects as leaf engineers: can leaf-miners alter leaf structure for birch aphids? Funct Ecol 16:575–584

Jones CG, Lawton JH (eds) (1995) Linking species and ecosystems. Chapman and Hall, New York

Karban R, Baldwin IT (1997) Induced responses to herbivory. University of Chicago Press, Chicago

Kessler A, Baldwin IT (2001) Defensive function of herbivore-induced plant volatile emissions in nature. Science 291:2141–2144

Khan ML, Tripathi RS (1991) Seedling survival and growth of early and late successional tree species as affected by insect herbivory and pathogen attack in sub-tropical humid forest stands of north-east India. Acta Ecol 12:569–579

Kinzig AP, Pacala SW, Tilman D (eds) (1991) The functional consequences of biodiversity. Princeton University Press, Princeton

Kosola KR, Dickmann DI, Paul EA, Parry D (2001) Repeated insect defoliation effects on growth, nitrogen acquisition, carbohydrates, and root demography of poplars. Oecologia 129:65–74

Lamb D (1985) The influence of insects on nutrient cycling in eucalypt forests: a beneficial role? Aust J Ecol 10:1–5

Lee KE (1979) The role of invertebrates in nutrient cycling and energy flow in grasslands. In: (Crosby TK, Pottinger RP (eds) Proc 2nd Australian Conf on Grassland Invertebrate Ecology. Government Printer, Wellington, pp 26–29

Lee KE, Wood TG (1971) Termites and soil. Academic Press, London

Lerdau M (1996) Insects and ecosystem function. TREE 11:151–152

Lewinsohn TM, Price PW (1996) Diversity of herbivorous insects and ecosystem processes. In: Solbrig OT, Medina E, Silva JF (eds) Biodiversity and savanna ecosystem processes. Ecological studies, vol 121. Springer, Berlin Heidelberg New York, pp 143–157

Lightfoot DC, Whitford WG (1990) Phytophagous insects enhance nitrogen flux in a desert creosotebush community. Oecologia 82:18–25

Likens GE (1992) The ecosystem approach: its use and abuse. Ecology Institute, Oldendorf, Germany

Likens GE, Bormann FH, Pierce RS, Eaton JS, Johnson NM (1977) Biogeochemistry of a forested ecosystem. Springer, Berlin Heidelberg New York

Loreau M, Naeem S, Inchausti P, Bengtsson J, Grime JP, Hector A, Hooper DU, Huston MA, Raffaelli DG, Schmid B, Tilman D, Wardle DA (2001) Biodiversity and ecosystem functioning: current knowledge and future challenges. Science 294:804–808

Loreau M, Naeem S, Inchausti P (eds) (2002) Biodiversity and ecosystem functioning. Oxford University Press, Oxford

Louda SM, Keeler KH, Holt RD (1990) Herbivore influences on plant performance and competitive interactions. In: Grace JB, Tilman D (eds) Perspectives on plant competition. Academic Press, San Diego, pp 413–444

Lovett GM, Ruesink AE (1995) Carbon and nitrogen mineralization from decomposing gypsy moth frass. Oecologia 104:133–138

Lovett GM, Christenson LM, Groffman PM, Jones CG, Hart JE, Mitchell MJ (2002) Insect defoliation and nitrogen cycling in forests. Bioscience 52:335–341

Marquis RJ (1996) Plant architecture, sectoriality and plant tolerance to herbivores. Vegetatio 127:85–97

Mattson WJ, Addy ND (1975) Phytophagous insects as regulators of forest primary production. Science 190:515–522

McCullough DG, Werner RA (1998) Fire and insects in northern and boreal forest ecosystems of North America. Annu Rev Entomol 43:107–127

McGrady-Steed J, Harris PM, Morin PJ (1997) Biodiversity regulates ecosystem predictability. Nature (Lond) 390:162–165

McNaughton SJ (1993) Biodiversity and function of grazing ecosystems. In: Schulze E-D, Mooney HA (eds) Biodiversity and ecosystem function. Springer, Berlin Heidelberg New York

Mitchell CE (2003) Trophic control of grassland production and biomass by pathogens. Ecol Lett 6:147–155

Montoya JM, Rodriguez MA, Hawkins BA (2003) Food web complexity and higher-level ecosystem services. Ecol Lett 6:587–593

Mooney HA (1991) Emergence of the study of global ecology. Is terrestrial ecology an impediment to progress? Ecol Appl 1:2–5

Mulder CPH, Koricheva J, Huss-Danell K, Högberg P, Joshi J (1999) Insects affect relationships between plant species richness and ecosystem processes. Ecol Lett 2:237–246

Naeem S, Li S (1997) Biodiversity enhances ecosystem reliability. Nature (Lond) 390:507–509

Naeem S, Thompson LJ, Lawler JH, Lawton JH, Woodfin RM (1994) Declining biodiversity can alter the performance of ecosystems. Nature (Lond) 368:734–737

Odum EP (1953) Fundamentals of ecology. WB Saunders, Philadelphia

Petrusewicz K (ed) (1967) Secondary productivity of terrestrial ecosystems (principles and methods), vol II. Panstwowe Wydawnictwo Naukowe, Warszawa, Poland

Polis GA (1999) Why are parts of the world green? Multiple factors control productivity and the distribution of biomass. Oikos 86:3–15

Price PW (1997) Insect ecology. Wiley, New York

Sallabanks R, Courtney SP (1992) Frugivory, seed predation, and insect–vertebrate interactions. Annu Rev Entomol 37:377–400

Schmitz OJ (2003) Top predator control of plant biodiversity and productivity in an old-field ecosystem. Ecol Lett 6:156–163

Schmitz OJ, Hambäck PA, Beckerman AP (2000) Trophic cascades in terrestrial systems: a review of the effects of carnivore removals on plants. Am Nat 155:141–153

Schowalter TD (2000a) Insect ecology: an ecosystem approach. Academic Press, San Diego

Schowalter TD (2000b) Insects as regulators of ecosystem development. In: Colema PC, Hendrix PF (eds) Invertebrates as webmaster in ecosystem. CAB International, Wallingford, UK, pp 99–114

Schowalter TD, Hargrove WW, Crossley JDA (1986) Herbivory in forested ecosystems. Annu Rev Entomol 186:177–196

Schulze ED, Mooney HA (eds) (1993) Biodiversity and ecosystem function. Springer, Berlin Heidelberg New York

Seastedt TR, Crossley DAJ (1984) The influence of arthropods on ecosystems. Bioscience 34:157–161

Stadler B, Solinger S, Michalzik B (2001) Insect herbivores and the nutrient flow from the canopy to the soil in coniferous and deciduous forests. Oecologia 126:104–113
Stork NE (1988) Insect diversity: facts, fiction and speculation. Biol J Linn Soc 35:321–337
Stowe KA, Marquis RJ, Hochwender CG, Simms EL (2000) The evolutionary ecology of tolerance to consumer damage. Annu Rev Ecol Syst 31:565–595
Strauss SY, Agrawal AA, Strand MR (1999) The ecology and evolution of plant tolerance to herbivory. TREE 14:179–185
Tilman D, Downing JA (1994) Biodiversity and stability in grasslands. Nature (Lond) 367:363–365
Tilman D, Wedin D, Knops J (1996) Productivity and sustainability influenced by biodiversity in grassland ecosystems. Nature (Lond) 379:718–720
Tilman D, Knops J, Wedin D, Reich P, Ritchie M, Siemann E (1997) The influence of functional diversity and composition on ecosystem processes. Science 277:1300–1302
Trumble JT, Kolodny-Hirsch DM, Ting IP (1993) Plant compensation for arthropod herbivory. Annu Rev Entomol 38:93–119
Turlings TCJ, Tumlinson JH, Lewis WJ (1990) Exploitation of herbivore-induced plant odors by host-seeking parasitic wasps. Science 250:1251–1253
Urbanek RP (1988) The influence of fauna on plant productivity. In: Majer JD (ed) Animals in primary succession. Cambridge University Press, Cambridge, pp 71–106
van der Heijden MGA, Klironomos JN, Ursic M, Moutoglis P, Streitwolf-Engel R, Boller T, Wiemken A, Sanders IR (1998) Mycorrhizal fungal diversity determines plant biodiversity, ecosystem variability and productivity. Nature (Lond) 396:69–72
Wallace JB, Webster JR (1996) The role of macroinvertebrates in stream ecosystem function. Annu Rev Entomol 41:115–139
Wardle DA (2002) Communities and ecosystems: linking the above- and belowground components. Princeton University Press, Princeton
Whelan RJ (1989) The influence of fauna on plant species composition. In: Majer JD (ed) Animals in primary succession. Cambridge University Press, Cambridge, pp 107–143
Wiegert RG, Evans FC (1967) Investigations of secondary productivity in grasslands. In: Petrusewicz K (ed) Secondary productivity of terrestrial ecosystems (principles and methods), vol 1. Panstwowe Wydawnictwo Naukowe, Warszawa, Krakow, pp 499–518
Wilson EO (ed) (1988) Biodiversity. National Academy Press, Washington
Willson MF (1992) The ecology of seed dispersal. CAB International, Wallingford, UK
Zlotin RI, Khodashova KS (1980) The role of animals in biological cycling of forest-steppe ecosystems. Dowden, Hutchinson and Ross., Stroudsburg, Pennsylvania

Section II
Insects and the Belowground System

2 Insect Herbivores, Nutrient Cycling and Plant Productivity

S.E. Hartley and T.H. Jones

2.1 Summary

We review the various ways in which insect herbivores affect ecosystem function, focusing particularly on their impacts on decomposition, nutrient cycling and plant productivity. Many of the most profound effects of insects on these processes occur below ground and until recently have been considered relatively inaccessible to study. However, new approaches, particularly the development of stable isotope techniques and more inventive uses of controlled environment studies, have allowed significant advances in our understanding of the role of soil biota in both below- and aboveground processes. Undoubtedly, one of their most important roles is to physically break up organic matter within the soil and make it accessible to the microbial component, but these new techniques mean we now know far more about the more indirect effects of soil biota as well, such as the impact of root-feeding herbivores on the rhizosphere. In addition, the development of more quantitative techniques for assessing canopy herbivory and the impacts of defoliation on nitrogen cycling in forests has provided new evidence for the impact of aboveground foliovores on soil processes. We draw on the literature and on our own work to summarize the 'state of play'. We highlight some of the progress that has been made in the study of the role of insect herbivores in ecosystem function, examine some intriguing interactions between insect herbivores and other organisms, and draw attention to some neglected impacts on ecosystem processes. Thus we highlight gaps in our current understanding and hence areas that future research might profitably examine.

Ecological Studies, Vol. 173
W.W. Weisser and E. Siemann (Eds.)
Insects and Ecosystem Function

2.2 Introduction

In this chapter, we review the various ways insects affect ecosystem function, particularly in relation to processes important in carbon and nitrogen cycling, namely decomposition, mineralization and plant productivity. It has been 20 years since the last major review of this subject (Seastedt and Crossley 1984). Since then, developments in experimental approach and the availability of new techniques have advanced the field considerably. For example, we are now in a much better position to understand the links between above- and belowground processes (Van der Putten et al. 2001) because of new conceptual models explaining the impact of foliar feeders on root feeders and vice versa (Seastedt et al. 1988; Masters et al. 1993; Bardgett et al. 1998a) and the use of ^{15}N and ^{13}C stable isotope labelling to follow the fate of nutrients through ecosystems (Eggers and Jones 2000; Scheu and Falca 2000). These new approaches have revealed novel insights into the importance of insect herbivores in ecosystem function.

Insects play as important a role below ground as they do above ground: Collembola, termites and ants are major modifiers of soil processes, thus affecting plant nutrient availability, and in food webs dominated by invertebrates most plant productivity goes to soil biota rather than to herbivores (Oksanen et al. 1997; ; Wardle and Bardgett, Chap. 3, this Vol.; Bonkowski and Scheu, Chap. 4, this Vol.; Masters, Chap. 5, this Vol.). The effects of belowground insects on ecosystem processes could thus exceed those of aboveground ones, particularly in low productivity systems, although this is disputed by some authors who argue that herbivores have considerably more influence on the ecosystem than do detritivores (Mattson and Addy 1975). Hence we begin our assessment of the impact of insects on nutrient cycling and ecosystem function below ground, before moving on to consider the influence herbivores have above ground on plant productivity and plant community dynamics. Throughout, by attempting to summarize what is known from existing studies, we hope to highlight gaps in our current understanding of the impact insects have on ecosystem processes and hence signpost new research directions.

2.3 Decomposition

2.3.1 The Resources Available

The quantity and quality of dead and decaying plant tissue entering soils to be utilized by decomposers clearly depend on which plant species are present in the community. In addition, aboveground consumers of plant material can exert important effects on decomposer communities and processes, and ultimately plant-available nutrient supply, through a variety of mechanisms. These mechanisms operate over a range of spatial scales, from individual plant level to that of the plant community, and can involve either positive or negative effects on both the quantity and quality of resources that enter the decomposer subsystem (Fig. 2.1; see also reviews by Bardgett et al. 1998a; Wardle 2002; Wardle and Bardgett, Chap. 3, this Vol.).

Data from diverse ecosystems indicate that more than 50 % of net primary production is commonly allocated to belowground plant parts (Coleman 1976), whilst values for particular plant species may approach 90 % (Andersen 1987). There is calculated to be between 330 and 4,000 g of root biomass to each square meter of soil, which, although of lower nutritional value than foliar tissue, would seem to be an important resource for insect herbivores. Although roots are the principal energy source for the majority of belowground consumers, a significant proportion of carbon allocated below ground may be directed to mycorrhizal fungi (Read 1991). Indeed, some workers have suggested that as much as 40–50 % of total plant photosynthate may be directed to ectomycorrhizal fungi (Fogel and Hunt 1979). Potentially more realistic figures of 10–20 % have been recorded for arbuscular mycorrhizal (AM) fungi (Jakobsen and Rosendahl 1990), and although some of this feeds the higher respiration rate in infected roots, it represents a significant energy source for subterranean mycophagous insects.

Dead plant material poses special problems to decomposers because it contains high proportions of cellulose and lignin, neither of which is readily digested by animals. In fact, amongst the insects, some cockroaches and higher termites in the subfamily *Nasutitermitinae* (Isoptera) are the only taxa known to synthesize enzymes capable of degrading cellulose (Martin 1991). Termites are perhaps the most impressive decomposers in the insect world (Whitford et al. 1988; Moorhead and Reynolds 1991) due to their symbioses with micro-organisms that live in their guts and produce celluloytic (cellulose-digesting) enzymes (Basaglia et al. 1992; Yoshimura et al. 1993). These enzymes mean they are highly efficient digesters of litter and wood in desert (Whitford et al. 1988), savannah (Wood and Sands 1978) and forest ecosystems (Bignell et al. 1997), and they may consume 55 % of the surface litter (Wood and Sands 1978). This makes them the major regulators of the dynamics of litter and soil organic matter in many ecosystems (Lavelle 1997).

Collembola are extremely abundant in soil and leaf litter, and are one of the principal fungal feeders in many soil ecosystems (Klironomos and Kendrick 1995; Thimm and Larink 1995). In most terrestrial ecosystems they occur at densities of 10^4–10^5 m^{-2} (Petersen and Luxton 1982) and are extremely important in influencing the structure of some soils (Rusek 1998). For example, most soils contain millions of Collembola faecal pellets per square meter and these must be beneficial in slowly releasing essential nutrients to plant roots as the pellets are broken down by microbes. One of the

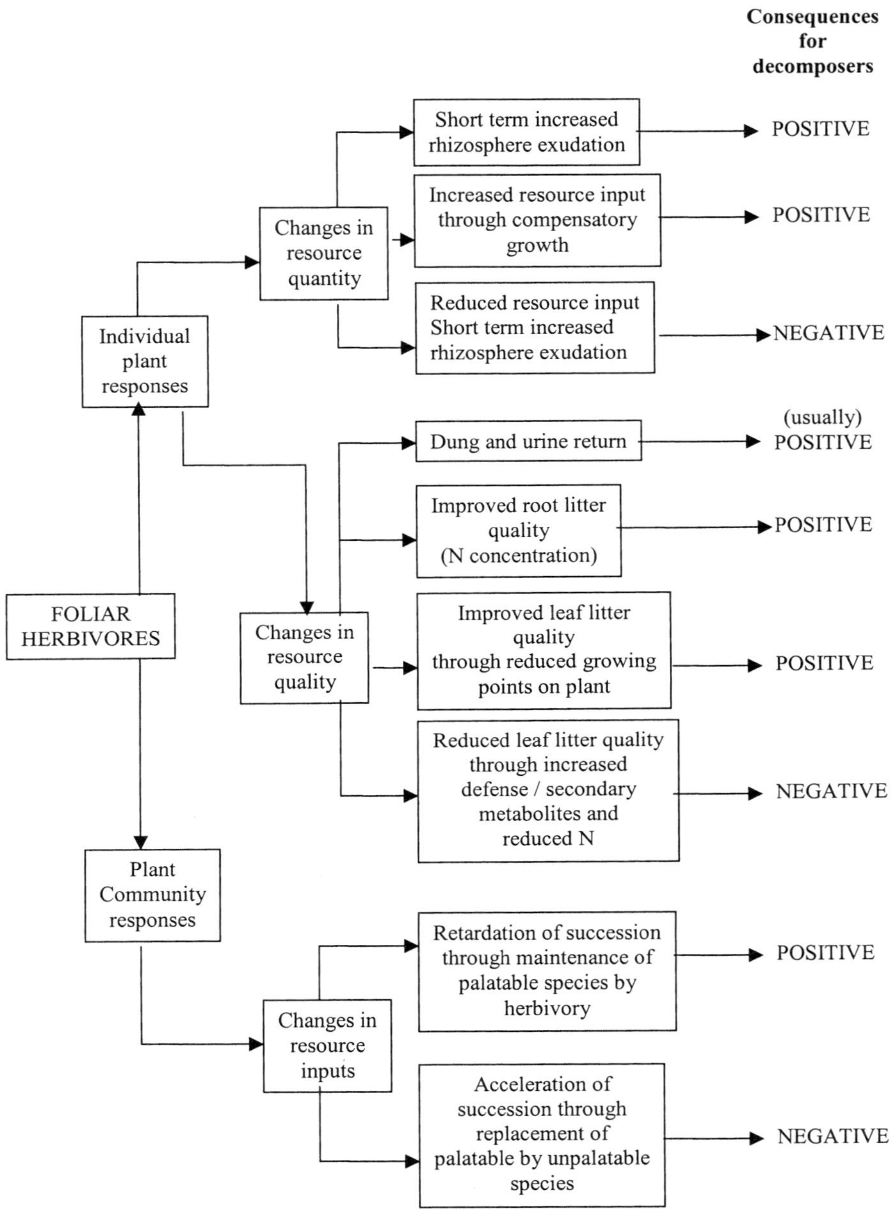

Fig. 2.1. Mechanisms by which foliar herbivores may influence soil decomposer organisms, both positively and negatively. (Loreau et al. 2002)

main effects of Collembola on decomposition and 'soil respiration' is via their feeding on fungal hyphae. Some studies have shown that Collembola grazing of mycorrhizae on roots can simulate growth of the symbiont and improve plant growth (Lussenhop 1996). However, at higher densities Collembola grazing may also reduce the functioning of mycorrhizae, with serious consequences for the host plant. For example, Warnock et al. (1982) grew seedlings of leek (*Allium porrum*) with or without mycorrhiza (*Glomus fasciculatum*), with or without soil leachings (an aqueous suspension of the soil micro-flora, without AM fungi), and with or without the collembolan *Folsomia candida*, in a factorial design. Both mycorrhizal infection and leachate increased plant growth, but the addition of *F. candida* to infected plants meant that these grew little better than uninfected individuals because grazing on the external hyphae rendered the infection ineffective. Whether Collembola will choose to graze on mycorrhiza when provided with a choice of resources is a question that has been raised (Klironomos and Ursic 1998); studies such as those of Warnock et al. (1982) must therefore be considered in that context.

There are also numerous insects associated with the breakdown of organic material from animals, such as blow flies and flesh flies (Diptera: Calliphoridae and Sarcophagidae, respectively), carrion beetles (Coleoptera: Silphidae) and dung beetles (Coleoptera: Scarabaeinae). Dung beetles have actually been used as 'decomposition control agents' in Australia (Waterhouse 1977). With 30 million cattle, dung pads could put about 2.5 million ha of pasture land out of service each year, so in 1967, African dung beetles were introduced for the control of cattle dung. Thus we have compelling evidence of the importance of certain insects to decomposition processes in ecosystems!

2.3.2 Effects of Insect Herbivory on Decomposition

2.3.2.1 Herbivory and Litter Quality

Secondary compounds in leaves are a major influence on rates of litter decomposition (Wardle and Bardgett, Chap. 3, this Vol.). Thus, although lignin and fibre content are two of the most important determinants of decomposition rate, concentrations of phenolic compounds and condensed tannins are also significantly correlated with decomposition rates in litter from a range of plant species (Wardle et al. 2002). Complexes of tannins bound to protein are some of the most difficult compounds for soil decomposers to digest (Lavelle 1997). This explains why leaf litter decomposition rate may be correlated with palatability to generalist herbivores (Cornelissen 1996; Wardle et al. 1998), because the tannins and phenolic compounds, which slow decomposition, are also important deterrents to insect feeding (Hartley and Jones 1997). Furthermore, insect herbivory may cause increases in the level of phenolics and tannins, so-called induced defences (Hartley and Firn 1989; Karban and Baldwin 1997), and this decline in plant quality could persist in the litter, with potential consequences for decomposition rates. For example, it has been demonstrated that the tannin content of *Pinus muricata* litter controls the proportion of nitrogen released in dissolved organic forms relative to mineral forms (Northrup et al. 1995).

Litter decomposition rates are related to plant life-form and leaf life-time (i.e. evergreen vs deciduous habit), in part reflecting between-species differences in lignin, and secondary compound content (Cornelissen 1996). Thus woody shrubs contain more lignin and more tannins than herbaceous plants, and hence their litter takes longer to decay. The selective feeding of insect herbivores could affect decomposition rates if it changes the dominant species in a community from one with recalcitrant litter which decomposes slowly, such as a woody species with high tannin content, to one with litter which decomposes more easily, such as a herbaceous species with lower levels of secondary compounds. One well-known example of insect herbivory causing just such an effect is outbreaks of heather beetle (*Lochmaea sutralis*) (Bredowski and Zeilinga 1987) where *Calluna vulgaris*, a woody shrub with high levels of tannins and phenolics (Hartley and Gardner 1995), is replaced by grasses, which have much lower secondary compound content (Hartley et al. 2003).

2.3.2.2 Herbivory, Root Exudation and Root Biomass

As well as its direct effects on decomposition and on litter quality, insect herbivory, both above and below ground, can affect decomposition processes because of its impacts on root biomass and root exudation (Bonkowski and Scheu, Chap. 4, this Vol.). For example, Bokhari and Singh (1974) found that artificially defoliated plants of *Agropyron smithii* grown in hydroponic solution released nearly 10% more carbon into the growth medium than did intact plants, and also had a greater rate of respiration associated with the rooting zone. Further, ^{14}C pulse labelling studies of monocotyledonous herbaceous plants provide evidence that herbivory by grasshoppers causes an allocation shift of carbon from shoots to roots (Dyer et al. 1991) and that this leads to both greater exudation of ^{14}C into the rhizosphere and enhanced rooting zone respiration (Holland et al. 1996). This enhanced exudation of carbon in the rhizosphere stimulates soil organisms. For example, Holland (1995) found that microbial biomass associated with *Zea mays* roots was maximized when plants were subjected to intermediate levels of herbivory by grasshoppers in a no-tillage cropping system, and proposed that increased root exudation due to defoliation was responsible.

Any reduction in root biomass in response to defoliation may, in turn, lead to more dead roots for the soil microbial community to decompose (Seastedt et al. 1988). However, experimental evidence for the impacts of foliar insect herbivory on belowground processes remains relatively scarce; most studies have involved either artificial defoliation or mammalian grazers. For example, in pot trials based on individual plants, experimental defoliation usually reduces root biomass (Ruess 1998; Mikola et al. 2001). However, evidence from field studies is more mixed. McNaughton et al. (1998) found no evidence for greater root productivity in plots in the Serengeti where grazers had been excluded. In contrast, Ruess et al. (1998) found, through the use of mini-rhizotrons, that mammalian browsing in central Alaskan forest caused a reduction of fine root productivity, an increase in root mortality and an increase in root turnover rate. The most convincing evidence that damage by insect herbivores can also affect root biomass comes from studies of aphids. Phloem-feeding insects can create an additional aboveground sink for plant carbon, and hence a reduction in root biomass is a common consequence of heavy aphid infestations (Choudhury 1984; Inbar et al. 1995).

Clearly root-feeding herbivores will affect root biomass, but there is also evidence that they alter root exudation in a similar way to foliar-feeding herbivores (Masters, Chap. 5, this Vol.). For example, Yeates et al. (1998) used $^{14}CO_2$ pulse-labelling techniques to show that infection of white clover (*Trifolium repens*) roots by low numbers of clover cyst nematodes (*Heterodera trifolii*) increased 'leakage' of carbon from roots, resulting in an enhanced microbial biomass in the rhizosphere. The highest carbon flux (measured as microbial biomass) occurred at low levels of herbivory; above a certain threshold any positive effects were lost (Denton et al. 1999). The difficulties of manipulating and observing root herbivores mean that studies are technically more difficult than those on foliar feeders. Their impacts are also more difficult to measure because root-feeding insects are often slow-growing and have a long life cycle. Despite these methodological difficulties, there is increasing evidence that belowground grazers do influence nutrient cycling. In particular, there is increasing appreciation that the responses of belowground feeders to increases in carbon supply from root exudation have 'knock-on' effects for the aboveground plant community. We deal with both these issues in the next section.

2.4 Nutrient Cycling and Plant Productivity

Given their ability to decompose various organic materials (see previous section), it is not surprising that insects play a major role in nutrient cycling. Although the impact of insect herbivores is most associated with nitrogen cycling (e.g. Lovett et al. 2002), it is increasingly recognized that they are able to influence the dynamics of other nutrients too. They also affect soil structure, breaking up plant residues and vastly increasing the surface area accessible to microbial attack, with obvious consequences for nutrient cycling

(Boddy et al. 1999). The advent of new methods for tracking the movement of assimilation of nutrients has led to better information on the constraints on nutrient cycling by insects (e.g. their ability to conserve nitrogen in their guts) and has also illustrated the fate of insect-derived nitrogen inputs (e.g. from frass) in ecosystems. Interestingly, labelling techniques are now sufficiently sophisticated to show that nutrients traditionally moving from plant to insect can move back to the plant. For example, in plant–ant mutualisms, nitrogen has now been shown to move in both directions between partners (Fischer et al. 2002, 2003).

2.4.1 Effects on Carbon and Nitrogen Cycling

2.4.1.1 Methane and Carbon Dioxide

Insects play their major role in the carbon cycle during the decomposition process. Perhaps the most important insects, at least in terms of potential impact on the global carbon cycle, are termites. Termite guts contain anaerobic microsites, hence their decomposition of plant material produces methane (CH_4) as well as carbon dioxide (CO_2). They therefore have the potential to recycle significant amounts of carbon to the atmosphere in two gaseous forms. CH_4 is one of the principal greenhouse gases, contributing about 18 % to the effects of such gases on climatic variation (Intergovernmental Panel on Climate Change 2001). With CH_4 a far more efficient absorber of energy than CO_2, the relative importance of termites to global fluxes of greenhouse gases is a subject of considerable current debate (Houghton et al. 1990; Lawton 2000). Forests contain more organic carbon than all other terrestrial systems and, at present, account for about 90 % of the annual carbon flux between the atmosphere and the Earth's land surface (Groombridge 1992). The huge abundance of termites, particularly in forest systems, makes them a key player in the terrestrial carbon cycle. In fact, termites may be an order of magnitude more abundant that the next most abundant arthropod group, the ants. The highest biomass reported for termites, 50–100 g m^{-2} in southern Cameroon forest (Eggleton et al. 1996), is greater than any other component of the invertebrate (or vertebrate) biota, and may constitute as much as 95 % of all soil insect biomass (Bignell et al. 1997). If direct carbon fluxes by insects are significant components of ecosystem processes, termites will make by far the greatest contribution. Some estimates consider they may be responsible for 2 % of global CO_2 production and 20 % of global CH_4 production (Speight et al. 1999)!

2.4.1.2 Nitrogen and Phosphorus

In contrast to their role as major players in CH_4 and CO_2 fluxes, termites appear to be far less important in the nitrogen cycle. Although in some ecosystems, such as the semi-arid grasslands of Africa, they remove similar amounts of herbage as grazing mammals and consume up to 49 % of annual plant production (Lavelle 1997), they do not appear to release important amounts of assimilable nutrients in their faeces. This reflects the fact that they have developed highly efficient mechanisms for nitrogen conservation during digestion. For example, nitrogen-fixing organisms have been found in their guts (Breznak 1982). However, they do have significant effects on the release of mineral nitrogen in the structures they create, particularly those made by macro-termitine termites, which cultivate fungi on leaf or wood material. In addition, some experiments have demonstrated that soil from termite nests is much better for plant growth than the surrounding soil – an effect long recognized by indigenous populations (Lavelle 1997), although quantitative data are lacking.

Generally, however, it is recognized that insects have a vital role to play in nitrogen cycling. Most plant communities occur on soils that have low levels of available nutrients, and nutrient mineralization is often the main limit on primary production. In most ecosystems, particularly where there are no macro-herbivores (e.g. in the Arctic or sub-Arctic where plant productivity is too low to support them; Oksanen et al. 1997), most nutrients pass through the detritivore cycle. Insects are a significant part of the detritivore community, and their ability to digest litter means they can have a major influence on the availability of nitrogen and phosphorus to higher plants. One excellent example is Marion Island in the oceanic sub-Antarctic, a species-poor plant community that has high primary production. Larvae of a flightless moth (*Pringleophaga marioni*), as a result of digestion of plant litter, stimulate nitrogen mineralization ten-fold and phosphorus mineralization three-fold. Marion Island supports a large population of introduced house mice (*Mus musculus*) which eat 0.7 % of the standing crop of these arthropods every day (Burger 1978). The recent increase in mouse population has reduced the annual turnover of litter by moth larvae from 2,500 to 1,500 kg ha^{-1}, and Smith and Steenkamp (1990) argue that the predation pressure on soil invertebrates has decreased rates of nutrient cycling, causing further imbalances between primary production and decomposition.

Collembola grazers are also reported to have potentially marked effects on nutrient concentrations in the soil (Bonkowski and Scheu, Chap. 4, this Vol.). For example, Collembola in the soil play a major role as comminutors (cf. earthworms) breaking up plant residues and vastly increasing the surface area accessible to microbial attack (Hopkin 1998). This increases nitrogen flux. Verhoef and de Goede (1985) show that the presence of Collembola in a soil can affect its nutrient concentration. In a field experiment, removal of

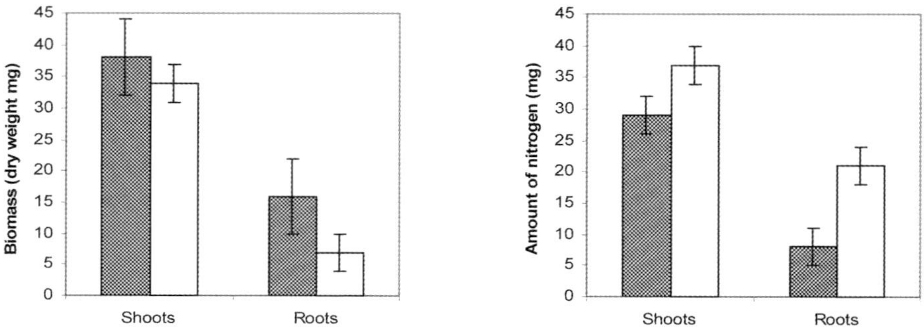

Fig. 2.2. Effect of Collembola (*unshaded bars*) on shoot and root mass (mean ± SD) and concentration of nitrogen in shoots and roots (mean ± SD) in *Poa annua*. *Shaded bars* are values in the absence of Collembola. (Adapted from Scheu et al. 1999)

Collembola by non-chemical means resulted in a decrease in the nitrogen concentration: plots with Collembola had 2.3 times the concentration of nitrogen compared with those without. A subsequent laboratory experiment showed a non-linear response, with plant nitrogen concentration peaking at intermediate densities of Collembola. One interpretation of these results is that at low densities, the positive effects of Collembola on mineralization rates of both nitrogen and phosphorus outweigh any negative impacts they have on plant nutrient availability due to their feeding on mycorrhizae, while at high densities, the increase in mineralization is insufficient to offset the mycorrhizal loss. Jones and colleagues (Scheu et al. 1999; Wurst and Jones 2003) have found that while the presence of Collembola caused a reduction in plant biomass, particularly that of *Poa annua* roots, plant tissue nitrogen was increased (Fig. 2.2). This in turn increased aphid (*Myzus persicae*) populations on *P. annua* three-fold (Bonkowski and Scheu, Chap. 4, this Vol.).

Belowground insect herbivores can also remove nitrogen from the soil system: in tallgrass prairie, the emergence of annual cicadas can represent a significant flux of non-gaseous nitrogen from below to above-ground. In this system, total nitrogen inputs were estimated at 11–25 kg N ha^{-1} $year^{-1}$, but the emergence of cicadas represented a redistribution of 4 kg N ha^{-1} $year^{-1}$, i.e. 16–36 % of the total annual input (Callaham et al. 2000)!

2.4.1.3 Inputs from Aboveground Herbivores

Large-scale defoliation by outbreaking forest insects can lead to a larger input of foliar litter and insect frass to the forest floor, potentially increasing nutrient cycling rates (Seastedt and Crossley 1980, 1984). Studies measuring the impact of oak forest defoliation by gypsy moth (*Lymantria dispar*) on litter fall have found that while the total quantity remained unaltered, the composition and seasonal distribution of litter fall was affected (Grace 1986). The nitrogen, phosphorus and potassium in the litter increased, whilst calcium levels decreased in response to defoliation. The most marked change was that on defoliated plots 56 % of the litter fell during the growing season, whereas on control plots 90 % of the litter was deposited in the autumn. Only 30 % of the nutrients on defoliated plots were derived from tree litter (compared to more than 85 % on control plots). Effects on litter quality might also be expected and indeed insect herbivory has been reported to increase the nitrogen in litter fall from 31–52 kg N ha^{-1} (Grace 1986).

Early studies (Mattson and Addy 1975; Schowalter 1981) assumed that defoliation of forests by insects would increase available nitrogen due to the deposition, and subsequent decomposition, of nitrogen-rich frass pellets. For example, outbreaks of the California oak moth (*Pytygandia californica*) were thought to lead to increased inputs of nitrogen and phosphorus into soils because short-term leaching experiments showed nitrogen was more rapidly lost from frass than from leaf litter (Hollinger 1986). However, when Lovett and Ruesink (1995) measured the rates of carbon and nitrogen mineralization from the frass of gypsy moth caterpillars in laboratory incubations they found that the frass contained much labile carbon as well as nitrogen, and that it stimulated microbial growth. This in turn increased microbial immobilization of nitrogen. Hence there was not the expected pulse of nitrate and ammonium; instead microbial immobilization appeared to be an efficient mechanism for conserving nitrogen within the forest ecosystem after a defoliation event.

While laboratory studies seemed clear cut, in the field the effects of defoliation on the nitrogen cycle in forest ecosystems remained uncertain for rather longer: some studies (Swank et al. 1981; Webb et al. 1995; Eshleman et al. 1998) did demonstrate increased export of nitrate in streams after large-scale defoliation events, whilst others, particularly those studies in ecosystems where baseline concentrations of nitrate were relatively high, did not (Bormann and Likens 1979). The advent of better labelling techniques has now enabled resolution of this issue, at least in forest ecosystems in the northern USA. Christenson et al. (2002) followed the fate of ^{15}N in gypsy moth frass deposited on an oak forest floor and compared it to that of ^{15}N in leaf litter. They found that the nitrogen in frass had a different fate to that in leaf litter, but in neither case did much of the nitrogen get taken up by oak seedlings: 40 % of the ^{15}N in the frass became incorporated in the soils and only 1 % was found in seedlings, whilst 80 % of the ^{15}N in leaves remained as undecomposed leaf material after 2 years. The data indicate that although the nitrogen in frass was mobilized more quickly than that in leaf litter, it remained largely unavailable to plants and micro-organisms because little of it was found in the extractable, microbial or readily mineralizable pools. Thus it seems that although insect defoliation represents a major perturba-

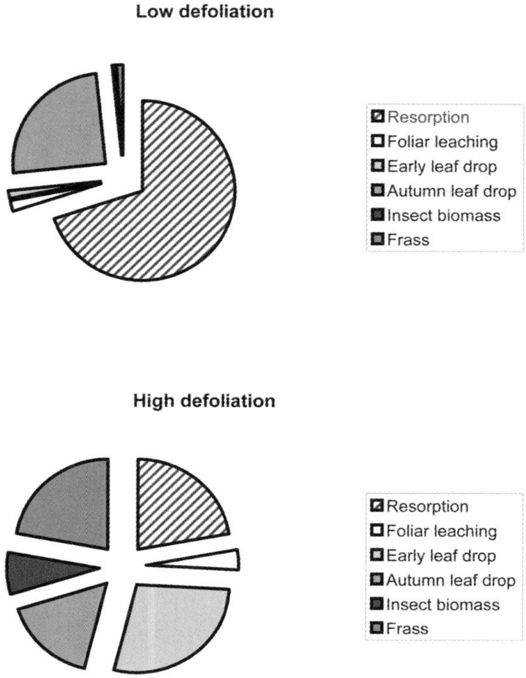

Fig. 2.3. Fate of nitrogen in foliage in years of high and low defoliation in an oak forest in the eastern United States (from Lovett et al. 2002). *Pie charts* show the relative proportion of nitrogen in oak leaves which is resorbed, lost through leaching, lost through early or autumn leaf drop, or converted into insect biomass or frass, in a forest subject to low defoliation by insect herbivores (upland mixed-oak forest in Millbrook, New York) and those subject to high levels of defoliation (oak forests in Pennsylvania; Grace 1986)

tion to the internal nitrogen cycle of the forest (Fig. 2.3), this perturbation primarily causes a redistribution of nitrogen within the ecosystem rather than a large loss of nitrogen (Lovett et al. 2002).

As well as frass, the activity of canopy herbivores includes other inputs, notably throughfall (precipitation which has fallen through the canopy and been modified by the activity of canopy herbivores) and greenfall (portions of green leaves that fall due to insect herbivores being messy eaters!). Most studies to date have focused on frass, but one recent study (Reynolds and Hunter 2001) has attempted to examine all three factors separately, namely frass, throughfall and greenfall, and to determine their relative contributions to soil respiration, soil nutrients and litter decomposition. Frass did not affect soil respiration, possibly because the amount added was typical of field densities, whereas Lovett and Ruesink's (1995) study (which did find increased carbon mineralization rates) was a laboratory microcosm study with an added amount of frass equivalent to that during outbreak conditions. Throughfall reduced soil respiration, possibly because soil microbes were outcompeted for the added mineral nitrogen by mycorrhizal fungi. Greenfall also decreased soil respiration. Surprisingly, none of the treatments affected litter decomposition, despite the effects of all three treatments on soil respiration and soil micro-arthropods (Reynolds and Hunter 2001; Reynolds et al. 2003).

2.4.1.4 The Importance of Belowground Biota: Evidence from Controlled Environment Studies

Lovett and Ruesink's (1995) study above provides an excellent example of how laboratory microcosm studies, and controlled environment studies in general, provide us with a means of obtaining detailed understanding of the ecosystem consequences of environmental change. Many such studies have recently been used to highlight salient features of ecosystem interactions (see Beyers and Odum 1993; Lawton 2000); here we consider some recent findings from the Ecotron controlled environment facility based at Imperial College London, UK.

^{14}C-labelling techniques have demonstrated that foliar herbivory stimulates the translocation of materials to the roots and hence increases root exudation (Dyer et al. 1991; Holland et al. 1996; Bardgett et al. 1998b) . However, the impact of these root exudates on decomposition and nutrient cycling depends on the response of soil organisms such as Collembola and soil fungi to this increase in carbon supply. The responsiveness of soil decomposers to carbon inputs can be seen from experiments on the impacts of rising atmospheric CO_2 on model terrestrial ecosystems maintained in the Ecotron facility (Jones et al. 1998). In this study, elevated CO_2 led to increased photosynthesis, and to increased belowground transport of carbon, where the increased level of dissolved organic carbon (DOC) changes soil fungal assemblages with knock-on effects on Collembola abundance and species composition. Studies on fungal enzymatic activities in Swiss grasslands show very similar changes in fungal community composition under conditions of elevated CO_2 (Jones et al. 2000). These belowground changes lead to increased rates of decomposition [in Jones et al. (1998) cotton decomposition almost doubled under conditions of elevated CO_2] and hence have the potential to alter nutrient availability to plants. More recently, again in the Ecotron facility, Bradford et al. (2002), in a study that explored the effects of changing soil biota composition on a model grassland community, have found that decomposition rate was significantly enhanced in the most complex faunal treatment (containing Collembola, Diptera and Coleoptera grazers). As decomposition rate is generally positively correlated, within a system, to nutrient availability (Swift et al. 1979), it was surprising that Bradford et al. did not also find a net primary production increase in these communities. Both mycorrhizal colonization and root biomass were lower in the more complex communities and it is possible that these decreases may explain why plants in these communities were unable to capitalize on the potentially higher nutrient availability. What is obvious from this limited number of studies is that these sort of complex feedback effects, linking above- and belowground processes, are increasingly being identified – insect herbivores have a key role to play in these links whether they are feeding above or below the soil surface.

Aboveground herbivores often remove the most nutrient-rich species as these are usually the most palatable (Ritchie et al. 1998; Wardle and Bardgett, Chap. 3, this Vol.), but below ground, the opposite effect can be observed. For example, the Bradford et al. (2002) study (see above) also found that plant functional group (grass, forb, legume) and species composition were markedly affected by soil biota treatments. Foliar biomass of both forbs and legumes decreased in the most complex communities (containing highest numbers and variety of soil grazers), whereas in the other two treatment communities [microbiota only, and microbiota and mesofauna (that did contain Collembola)] their biomass increased over time. This shift towards the more nitrogen-rich plant functional groups was reflected in shifts in grass species composition – *Holcus mollis*, the most nitrogen-rich graminoid species, increased in biomass in both the community containing only microbiota and in the community with both microbiota and mesofauna added. In all these studies, the important issue that remains unanswered is whether such controlled environment findings are representative of what actually happens in the field.

2.4.1.5 Insect Herbivory and Spatial Variation in Nutrient Availability

Insect herbivores are important contributors to spatial heterogeneity in nutrient availability for plants. These effects can be very obvious visually, as when termite and ant mounds support different vegetation types to those in the surrounding landscape (Blomqvist et al. 2000). Trees growing in mounds made by wood-ants have been shown to have 20 % more foliar nitrogen than surrounding trees more distant from any mounds (Karhu and Neuvonen 1998). This had beneficial effects on insect herbivores, which grew 30 % more rapidly on foliage from trees on ant mounds than on other foliage. Ant nests are rich in organic matter, nitrogen and phosphorus (Folgarait 1998) and they may contain six times the amount of ammonium and nitrate than off-mound soil (McGinley et al. 1994).

These effects on nutrient movement are at their most dramatic in the Neotropics, where leaf-cutter ants from the tribe Attini harvest leaves to feed their underground fungus gardens. Their nests may be 6 m deep, contain several million ants and require the same daily plant biomass as a cow! A typical colony moves 40 tons (23 m^3) of soil to the surface, with consequences for a range of soil processes; globally, ants may move more soil than earthworms. They are also the primary consumers within New World terrestrial ecosystems, where they remove up to 20 % of total leaf production (Beattie and Hughes 2002). Carrying this material to their nests, they effectively redistribute nutrients from the canopy to the forest floor, as do other canopy-dwelling insects which drop leaf fragments and frass, so returning nutrients from the tops of trees to the litter and soil.

It is not just in terms of their nest-building, soil-moving and fungal gardens that ants have the potential to influence plant nutrient supply. It is now known that ant–plant mutualisms can involve direct nutrient exchange between ants and plants. Several species of *Piper* (Piperaceae) live in symbiosis with *Pheidole bicornis* (Formicidae: Myrmicinae) on the southern Pacific slope of Costa Rica. These plants produce small single-celled food bodies in leaf domatia, formed by the petiole bases and rolling leaf sheaths, which are the main food source for *P. bicornis* ants, as shown by labelling studies comparing the natural abundance of ^{13}C and ^{15}N in the ants and the food bodies (Fischer et al. 2002). The flux from plants to ants is directly evident if plants provide food for ants, such as extra-floral nectar or food bodies. However, nutrient fluxes from ants to plants are less obvious, but by using ^{15}N-labelled glycine Fischer et al. (2003) discovered the presence of a nitrogen flux from ants to plants in the symbiosis of *P. bicornis* and *Piper fimbtiulatum* and *P. obliquum*. Nutrient transfer from ants to plants occurred remarkably quickly. Within 6 days up to 25 % of the nitrogen ingested by the ants was incorporated by the plants and accounted for a minimum daily input rate of 0.8 % of the plants' above-ground nitrogen uptake.

Spatial transport of resources by insect herbivores can also, at least hypothetically, be important when habitats of high and low productivity are in close proximity to one another. Empirical evidence for this is limited, but examples include a variety of insects that are important vectors of resource transport into caves, maintaining a diverse assemblage of microbial and faunal decomposers in the absence of autotrophs (Howarth 1983).

2.4.2 Herbivory and Plant Biomass

Insects have an obvious impact on plant biomass – they remove it! In some systems, insect herbivores may remove more biomass than any other type of herbivore. For example, in the Neotropics leaf-cutter ants remove 17 % of leaf biomass, more than any other taxon (Cherret 1986), whilst in savannahs termites can remove up to 16 % of annual primary production (Lavelle 1997). One area of current debate is the relative impact of insect vs vertebrate herbivory on plant productivity. This has been calculated for the Serengeti system: on average, insect herbivores consume about 6 % of the annual plant production, vertebrate herbivores about 7 % and decomposers a massive 87 % (Scholes and Walker 1993). A controversial recent meta-analysis by Bigger and Marvier (1998) reviewed 246 experiments where herbivores had been excluded and looked at the effects on plants. They concluded that protected plants were significantly larger than those exposed to herbivores, and that insect herbivores were found, on average, to have a larger impact on plant growth than vertebrate ones. Some doubt the validity of this latter result because of the confounding effects of most of the insect studies being on one plant species and the vertebrate studies being on many species, some of which were positively affected and others negatively, hence effectively 'cancelling out' the vertebrate impacts. Many authors consider that the impact of vertebrate herbivores is greater, particularly in terms of their impact on plant population dynamics and community structure (Crawley 1989).

We may speculate whether the positive effects of insect herbivores on plant productivity, mediated by changes in nutrient cycling, outweigh their negative effects on plant productivity due to biomass removal. This debate was addressed by Dyer et al. (1993) who concluded that herbivores induce non-linear growth and development in plants, which means that defoliation can cause a biphasic community response, such that at low levels of herbivory plant communities show an increase in productivity, but extreme herbivory causes a reduction in productivity. They further argued that the transition between these two states depends on temporal and spatial variation within the particular system, as well as on the relative impacts of herbivory on carbon and nitrogen availability (Holland and Detling 1990).

On average, in terrestrial systems, only 18 % of plant biomass is consumed by herbivores, whether insect or vertebrate (Cyr and Pace 1993). However, all attempts to measure the impacts of herbivory on plant productivity are likely to be an underestimate because some of the effects of insect herbivores are relatively neglected (Weisser and Siemann, Chap. 1, this Vol.). For example, the impacts of root-feeding herbivores are easy to underestimate (Hunter 2001), but they are still able to reduce nutrient uptake and hence plant productivity, particularly in plants with nitrogen-fixing root nodules (Murray et al. 1996; Teixeira et al. 1996; Masters, Chap. 5, this Vol.). In fact, there is at least one case where removal of root tissue has been shown to have a greater effect on plant performance than the removal of shoot tissue, although this was a study using simulated herbivory (Reichman and Smith 1991). In grassland systems, the plant production consumed by herbivores below ground may be between 3 and 7 times the amount consumed above ground (Scott et al. 1979).

There is still a relative lack of quantitative data on the comparative impact of root and shoot herbivory, but there have been rather more studies comparing the relative impacts of different types of aboveground herbivores on plant productivity. For example, in a comparison of the impact of three species of insect herbivores on goldenrod *Solidago altissima*, Meyer (1993) found, perhaps surprisingly, that the xylem-sucking spittlebug *Philaenus spumarius* was the most damaging, the leaf-chewing beetle *Trirhabda* sp. was intermediate in impact and the phloem-sucking aphid *Uroleucon caligatum* was least damaging in terms of effects on reducing plant biomass. It is possible that the impacts of the less obvious and easy to study insect guilds, such as xylem feeders, have been relatively overlooked compared to leaf chewers (Stadler et al., Chap. 11, this Vol.).

Another relatively neglected aspect of herbivory that can impact on plant productivity is early leaf abscission. In tropical systems, where 75 % of leaf damage occurs on leaves before full expansion (Coley and Aide 1991; Coley and Barrone 1996) and competition for light is intense, plants often abscise young leaves that have become damaged. Early leaf abscission allows the plant to shed damaged leaves, which are less efficient at light capture and can often be the site of pathogen attack and water loss, and divert resources to undamaged leaves (Blundell and Peart 2000). A recent test of this using Dipterocarp seedlings in Malaysian Borneo (F.P. Massey, unpublished) demonstrated that approximately 75 % of unexpanded leaves that were damaged by insect herbivores were abscised. Thus the photosynthetic leaf area lost to herbivores may greatly exceed the biomass eaten. Another good example is that of leaf-miners: the mine itself removes very little green leaf area but the effect on plant productivity may be much greater because many mined leaves are abscised (Faeth et al. 1981; Simberloff and Stiling 1989). To detect these effects requires studies involving careful, frequent measurements on large samples of individually marked leaves, but these have been rare. One study on birch (*Betula pendula*) measured each leaf on six trees every day and found that 17 % of leaves produced between April and mid-August fell prematurely or were consumed entirely. All these leaves, and hence their contribution to any measure of biomass lost to herbivores, would have been missed in a single survey of 'damage' by herbivores (Zakaria 1989). Studies of the effects of herbivory on plant biomass are usually restricted to a single season, or at most a few seasons, so longer-term impacts of herbivory on plant productivity may be missed. A recent study using tree-ring analysis demonstrated that the growth rate of juvenile Pinyon pines determined their vulnerability to a stem-boring moth (*Dioryctria albovitella*) when mature. Faster growing juveniles had 11 times more shoots killed by this moth when they matured than more slowly growing trees (Ruel and Whitham 2002).

As well as their impact on total plant biomass, there are more subtle effects of insect herbivores on plant productivity. For example, many types of herbivory, particularly sap-sucking, alter root:shoot ratios (Crawley 1997; Voelckel and Baldwin, Chap. 17, this Vol.). Many other examples of effects on plant allocation patterns exist. Herbivory can alter the balance between sources and sinks within a plant that in turn affects photosynthetic rates. Again sap-sucking insects often produce marked changes in source-sink ratios (e.g. Meyer and Whitlow 1992), but one of the most striking impacts in this regard is the effect of gall-formers. These herbivores cause the plant to produce what are often very large structures. The galls are effective sinks that attract and consume a large proportion of the plant's resources (Abrahamson and Weis 1986); one of the benefits suggested for the galling habit is that gall tissue is higher in nutrients than the surrounding plant tissue (Price et al. 1987; but see Hartley and Lawton 1992). Gall-insects appear to be able to manipulate the levels of nutrients within the gall. For example, galls that form on the vascular system of leaves can alter the flow of resources within the leaf, although their ability to do this appears to depend on their precise location with respect to the midrib (Hartley 1998). Leaf-miners can also alter the vascular flow within leaves because they frequently cut the midrib – this can have severe consequences for sap-feeding insects sharing the same leaf (Johnson et al. 2002). Although insect herbivores, even gall-formers and leaf-miners, have less impact on plant morphology and allocation patterns than vertebrate herbivores, they can alter plant structure quite markedly if they damage the apical bud. Gall-formers often attack buds and can have striking effects on plant morphology (Shorthouse and Rofritsch 1992). Aphids which damage apical meristems cause increased branching in their host plants, which in turn leads to an increase in shoots which can be attacked by the aphids (Pilson 1992). Shoot borers can also have a major impact on plant morphology, and indeed mortality, by causing damage to the leading shoot and/or an increase in secondary branching (Sullivan 2003).

Mattson and Addy (1975) tried to take a 'forest-wide' perspective and reviewed the role of phytophagous insects as regulators of primary productivity in forests. They concluded that the impact of insect herbivores on primary productivity depended on the intensity of defoliation, the quantity and distribution of photosynthetic biomass and variations in net photosynthetic rate under different levels of damage and under different environmental conditions. Most attempts to quantify the impact of insect herbivores on above-ground plant productivity do not come close to measuring all these parameters, and effects on belowground productivity are usually ignored altogether. Despite the lack of appropriate data, Mattson and Addy (1975) felt that, on average, insect herbivores consumed around 10 % of the total net primary production in forests. However, they also noted that there were periodic outbreaks of certain forest Lepidoptera which removed 100 % of leaf biomass. Even this substantial loss of plant productivity did not usually cause mass mortality of trees unless outbreaks persisted over 3–5 years. More usually, the insect population crashed during the second or third year of an outbreak and the trees recovered without suffering any substantial mortality (Stalter and Serrao 1983). In a similar system, outbreaks of Douglas-fir tussock moths (*Orgyia pseudotsugata*) in California white-fir (*Abies concolor*) stands did at first lead to a decrease in tree growth, but in the longer term (over several decades) the outbreaks led to an 80 % increase in growth (Wickman and Starr 1990). The authors attributed this to tree thinning and increased nutrient availability. Thus in these forest systems at least, even severe defoliation by insect herbivores may have no long-term adverse effects on plant growth or mortality.

2.5 Conclusions

In this chapter, we have highlighted the numerous ways that insect herbivores can affect ecosystem process, namely decomposition, nutrient cycling and plant productivity. Insect herbivores have both direct and indirect effects on these processes. Perhaps one of their most important roles is to physically break up organic matter within the soil and make it accessible to fungi/bacteria. This means that Collembola, termites and ants have a major impact on belowground processes and nutrient cycling, usually by changing soil structure and decomposition and mineralization rates (Jones and Bradford 2002). Although these herbivores are very small, their huge abundance means they have an impact out of all proportion to their size – hence termites can make a contribution to the greenhouse effect! Above ground, the potential roles of insects in ecosystems are less well characterized, but in some systems at least they may have dramatic, although often temporary, effects on primary production (Schowalter et al. 1986; Dyer et al. 1993) by intense grazing pressure. In addition, by dropping leaf fragments and faeces to the forest floor, canopy-dwelling insects may return nutrients from the tops of trees to the litter and soil (Seastedt and Crossley 1980; Lovett and Ruesink 1995). Thus, insect herbivores not only influence the total amount of nutrients available to plants, but also cause temporal and spatial variation in nutrient availability. Insect herbivores also influence the growth, recruitment and mortality of plants. Again, the few studies that have attempted to assess these aspects suggest that the impact of insect herbivores is as large as vertebrates, if not larger (Bigger and Marvier 1998); in fact, given that insects make up some of the decomposer community and that much of the plant biomass ends up with them and not with herbivores, it seems likely that the impact of insect herbivores on ecosystem processes far exceeds that of vertebrate herbivores (Scholes and Walker 1993).

Generally, the importance of insect herbivores to ecosystem function seems likely to have been underestimated. Many of their impacts, particularly below ground, are difficult to measure. However, recent advances in labelling techniques have led to a better understanding of some of the impacts of insect herbivores on processes such as decomposition and nutrient cycling. For example, it has now been shown that ants feed plants as well as the other way around. Greater interest in ecosystem function, soil biodiversity and the impacts of belowground herbivores has led to new conceptual models, and to better tests of these models. We have much more data with which to assess a model of the interactions between foliar- and root-feeding herbivores (Masters et al. 1993; Masters, Chap. 5, this Vol.) and we now suspect that one prediction of the model, namely that foliar herbivory reduces the populations of belowground herbivores, may be wrong (Blossey and Hunt-Joshi 2003).

We have also highlighted some of the progress that has been made in the study of the role of insect herbivores in ecosystem function. We have examined some intriguing interactions between insect herbivores and other organisms, and drawn attention to some neglected impacts on ecosystem processes. In addition, we have highlighted some key issues which remain unresolved. For example, have practical developments like better labelling techniques advanced the field further than new conceptual models? Do herbivores or detritivores have the biggest impact on ecosystem function – are we any nearer to answering this question than when Mattson and Addy suggested herbivores were the most important in 1975? Most plants have most of their biomass below ground but we don't know what happens to it – what are the relative impacts of shoot herbivory vs root herbivory on plants? We still do not have adequate data to understand how the impact of insect defoliation on nitrogen cycling varies according to insect species or insect abundance – we cannot assume that all forest ecosystems are like the north-eastern USA! We do know that insect herbivores are vitally important to ecosystem functioning, but there is still a lot that we do not know.

References

Abrahamson WG, Weis AE (1986) Nutritional ecology of arthropod gall-makers. In: Slansky F, Rodriguez JG (eds) Nutritional ecology of insects mites and spiders. Academic Press, San Diego, pp 235–258

Andersen DC (1987) Below-ground herbivory in natural communities: a review emphasizing fossorial animals. Q Rev Biol 62:261–286

Bardgett RD, Keiller S, Cook R, Gilburn AS (1998a) Dynamic interactions between soil animals and microorganisms in upland grassland soils amended with sheep dung: a microcosm experiment. Soil Biol Biochem 30:531–539

Bardgett RD, Wardle DA, Yeates GW (1998b) Linking above-ground and below-ground interactions: how plant responses to foliar herbivory influence soil organisms. Soil Biol Biochem 30:1867–1878

Basaglia M, Concheri G, Cardinali S, Pasti-Grigsby MB, Nuti MP (1992) Enhanced degradation of ammonium-pretreated wheat straw by lignocelluloytic *Streptomycestes* spp. Can J Microbiol 38:1022–1025

Beattie AJ, Hughes LH (2002) Ant–plant interactions. In: Herrera CM, Pellmyr O (eds) Plant–animal interactions: an evolutionary approach. Blackwell, Oxford, pp 211–235

Beyers RJ, Odum HT (1993) Ecological microcosms. Springer, Berlin Heidelberg New York

Bigger PS, Marvier MA (1998) How different would a world without herbivory be? A search for generality in ecology. Integrative Biol 1:60–67

Bignell DE, Eggleton P, Nunes L, Thomas KL (1997) Termites as mediators of carbon fluxes in tropical forests: budgets for carbon dioxide and methane emissions. In: Watt AD, Stork NE, Hunter MD (eds) Forests and insects. Chapman and Hall, London, pp 109–134

Blomqvist MM, Olff H, Blaauw MB, Bongers T, Van der Putten WH (2000) Interactions between above- and belowground biota: importance for small-scale vegetation mosaics in a grassland ecosystem. Oikos 90:582–598

Blossey B, Hunt-Joshi TR (2003) Belowground herbivory by insects: influence on plants and aboveground herbivores. Annu Rev Entomol 48:521–547

Blundell AG, Peart DR (2000) High abscission rates of damaged expanding leaves: field evidence from seedlings of a Bornean rain forest tree. Am J Bot 78:1693–1698

Boddy L, Wells JM, Culshaw C (1999) Fractal analysis in studies of mycelium in soil. Geoderma 88:301–328

Bokhari UG, Singh JS (1974) Effects of temperature and clipping on growth, carbohydrate reserves and root exudation of western wheatgrass in hydroponic culture. Crop Sci 14:790–794

Bormann FH, Likens GE (1979) Patterns and process in a forested ecosystem. Springer, Berlin Heidelberg New York

Bradford MA, Jones TH, Bardgett RD, Black HIJ, Boag B, Bonkowski M, Cook R, Eggers T, Gange AC, Grayston SJ, Kandeler E, McCaig AE, Newington JE, Prosser JI, Setala H, Staddon PL, Tordoff GM, Tscherko D, Lawton JH (2002) Impacts of soil faunal community composition on model grassland ecosystems. Science 298:615–618

Bredowski JJM, Zeilinga R (1987) Transition from heathland to grassland: damaging effects of the heather beetle. J Ecol 75:159–175

Breznak JA (1982) Intestinal microbiota of termites and other xylophagous insects. Annu Rev Microb 36:323–343

Burger AE (1978) Terrestrial invertebrates: a food resource for birds at Marion Island. S Afr J Ant Res 8:87–99

Callaham MA, Whiles MR, Meyer CK, Brock BL, Charlton RE (2000) Feeding ecology and emergence production of annual cicadas (Homoptera: Cicadidae) in tallgrass prairie. Oecologia 123:535–542

Cherret JM (1986) History of the leaf-cutting ant problem. In: Lofgren CS, Vander Meer RK (eds) Fire ants and leaf-cutting ants: biology and management. Westview Press, Boulder, Colorado, pp 10–17

Choudhury D (1984) Aphids and plant fitness: a test of Owen and Weigert's hypothesis. Oikos 43:401–402

Christenson LM, Lovett GM, Mitchell MJ, Groffman PM (2002) The fate of nitrogen in gypsy moth frass deposited to an oak forest floor. Oecologia 131:444–452

Coleman DC (1976) A review of root production processes and their influence on soil biota in terrestrial ecosystems. In: Anderson JM, Macfadyen A (eds) The role of terrestrial and aquatic organisms in decomposition processes. Blackwell, London, pp 417–434

Coley PD, Aide TM (1991) A comparison of herbivory and plant defences in temperate and tropical broad-leaved forest. In: Price PW, Lewinsohn TM, Fernandes GW, Benson WW (eds) Plant–animal interactions: evolutionary ecology in tropical and temperate regions. Wiley, New York, pp 25–49

Coley PD, Barrone JA (1996) Herbivory and plant defenses in tropical forests. Annu Rev Ecol Syst 27:305–335

Cornelissen JHC (1996) An experimental comparison of leaf decomposition rates in a wide range of temperate plant species and types. J Ecol 84:573–582

Crawley MJ (1989) Insect herbivores and plant population dynamics. Annu Rev Entomol 34:531–564

Crawley MJ (1997) Plant ecology, 2nd edn. Blackwell Science, Oxford

Cyr H, Pace ML (1993) Magnitude and patterns of herbivory in aquatic and terrestrial ecosystems. Nature (Lond) 361:148–150

Denton CS, Bardgett RD, Cook R, Hobbs PJ (1999) Low amounts of root herbivory positively influence the rhizosphere microbial community in a temperate grassland soil. Soil Biol Biochem 31:155–165
Dyer ML, Acra MA, Wang GM, Coleman DC, Freckman DW, McNaughton SJ, Strain BR (1991) Source-sink carbon relations in two *Panicum coloratum* ecotypes in response to herbivory. Ecology 72:1472–1483
Dyer ML, Turner CL, Seastedt TR (1993) Herbivory and its consequences. Ecol Appl 3:10–16
Eggers T, Jones TH (2000) You are what you eat...or are you? TREE 15:265–266
Eggleton P, Bignell DE, Sands WA, Mawdsley NA, Lawton JH, Wood TG, Bignell NC (1996) The diversity, abundance and biomass of termites under differing levels of disturbance in the Mbalmayo Forest Reserve, southern Cameroon. Philos Trans R Soc Lond Ser B 351:51–68
Eshleman KN, Morgan RP, Webb JR, Deviney FA, Galloway JN (1998) Temporal patterns of nitrogen leakage from mid-Appalachian forested watersheds: role of insect defoliation. Water Resour Res 34:2005–2016
Faeth SH, Conner EF, Simberloff D (1981) Early leaf abscission: a neglected source of mortality for foliovores. Am Nat 177:409–415
Fischer RC, Richter R, Wanek W, Mayer V (2002) Plants feed ants: food bodies of myrmecophytic *Piper* and their significance for the interactions with *Pheidole bicornis* ants. Oecologia 133:186–192
Fischer RC, Wanek W, Richter R, Mayer V (2003) Do ants feed plants? A ^{15}N labelling study of nitrogen fluxes from ants to plants in the mutualism of *Pheidole* and *Piper*. J Ecol 91:126–134
Fogel R, Hunt G (1979) Fungal and arboreal biomass in a western Oregon Douglas fir ecosystem. Can J Forest Res 9:245–256
Folgarait PJ (1998) Ant biodiversity and its relationship to ecosystem functioning: a review. Biodiv Conserv 7:1221–1244
Grace JR (1986) The influence of gypsy moth on the composition and nutrient content of litter fall in a Pennsylvania oak forest. For Sci 32:855–870
Groombridge B (ed) (1992) Global biodiversity. status of the earth's living resources. Chapman and Hall, London
Hartley SE (1998) The chemical composition of plant galls: are levels of nutrients and secondary compounds controlled by the gall-former? Oecologia 113:492–501
Hartley SE, Firn RD (1989) Phenolic biosynthesis, leaf damage, and insect herbivory in birch (*Betula pendula*). J Chem Ecol 15:275–283
Hartley SE, Gardner SM (1995) The response of *Philaenus spumarius* (Homoptera: Cercopidae) to fertilising and shading a moorland host plant (*Calluna vulgaris*). Ecol Entomol 20:396–399
Hartley SE, Jones CG (1997) Plant chemistry and herbivory; or why the world is green. In: Crawley MJ (ed) Plant ecology, 2nd edn. Blackwell, Oxford, pp 284–324
Hartley SE, Lawton JH (1992) Host-plant manipulation by gall insects: a test of the nutrition hypothesis. J Anim Ecol 61:113–119
Hartley SE, Gardner SM, Mitchell RJ (2003) Indirect effects of grazing and nutrient addition on the Hemiptera community of heather moorlands: the role of soil type, vegetation structure and plant species richness. J Appl Ecol 40:793–803
Holland EA, Detling JK (1990) Plant response to herbivory and belowground nutrient cycling. Ecology 71:1040–1049
Holland JN (1995) Effects of above ground herbivory on soil microbial biomass in conventional and no-tillage agroecosystems. Appl Soil Ecol 2:275–279
Holland JN, Cheng WX, Crossley DA (1996) Herbage-induced changes in plant carbon allocation: assessment of below ground carbon fluxes using carbon-14. Oecologia 107:87–94

Hollinger DY (1986) Herbivory and the cycling of nitrogen and phosphorus in isolated California oak tree. Oecologia 70:219–297
Hopkin SP (1998) Biology of the springtails. Insecta: Collembola. Oxford University Press, Oxford, 330 pp
Houghton JT, Jenkins GJ, Ephraums JJ (1990) Climate change: the IPCC scientific assessment. Cambridge University Press, Cambridge
Howarth FG (1983) Ecology of cave arthropods. Annu Rev Entomol 28:365–389
Hunter MD (2001) Out of sight, out of mind: the impacts of root-feeding insects in natural and managed systems. Agric Forest Entomol 3:1–8
Inbar M, Eshel A, Wool D (1995) Interspecific competition among phloem-feeding insects mediated by induced host plant sinks. Ecology 76:1506–1515
Intergovernmental Panel on Climate Change (2001) Climate change 2001: the scientific basis. Cambridge University Press, Cambridge
Jakobsen I, Rosendahl L (1990) N_2 fixation and root respiration in pea – effects of VAM and phosphorus. Agric Eco Environ 29:205–209
Johnson SN, Mayhew PJ, Douglas AE, Hartley SE (2002) Insects as leaf engineers – can leaf-miners alter leaf structure for birch aphids? Funct Ecol 16:575–584
Jones TH, Bradford MA (2002) Assessing the functional implications of soil biodiversity in ecosystems. Ecol Res 16:845–858
Jones TH, Thompson LJ, Lawton JH, Bezemer TM, Bardgett RD, Blackburn TM, Bruce KD, Cannon PF, Hall GS, Hartley SE, Howson G, Jones CG, Kampichler C, Kandeler E, Ritchie DA (1998) Impacts of rising atmospheric carbon dioxide on model terrestrial ecosystems. Science 280:221–223
Jones TH, Bezemer TM, Korner C, Lawton JH, Thompson LJ (2000) Comparing studies of artificial and natural ecosystem responses to CO_2 enrichment. Biotronics 29:1–7
Karban R, Baldwin IT (1997) Induced responses to herbivory. University of Chicago Press, Chicago
Karhu KJ, Neuvonen S (1998) Wood ants and a geometrid defoliator of birch: predation outweighs beneficial effects through the host plant. Oecologia 113:509–516
Klironomos JN, Kendrick WB (1995) Stimulative effects of arthropods on endomycorrhizas of sugar maple in the presence of decaying litter. Funct Ecol 90:528–536
Klironomos JN, Ursic M (1998) Density-dependent grazing on the extraradical hyphal network of the arbuscular mycorrhizal fungus, *Glomus intraradices*, by the collembolan, *Folsomia candida*. Biol Fert Soils 26:250–253
Lavelle P (1997) Faunal activities and soil processes: adaptive strategies that determine ecosystem function. Adv Ecol Res 27:93–132
Lawton JH (2000) Community ecology in a changing world. In: Kinne O (ed) Excellence in ecology. Ecology Institute, Oldendorf, Germany
Loreau M, Naeem S, Inchausti P (eds) (2002) Biodiversity and ecosystem functioning. Synthesis and perspectives. Oxford University Press, Oxford
Lovett GM, Ruesink AE (1995) Carbon and nitrogen mineralization from decomposing gypsy moth frass. Oecologia 104:133–138
Lovett GM, Christenson LM, Groffman PM, Jones CG, Hart JE, Mitchell MJ (2002) Insect defoliation and nitrogen cycling in forests. Bioscience 52:335–341
Lussenhop J (1996) Collembola as mediators of microbial symbiont effects upon soybean. Soil Biol Biochem 28:363–369
Martin MM (1991) The evolution of cellulose digestion in insects. Phil Trans R Soc Lond Ser B: Biol Sci 333:281–288
Masters GJ, Brown VK, Gange AC (1993) Plant mediated interactions between above- and below-ground herbivores. Oikos 66:48–151
Mattson WJ, Addy ND (1975) Phytophagous insects as regulators of forest primary production. Science 190:515–522

McGinley MA, Dhillion SS, Neumann JC (1994) Environmental heterogeneity and seedling establishment: ant-plant-microbe interactions. Funct Ecol 8:607–615
McNaughton SJ, Banyikwa FF, McNaughton MM (1998) Root biomass and productivity in a grazing system: the Serengeti. Ecology 79:587–592
Meyer GA (1993) A comparison of the impacts of leaf feeding and sap feeding insects on the growth and allocation of goldenrod. Ecology 74:1101–1116
Meyer GA, Whitlow TH (1992) Effects of leaf and sap feeding insects on photosynthetic rates of goldenrod. Oecologia 92:480–489
Mikola J, Yeates GW, Barker GM, Wardle DA, Bonner KI (2001) Effects of defoliation intensity on soil food web properties in an experimental grassland community. Oikos:92 333–343
Moorhead DL, Reynolds JF (1991) A general model of litter decomposition in the Northern Chihuahuan Desert. Ecol Model 56:197–220
Murray PJ, Hatch DJ, Cliquet JB (1996) Impact of insect herbivory on the growth and nitrogen and carbon contents of white clover (*Trifolium repens* L.) seedlings. Can J Bot 74:1591–1595
Northrup RR, Yu Z, Dahlgren RA, Vogt KA (1995) Polyphenol control of nitrogen release from pine litter. Nature (Lond) 377:227–229
Oksanen L, Aunapuu M, Oksanen T, Scneider M, Ekerholm P, Lundberg PA, Armulik T, Aruoja V, Bondestad L (1997) Outlines of food webs in a low arctic tundra landscape in relation to three theories of trophic dynamics. In: Gange A, Brown VK (eds) Multitrophic interactions in terrestrial ecosystems. Blackwell, Oxford, pp 351–374
Petersen H., Luxton M (1982) A comparative analysis of soil fauna populations and their role in decomposition processes. Oikos 39:287–388
Pilson D (1992) Aphid distribution and the evolution of goldenrod resistance. Evolution 46:1358–1372
Price PW, Fernandes GW, Waring GL (1987) Adaptive nature of insect galls. Environ Entomol 16:15–24
Read DJ (1991) Mycorrhizas in ecosystems. Experientia 47:376–391
Reichman OJ, Smith SC (1991) Responses to simulated leaf and root herbivory by a biennial *Tragopogon dubius*. Ecology 72:116–124
Reynolds BC, Hunter MD (2001) Responses of soil respiration, soil nutrients and litter decomposition to inputs from canopy herbivores. Soil Biol Biochem 33:1641–1652
Reynolds BC, Crossley DA, Hunter MD (2003) Response of soil invertebrates to forest canopy inputs along a productivity gradient. Pedobiologia 47:127–139
Ritchie ME, Tilman D, Knoops JMH (1998) Herbivore effects on plant and nitrogen dynamics in oak savanna. Ecology 79:165–177
Ruel J, Whitham TG (2002) Fast growing juvenile pinions suffer greater herbivory when mature. Ecology 83:2691–2699
Ruess RW (1998) The interaction of defoliation and nutrient uptake in *Sporobolus kentrophyllus*, a short-grass species from the Serengeti plains. Oecologia 77:550–556
Ruess RW, Hendrick RL, Bryant JP (1998) Regulation of fine root dynamics by mammalian browsers in early successional Alaskan taiga forests. Ecology 79:2706–2720
Rusek J (1998) Biodiversity of Collembola and their functional role in the ecosystem. Biodiv Conserv 7:1207–1219
Scheu S, Falca M (2000) The soil food web of two beech forests (*Fagus sylvatica*) of contrasting humus type: stable isotope analysis of a macro- and a mesofauna dominated community. Oecologia 123:285–296
Scheu S, Theenhaus A, Jones TH (1999) Links between the detritivore and herbivore system: effects of earthworm and Collembola on plant growth and aphid development. Oecologia 119:541–551

Scholes RJ, Walker BH (1993) An African savanna: synthesis of the Nylsvley study. Cambridge University Press, Cambridge
Schowalter TD (1981) Insect herbivore relationship to the state of the host-plant: biotic regulation of ecosystem nutrient cycling through ecological succession. Oikos 37:126–130
Schowalter TD, Hargrove WW, Crossley DA (1986) Herbivory in forested ecosystems. Annu Rev Entomol 31:177–196
Scott JA, French NR, Leetham JW (1979) Patterns in consumption in grasslands. In: French NR (ed) Perspectives in grassland ecology. Springer, Berlin Heidelberg New York, pp 89–105
Seastedt TR, Crossley DA (1980) Effects of microarthropods on the seasonal dynamics of nutrients in forest litter. Soil Biol Biochem 12:337–342
Seastedt TR, Crossley DA (1984) The influence of arthropods on ecosystems. Bioscience 34:157–161
Seastedt TR, Ramundo RA, Hayes DC (1988) Maximisation of densities of soil animals by foliage herbivory – empirical evidence, graphical and conceptual models. Oikos 51:243–248
Shorthouse JD, Rofritsch O (1992) Biology of insect-induced galls. Oxford University Press, New York
Simberloff D, Stiling P (1989) Leaf abscission: induced defence against pests or response to damage? Oikos 55:43–49
Smith VR, Steenkamp M (1990) Climate change and its ecological implications at a sub Antarctic island. Oecologia 85:14–24
Speight MR, Hunter MD, Watt AD (1999) Ecology of insects. concepts and applications. Blackwell, Oxford
Stalter R, Serrao J (1983) The impact of defoliation by gypsy moths on the oak forest at Greenbrook Sanctuary, New Jersey. Bull Torrey Bot Club 110:526–529
Sullivan JJ (2003) Density-dependent shoot-borer herbivory increases the age of first reproduction and mortality of Neotropical tree seedlings. Oecologia 136:96–106
Swank WT, Waide JB, Crossley DA, Todd RL (1981) Insect defoliation enhances nitrate export from forest ecosystems. Oecologia 51:297–299
Swift MJ, Heal OW, Anderson JM (1979) Decomposition in terrestrial ecosystems. Studies in ecology volume 5. Blackwell, Oxford
Teixeira MLF, Coutinho HLC, Franco AA (1996) Effects of *Cerotoma arcuata* (Coleoptera Cyrosmelidae) on predation of nodules and on N_2 fixation of *Phaseolus vulgaris*. J Econ Entomol 89:165–169
Thimm T, Larink O (1995) Grazing preferences of some Collembola for ectomycorrhizal fungi. Biol Fertil Soils 19:266–268
Van der Putten WH, Vet LEM, Harvey JA, Wackers FL (2001) Linking above- and belowground multitrophic interactions of plants, herbivores, pathogens and their antagonists. TREE 16:547–554
Verhoef HA, de Goede RGM (1985) Effects of collembolan grazing on nitrogen dynamics in a coniferous forest. In: Fitter AH, Atkinson D, Read DJ, Usher MB (eds) Ecological interactions in the soil: plants, microbes and animals. Blackwell, Oxford, pp 367–376
Wardle DA (2002) Communities and ecosystems. linking the aboveground and belowground components. Monographs in population biology 34. Princeton University Press, Princeton
Wardle DA, Barker GM, Bonner KI, Nicholson KS (1998) Can comparative approaches based on ecophysiological traits predict the nature of biotic interactions and individual plant species effects in ecosystems? J Ecol 86:405–420

Wardle DA, Bonner KI, Barker GM (2002) Linkages between plant litter decomposition, litter quality, and vegetation responses to herbivores. Funct Ecol 16:585–595

Warnock AJ, Fitter AH, Usher MB (1982) The influence of a springtail *Folsomia candida* (Insecta, Collembola) on a mycorrhizal association of leek *Allium porrum* and the vesicular-arbuscular mycorrhizal endophyte *Glomus fasciculatus*. New Phytol 90:285–292

Waterhouse DF (1977) The biological control of dung. In: Eisner T, Wilson EO (eds) The insects. Scientific American, WH Freeman, San Francisco, pp 314–322

Webb JR, Cosby BJ, Deviney FA, Eshleman KN, Galloway JN (1995) Change in the acid-base status of an Appalachian catchment following forest defoliation by the gypsy moth. Water Air Soil Pollut 85:535–540

Whitford WG, Stinnett K, Andersen J (1988) Decomposition of roots in a Chihuahuan Desert ecosystem. Oecologia 75:8–11

Wickman BE, Starr MR (1990) Mammoth Lakes revisited – 50 years after a Douglas-fir tussock moth outbreak. US Dep Agric Pac N W Res Note 498

Wood TG, Sands WA (1978) The role of termites in ecosystems. In: Brian MV (ed) Production ecology of ants and termites. Cambridge University Press, Cambridge, pp 212–243

Wurst S, Jones TH (2003) Indirect effects of earthworms (*Aporrectodea caliginosa*) on an above-ground tritrophic interaction. Pedobiologia 47:91–97

Yeates GW, Shepherd TG, Francis GS (1998) Contrasting response to cropping of populations of earthworms and predacious nematodes in four soils. Soil Till Res 48:255–264

Yoshimura T, Azumi JI, Tsunoda K, Takahashi M (1993) Changes of wood-attacking activity of the lower termite, *Coptotermes formosanus* Shiraki in defaunation–refaunation process of the intestinal Protozoa. Mater Organ 28:153–164

Zakaria S (1989) The influence of previous insect feeding on the rate of damage of birch tree leaves. PhD Thesis, University of London

3 Indirect Effects of Invertebrate Herbivory on the Decomposer Subsystem

D.A. Wardle and R.D. Bardgett

3.1 Summary

Invertebrate herbivores can exert important effects on the decomposer subsystem through a range of mechanisms. In this chapter, we review the mechanistic bases through which invertebrate herbivory may affect the quantity and quality of plant-derived resources entering the soil. We identify four main types of mechanisms through which this may occur: (1) herbivores can influence resource quantity, both in the short term through promoting rhizosphere exudation and in the long term through optimizing or reducing net primary productivity (NPP); (2) herbivores may affect litter quality, either positively through causing greater tissue nutrient concentrations or negatively through inducing plants to produce secondary defence compounds; (3) herbivores sometimes return a significant proportion of NPP to the soil as fecal material, which can have very different consequences to plant litter for decomposers; (4) in the longer term herbivores can significantly alter the functional composition of vegetation which can in turn determine the quality of litter returned to the soil. There are therefore numerous ways in which invertebrate herbivores can affect decomposers either positively or negatively, and these can exert important aboveground feedbacks. Some of the most significant effects of invertebrate herbivores in ecosystems occur during periodic population outbreaks, and the likely consequences of this for the decomposer subsystem are discussed. Usually, herbivores occur in multiple species communities, which leads to the question of how herbivore diversity affects decomposer processes; while there is a dearth of information available on the topic, there are plausible mechanisms whereby such effects could theoretically occur. It is concluded that, since all ecosystems depend upon both the producer and decomposer subsystems, a more complete understanding of ecosystem-level consequences of invertebrate herbivory can only be gained through the application of approaches that explicitly consider both subsystems, as well as the feedbacks between them.

Ecological Studies, Vol. 173
W.W. Weisser and E. Siemann (Eds.)
Insects and Ecosystem Function

3.2 Introduction

All terrestrial ecosystems consist of explicit aboveground and belowground components. Although these have often been treated in isolation from one another, both are required for the maintenance of ecosystem function. There is a growing recognition that a more complete understanding of how ecosystems function requires specific consideration of both of these subsystems, as well as the nature of interactions between them (Van der Putten et al. 2001; Wardle 2002). Animals that feed on producers, decomposers or other animals can exert important effects on producer and decomposer activity, and this may have far-ranging community- and ecosystem-level implications.

Herbivores, or the primary consumers of living plant material, may ingest from 1 to over 50 % of aboveground net primary productivity (NPP) depending on the ecosystem considered (McNaughton et al. 1989). They can therefore operate as powerful determinants of the amount of resources that enter the decomposer subsystem. Despite this, most work on herbivore effects in ecosystems has traditionally focused on aboveground responses, such as vegetation productivity, community structure and diversity. However, there is a growing recognition that the effects of both aboveground and belowground herbivores on plants can exert important belowground effects, with likely long-term aboveground consequences through altered supply rates of plant-available nutrients from the soil (reviewed by Bardgett et al. 1998; Wardle 2002; Bardgett and Wardle 2003; Hartley and Jones, Chap. 2, this Vol.).

In this chapter, we discuss how invertebrate herbivory may affect the decomposer subsystem through influencing the quality and quantity of resources entering the soil. In doing so, we assess mechanisms that function over various temporal and spatial scales, from individual plant parts to the whole plant community. We then evaluate the effects of multiple species invertebrate herbivore communities on the decomposer subsystem. Finally, we consider how the belowground effects of invertebrate herbivores may differ across different types of ecosystems.

3.3 Mechanistic Bases of Invertebrate Herbivore Effects

A variety of mechanisms have been identified concerning how herbivory may influence the decomposer organisms and processes, some of which are stimulatory and others inhibitory. The relative importance of these different mechanisms varies depending on context, and this explains why there are several examples of both promotion and reduction of decomposers and decomposition processes by herbivores reported in the literature (reviewed by Wardle 2002; Bardgett and Wardle 2003; Hartley and Jones, Chap. 2, this Vol.;

Table 3.1. Mechanistic bases through which foliar invertebrate herbivores may positively or negatively impact upon the quantity and quality of resources entering the decomposer subsystem. (Based on Table 1 of Wardle 2004)

Resource nature	Nature of effect	Positive effects of herbivores	Negative effects of herbivores
Quantity of resources	Short-term effects on rhizosphere carbon exudation	Enhancement of rhizosphere exudation	(None)
	Effects on net primary productivity (NPP)	Optimization of NPP by herbivory	Reduction of NPP through loss of plant tissues
Quality of resources	Effects on litter quality through physiological response by plant	Increased concentrations of nutrients and reduced secondary metabolites in shoot and root litter	Induced defences resulting in greater concentrations of secondary metabolites in litter; reduced litter nutrient concentrations
	Return of waste produces (e.g., faeces, honeydew)	Resources returned to soil in a form more labile than from plant litter	Reduced carbon to nitrogen ratio of faecal pellets; reduced porosity of pellets
	Effects on litter quality through influencing vegetation succession	Promotion of flora by herbivory preventing establishment by later successional plant species with poorer litter quality	Acceleration of succession through browsed plant species being disadvantaged and replacement by species with poorer litter quality

Bonkowski and Scheu, Chap. 4, this Vol.; Masters, Chap. 5, this Vol.). Here, we focus on the effects of invertebrate herbivores on the decomposer subsystem through the various mechanisms presented in Table 3.1.

3.3.1 Immediate Effects on Resource Quantity

Most studies relevant to understanding how invertebrate herbivores affect decomposers have focused on the effects of removal of plant parts, notably through defoliation. In the short term, loss of foliage from herbaceous plants can lead to large allocation of carbon to the rhizosphere (Hartley and Jones, Chap. 2, this Vol.). For example, pulse labelling studies by Dyer et al. (1991) and Holland et al. (1996) found that herbivory of monocotyledonous species by grasshoppers leads directly to a shift in carbon allocation from shoots to roots, resulting in both increased rhizosphere carbon exudation and enhanced total respiration in the rooting zone. This increased belowground allocation also appears to stimulate soil organisms, and explains positive responses of the soil microbial biomass to aboveground defoliation by grasshoppers (Holland 1995) and clipping (Mawdsley and Bardgett 1997). These effects appear to be multitrophic; Mikola et al. (2001b) showed clear positive effects of experimental defoliation on densities of both enchytraeids and microbe-feeding nematodes. Multitrophic effects are consistent with donor- rather than recipient-control of the decomposer food web.

Despite the importance of root herbivory in many ecosystems (notably grasslands), the effects of root herbivores on the decomposer subsystem have attracted little attention (Masters, Chap. 5, this Vol.). However, Yeates et al. (1998) found that plants of *Trifolium repens* grown in pots inoculated with each of five root-feeding nematode species showed greater release of ^{14}C into the rooting zone and uptake by the soil microbial biomass than when no nematodes were present. Further, root herbivory may facilitate the release of nutrients from plants, ultimately benefiting associated plant species. For example, Bardgett et al. (1999) found that infestations of the plant parasitic nematode *Heterodera trifolii* (which feeds only on leguminous roots) caused large increases in both soil microbial biomass and the transfer of ^{15}N from *T. repens* to neighbouring *Lolium perenne* plants.

3.3.2 Longer-Term Effects on Resource Quantity

In the longer term, the effects of physical removal of plant biomass can exert important effects on NPP, and hence the amount of material entering the decomposer subsystem as plant litter. Repeated removal of either foliar or root material usually has negative consequences for NPP. However, this is not a universal pattern and there are several instances in which intermediate levels of herbivory have been shown to enhance or optimize NPP, especially in grassland systems (McNaughton 1985; Holland et al. 1992). Most of the available evidence arises from studies involving browsing mammals, which are able to enhance NPP productivity through converting plant tissue to fecal material and thereby increasing nutrient availability for plants. Little is known about whether invertebrate herbivores may have similar effects. However, Belovsky and Slade (2000) provided evidence that grasshoppers at intermediate densities can promote grassland NPP through enhanced cycling of nutrients, compared with when they were absent or at high densities. Further, there is some evidence that invertebrate root herbivory can optimize plant productivity. Riedell (1989) found that root mass of *Zea mays* was greater when roots were infested with low densities of the root cutworm *Diabotrica* sp. than when the insect was absent. However, there is insufficient knowledge at present to assess the generality of this (see Brown and Gange 1990; Bardgett and Wardle 2003).

How altered NPP may affect the decomposer subsystem is not well understood; increasing NPP has been shown to exert either positive or negative effects on components of the soil food web depending on the situation considered (Wardle 2002). There are two possible mechanisms for these varied effects (Bardgett and Wardle 2003). First, the balance of top-down and bottom-up forces in structuring the decomposer food web is likely to be context dependent; only organisms regulated by bottom-up forces are likely to respond strongly to increases in resource availability. Some components of the decomposer food web are regulated by top-down forces while others are regulated by bottom-up forces (De Ruiter et al. 1995; Wardle 2002). Second, plants not only provide resources for soil microbes but also compete with them for nutrients, so an increase in plant growth could either increase or reduce the microflora depending on which of these mechanisms dominates. Therefore, any shift in NPP caused by invertebrate herbivores could exert either positive or negative effects on decomposers depending on the situation considered.

3.3.3 Effects of Changed Litter Quality

In most ecosystems the majority of NPP enters the soil as plant litter, and the quality of this litter (e.g., the concentrations of N, lignin and secondary metabolites) has a profound effect on the structure and functioning of the decomposer subsystem. Invertebrate herbivores induce physiological changes in plants which are in turn likely to influence the quality of litter that they produce, although this particular mechanism has received surprisingly little attention. However, it is likely that for grassland plants removal of live tissues by invertebrate herbivores would increase the concentration of nitrogen in the remaining tissues and therefore improve litter quality. The results of two previous studies support this. First, Seastedt et al. (1988) found that trimming of the shoots of the grass *Agropyron gerardii* caused an increase in root nitrogen concentrations, and suggested that this should in turn promote soil organisms. Second, Bardgett et al. (1999) found that root herbivory of grassland plant species by plant parasitic nematodes promoted the nitrogen content of foliar and root material.

Foliar herbivory by invertebrates may also induce plants to produce a range of antiherbivore defence compounds (Rhoades 1985; Agrawal 1998; Van der Putten et al. 2001), and while it is reasonable to expect that following their synthesis some of these compounds persist in the litter produced by these plants, much remains unknown about how this affects decomposers. Severe defoliation of deciduous trees by invertebrate herbivore outbreaks often results in greater concentrations of phenolics in subsequently produced tissues (Rhoades 1985), and given the role that phenolics can play in retarding decomposer-mediated processes (Northup et al. 1995; Wardle et al. 1997) it seems likely that litter produced by previously attacked trees would inhibit decomposers. Although direct evidence for such a mechanism is scarce, Findlay et al. (1996) found that damage by spider mites to *Populus deltoides* seedlings increased the concentration of phenolics in leaf tissues, and this in turn reduced the subsequent rate of decomposition of leaf litter produced by these plants.

3.3.4 Return of Invertebrate Waste Products

In ecosystems where herbivores remove a significant proportion of NPP, much of the plant-derived organic material is not added to the soil as leaf litter but instead as fecal material. The belowground effects of this are best understood for mammalian herbivores, and in most cases mammalian fecal material has been shown to promote soil organisms and processes (Bardgett et al. 1998). However, large amounts of fecal material may also be returned to the soil following outbreaks of foliar invertebrate herbivores. Lovett and Ruesink (1995) and Lovett et al. (2002) provide evidence that frass produced by the gypsy moth (*Lymantria dispar*) during outbreaks in deciduous forests can cause significant stimulation of microbial growth through having a high concentration of labile carbon. This microbial growth in turn leads to net immobilization of nitrogen by the soil microflora, suggesting a major role for gypsy moth frass in determining the availability of nitrogen to defoliated forests. It is expected that while insect frass would lead to net nitrogen immobilization in infertile ecosystems, net mobilization should instead occur in more fertile systems such as grasslands. For example, the positive effect of grasshoppers on NPP reported by Belovsky and Slade (2000) is likely to be due to positive effects of frass on net nitrogen mineralization.

While chewing insects return organic matter as frass, sucking insects such as aphids and scale insects often secrete surplus ingested carbohydrates as honeydew. Honeydew sugars are usually extremely labile and readily utilized by microbes when they reach the ground. For example, Dighton (1978) found that addition of synthetic honeydew to soils at a rate comparable to that produced by lime aphids (*Eucalipteris tilidae*) on trees of *Tilia* spp. caused an increase in fungal and bacterial populations of 1.3 and 4.0 times respectively. Further, honeydew-producing aphids in central European *Picea abies* forests have been shown to cause substantial increases in dissolved organic carbon in soils, and appear to promote net immobilization of nitrogen by the soil microflora (Stadler and Michalzik 1998). Given the abundance of sucking insects in many ecosystems, honeydew effects on decomposer organisms and processes may be of widespread importance.

3.3.5 Effects of Changes in Vegetation Composition

Invertebrate herbivores also have the capacity to influence the decomposer subsystem through driving changes at the level of the plant community. Recent studies have shown that across both plant species and functional groups, plants that produce foliage that is more palatable to generalist invertebrate herbivores tend also to be those that produce faster decomposing litter (Fig. 3.1), suggesting that similar suites of plant traits influence the activity of both herbivores and decomposers. Plant species that are the most favourable for herbivores and decomposers tend to be those that dominate earlier in succession (Cates and Orians 1975; Grime 2001; Wardle 2002). Herbivory is likely to favour later successional vegetation when earlier successional species are disadvantaged, and earlier successional vegetation whenever optimization of this vegetation by herbivores occurs (Davidson 1993; Augustine and McNaughton 1998).

In this light, invertebrate herbivory can operate as an important determinant of plant community structure and vegetation succession. Brown and Gange (1990, 1992) found through the use of selective insecticides that root herbivores promoted succession by reducing the success of early successional forb species, while foliar herbivores retarded succession by reducing the success of grasses. In contrast, Wardle and Barker (1997) found that both aboveground and belowground invertebrate herbivores reduced earlier successional forbs relative to later successional grasses when herbivores were added to experimental plant communities. Meanwhile in synthesized plant communities, Buckland and Grime (2000) found that invertebrate herbivores, notably the slug *Deroceros reticulatum*, suppressed early successional fast-growing plant species in infertile conditions. However, in fertile conditions early successional species dominated when herbivores were present, presumably because of their ability to compensate for tissue loss by defoliation. The consequences for the decomposer subsystem of the effects of invertebrate herbivory on succession have not been explicitly addressed to date. However, it

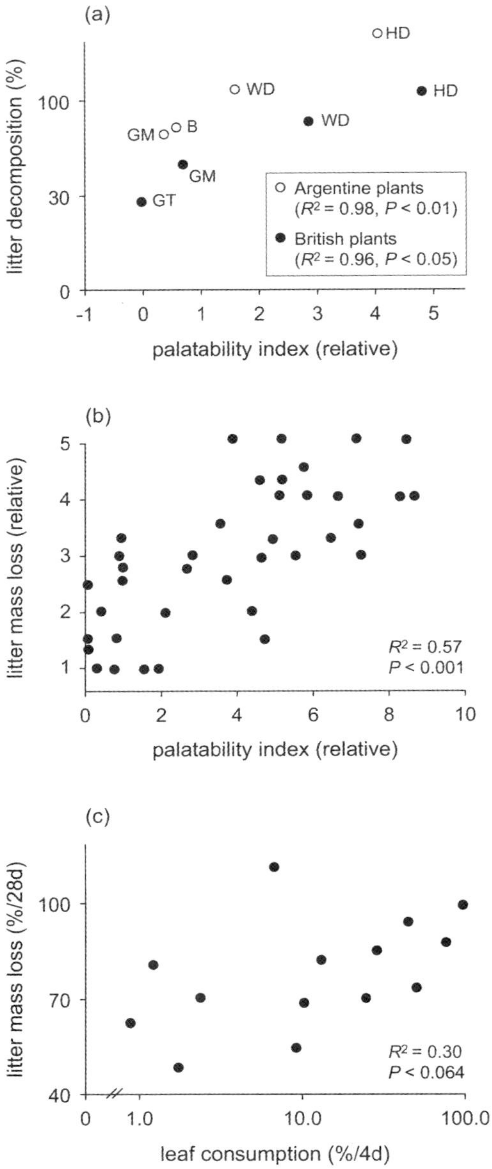

Fig. 3.1. Relationships between litter decomposability and palatability to generalist invertebrate herbivores across plant species. **a** Different plant life forms based on a data set involving 48 Argentinian and 72 British plant species. *HD* Herbaceous dicots; *WD* woody dicots; *GM* graminoid monocots; *B* bromeliads; *GT* gymnosperm trees. The herbivores used were generalist snails. (From Cornelissen et al. 1999.) **b** Forty-three plant species (each represented by a different *point*) representing a range of plant life forms. The herbivores used were *Helix aspersa* and *Acheta domestica*. (From Grime et al. 1996.) **c** Twenty dicotyledonous herbaceous species (each represented by a different *point*). The herbivore used was *Deroceros reticulatum*. (From Wardle et al. 1998.). (Figure reproduced from Wardle 2002, with permission from Princeton University Press)

appears reasonable that promotion of succession by invertebrates should adversely affect decomposers through reducing the quality of litter entering the soil (such as has been shown for mammalian herbivores; Pastor et al. 1988), while reversal of succession by invertebrates should have the opposite effect.

3.3.6 Feedbacks and Aboveground Consequences

Foliar herbivores affect plant nutrition and growth not just in the short term through physical removal of tissues, but also in the longer term by influencing feedbacks between the producer and decomposer subsystems. Herbivorous invertebrates have the capacity to determine the quality and quantity of organic materials produced by plants and added to the soil, which in turn affects decomposer activity and therefore the supply of nutrients to plants from the soil. Indeed, most studies indicating optimization of NPP by herbivores involve promotion by herbivores of nutrient cycling in the soil (McNaughton 1985; Belovsky and Slade 2000; De Mazancourt and Loreau 2000). As is apparent from the text this far, there are a range of mechanisms through which invertebrate herbivores may affect decomposers (some positive and some negative), so a range of possible effects of herbivores on plant growth, nutrient uptake and community structure may occur in the long term as a result of indirect herbivore effects on the decomposer subsystem. These feedbacks could in turn affect invertebrate herbivore performance, given recent findings that the positive effects of decomposer organisms on plant nutrient supply promote the growth and reproduction of foliar aphid species (Scheu et al. 1999; Bonkowski et al. 2001).

3.4 Significance of Invertebrate Herbivore Outbreaks

Invertebrate herbivores differ from vertebrate herbivores in that they generally have short life spans and rapid rates of reproduction, and therefore often undergo large fluctuations in abundance between seasons or between years. Under usual conditions, invertebrate populations may be expected to be low, partly because they can be regulated by top-down control (Hairston et al. 1960; Oksanen et al. 1981), and partly because of the inherently poor quality of their food resources (White 1978). However, populations of specific herbivorous invertebrate species can occasionally increase extremely rapidly, leading to population outbreaks, with severe effects on vegetation. These outbreaks are usually triggered directly or indirectly through changes in climatic conditions (e.g., Polis et al. 1997; Selas et al. 2001). It appears likely that normally, densities of invertebrate herbivores are insufficient for them to have a particularly important effect on the decomposer subsystem, and that significant belowground effects of these herbivores usually occur only during short, intense periods of population outbreak. Many of the mechanisms described in the preceding text are likely to be mainly manifested during outbreaks of these herbivores. For example, intense or repeated defoliations caused by outbreaks of the moths *L. dispar* in deciduous forests of eastern

North America or *Epirrita autumnata* in *Betula* forests of northern Europe result in death of tree stands, causing a pulse of inputs to the soil of recently killed plant material in the short term but reduced inputs through lower NPP in the longer term. Some outbreaks also result in a much higher proportion of NPP entering the decomposer subsystem as recently severed leaves or faecal pellets rather than as naturally senesced litter (Lovett et al. 2002). Further, non-lethal defoliation of *Betula* trees (such as often occurs during outbreaks of *E. autumnata*) leads to a poorer quality of subsequently produced foliage (and presumably leaf litter), with increased concentrations of secondary metabolites and decreased concentrations of nitrogen (Haukioja et al 1988). A better understanding of how invertebrate herbivores affect decomposition in many ecosystems would be gained by considering long-term temporal and spatial patterns of invertebrate herbivore abundance, and recognizing that the ecologically most important consequences of invertebrate herbivores in the long term could result from irregular and brief, and often spatially localized periods of significant herbivore outbreak (Carson et al., Chap. 10, this Vol.).

3.5 Multiple Species Herbivore Communities

While most studies on invertebrate herbivore effects on ecosystems have focused on single species, in real ecosystems several species of herbivore usually coexist. Therefore the issue arises as to whether increasing the diversity of invertebrate herbivores influences ecosystem processes, including those driven by the decomposer community. Although there has been considerable interest over the past decade in the ecosystem-level implications of biological diversity (Loreau et al. 2001), the issue of how invertebrate herbivore diversity affects terrestrial ecosystems remains essentially unexplored (but see Duffy et al. 2001 for a marine example). However, we would expect any effects of herbivore diversity on ecosystem processes to be promoted by the magnitude of differences among species in key functional traits and therefore the degree of interspecific resource partitioning, much as has been shown for plant diversity (Díaz and Cadibo 2001).

Although plant diversity can only influence production-driven processes through interspecific resource use complementarity in a positive direction, in the case of herbivores, diversity effects which are manifested through greater net resource use (i.e., consumption) can theoretically have a range of possible effects on ecosystem processes (Fig. 3.2a vs b; Bardgett and Wardle 2003). This arises because, as discussed earlier, there are a variety of ways through which invertebrate herbivores can affect the quality and quantity of resources entering the decomposer subsystem. Thus, an increasing net intensity of herbivory (such as may occur when herbivore diversity promotes greater resource uti-

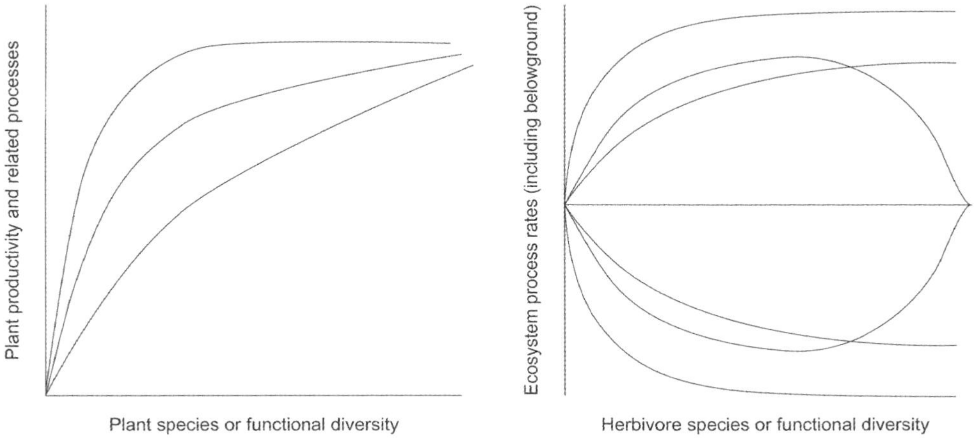

Fig. 3.2. *Left* Types of responses that may be expected from plant productivity and related ecosystem functions to increasing plant diversity. *Right* The response of ecosystem functions that can theoretically occur as a result of increasing herbivore diversity through resource use complementarity. Both positive and negative responses of belowground processes are possible because increasing net herbivory (resulting from greater resource use complementarity) can have either positive or negative effects on decomposer organisms depending upon which mechanisms are dominating. Further, unimodal responses can theoretically occur if plant productivity and nutrient cycling rates are optimized by intermediate levels of herbivory. (Figure reproduced from Bardgett and Wardle 2003, with permission from the Ecological Society of America)

lization) could conceivably have either positive or negative effects on decomposer processes depending on which mechanisms dominate. Further, herbivore diversity, and hence the amount of net consumption of NPP, could even result in unimodal effects on ecosystem processes, especially when these processes are optimized by intermediate levels of herbivory (Fig. 3.2b; Bardgett and Wardle 2003).

While Fig. 3.2 provides a theoretical framework for testable hypotheses about how invertebrate herbivores may affect the decomposer subsystem, little relevant data are available. However, it is well recognized that different invertebrate species in the same community partition resources, providing opportunities for the effects of diversity to manifest themselves. At the level of the whole plant, different herbivore species can partition resources both spatially (by specializing on different plant tissues) (Root 1973) and temporally (Masters et al 1993). At the plant community level, possible resource partitioning among invertebrate herbivore species can be inferred from studies that find herbivore species diversity correlates with plant species diversity

(Southwood et al 1979; Siemann 1998). The consequences of this partitioning for ecosystem functioning remain largely unexplored, although two studies provide relevant data. Mikola et al. (2001b) set up synthetic plant communities with three herbaceous species, and imposed treatments consisting of defoliating these species in all the possible one-, two- and three-way combinations. Although defoliation was found to influence belowground processes and organisms, there was no effect on the number of plant species that were defoliated. Wardle et al. (2000) amended synthesized plant communities with monocultures and two-species mixtures of foliar invertebrate herbivores and found that the above- and belowground effects of the two herbivore species mixtures did not differ from those of the corresponding one-species treatments.

Finally, we note that any effects of invertebrate herbivore diversity are likely to be context dependent. First, effects of herbivore diversity can only occur if resource partitioning occurs. This is most likely to take place when species comprising the herbivore community show a degree of specialization, and not when the community consists mainly of generalist feeders. Second, we anticipate that invertebrate herbivore diversity effects are most likely in systems where herbivores consistently affect the vegetation (see Schädler et al. 2003); for ecosystems in which herbivore effects are manifested mainly through intermittent population outbreaks, herbivore diversity effects may be less important given that outbreaks usually consist of a single species. However, we recognize that there are theoretical grounds for predicting that the diversity of herbivore communities may influence the chance of any one species reaching sufficient densities for an outbreak to occur.

3.6 Comparisons of Ecosystems

Given that different types of ecosystems differ tremendously in the functional composition of their vegetation, it is expected that the relative importance of herbivory as a driver of the decomposer subsystem should vary across ecosystems. This is borne out by the analyses of McNaughton et al. (1989) who found that for a global data set the proportion of NPP that is consumed by herbivores (and therefore does not enter the belowground subsystem as plant litter) increases with increasing productivity from less than 1 % to over 50 %. Further, Cebrian et al. (1998) and Cebrian (1999) showed, through a global synthesis of data from both terrestrial and aquatic systems, that ecosystems dominated by plants that are more productive and have higher tissue nitrogen and phosphorous concentrations also support greater rates of plant litter decomposition and have greater relative consumption of NPP by herbivores. Therefore, ecosystem effects of herbivory are expected to be greatest in productive ecosystems where a relatively high proportion of NPP enters the decomposer subsystem as fecal material, leading to optimization of NPP and promotion of grazing-tolerant plants with high litter quality. This would in turn lead to low net accumulation of organic matter and development of mull soils, together with domination of the soil food web by the bacterial-based energy channel and earthworms (Wardle 2002). In contrast, organic matter return in unproductive environments with low levels of herbivory would be in the form of poor litter quality, leading to the formation of thick moroid humus layers and domination of the soil food web by the fungal-based energy channel and microarthropods (Wardle 2002). Further, in unproductive environments, herbivory is likely to cause replacement of existing vegetation by less palatable plant species that produce litter of poorer quality, with negative implications for the decomposer subsystem.

It is unclear to what extent invertebrate vs vertebrate herbivores contribute to the total amount of herbivory encountered in most ecosystems. Although it is likely that in productive systems such as grasslands vertebrate herbivores usually consume more foliage than do invertebrates, Schädler et al. (2003) showed through meta-analysis of published manipulation experiments that invertebrate herbivores often exert important effects on vegetation in herbaceous plant communities. Further, in productive ecosystems, invertebrates often make a particularly significant contribution to the net amount of root herbivory. In ecosystems of intermediate productivity, such as temperate and boreal deciduous forests, invertebrates may exert particularly strong effects on ecosystem processes by causing large periodic shifts in vegetation composition as a result of single species outbreaks. Plant productivity and underlying determinants of NPP such as soil fertility and macroclimate are therefore key drivers of invertebrate herbivore densities in ecosystems, and thus the role that these herbivores play in influencing those processes that occur below ground.

3.7 Conclusions

It is apparent that there are a range of mechanisms by which invertebrate herbivores can indirectly influence the decomposer subsystem, through their effects on the quantity and quality of organic matter that plants produce. These effects operate at a range of spatial and temporal scales, and can have either positive or negative consequences for the decomposer subsystem; which of these mechanisms dominates will determine whether the net effect of herbivory on decomposers is positive or negative. The effects of invertebrate herbivory on ecosystem processes are inherently context dependent and vary tremendously across ecosystems, and it is reasonable to expect that local conditions are likely to determine herbivore effects on vegetation and therefore the magnitude and direction of their effects on the decomposer subsystem. Because the decomposer subsystem affects plant-available nutrient supply, and therefore the quality and quantity of foliage available for invertebrate herbivores, these herbivores can function as important participants in feedbacks between the aboveground and belowground subsystems.

Much remains unknown about how invertebrate herbivores affect the decomposer subsystem. For example, little is understood about the effects of different types of herbivores even though they may have vastly differing effects on the physiology of the host plant, i.e., root vs. foliar herbivores, different types of foliar invertebrate herbivores (e.g., chewing, sucking, rasping, leaf-mining), or invertebrate vs. vertebrate herbivores. Further, invertebrate herbivores exist in multiple species communities, but little is understood about whether ecosystem processes are driven by multiple herbivore species effects, or whether the most important effects emerge from single species outbreaks. One area that remains to be explored is the consequences for the decomposer subsystem of plant biochemical changes induced by invertebrate herbivore attack. These biochemical signals have the potential to alter the nature of multitrophic interactions including those that occur within the soil food web (van der Putten et al. 2001). Finally, since all ecosystems depend upon both the producer and decomposer subsystems, a more complete understanding of ecosystem-level consequences of invertebrate herbivory can only be gained through the application of approaches that explicitly consider both the aboveground and belowground subsystems, as well as the feedbacks between them.

References

Agrawal AA (1998) Induced responses to herbivory and increased plant performance. Science 279:1201–1202

Augustine DJ, McNaughton SJ (1998) Ungulate effects on the functional species composition of plant communities: herbivore selectivity and plant tolerance. J Wildl Manage 62:1165–1183

Bardgett RD, Wardle DA (2003) Herbivore-mediated linkages between aboveground and belowground communities. Ecology 84:2258–2268

Bardgett RD, Wardle DA, Yeates GW (1998) Linking above-ground and below-ground food webs: how plant responses to foliar herbivory influence soil organisms. Soil Biol Biochem 30:1867–1878

Bardgett RD, Denton CS, Cook R (1999) Below-ground herbivory promotes soil nutrient transfer and root growth in grassland. Ecol Lett 2:357–360

Belovsky GE, Slade JB (2000) Insect herbivory accelerates nutrient cycling and increases plant production. Proc Nat Acad Sci USA 97:14412–14417

Bonkowski M, Geoghegan IE, Birch ANE, Griffiths BS (2001) Effects of soil decomposer invertebrates (protozoa and earthworms) on an above-ground phytophagous insect (cereal aphid) mediated through changes in the host plant. Oikos 95:441–450
Brown VK, Gange AC (1990) Insect herbivory below ground. Adv Ecol Res 20:1–58
Brown VK, Gange AC (1992) Secondary plant succession: how is it modified by insect herbivory? Vegetatio 101:3–13
Buckland SM, Grime JP (2000) The effects of trophic structure and soil fertility on the assembly of plant communities: a microcosm experiment. Oikos 91:336–352
Cates RG, Orians GH (1975) Successional status and the palatability of plants to generalized herbivores. Ecology 56:410–418
Cebrian J (1999) Patterns in the fate of production in plant communities. Am Nat 154:449–468
Cebrian J, Williams M, McLelland J, Valeila I (1998) The dependence of heterotrophic consumption and C accumulation on autotrophic nutrient content in ecosystems. Ecol Lett 1:165–170
Davidson DW (1993) The effect of herbivory and granivory on terrestrial plant succession. Oikos 68:23–35
De Mazancourt C, Loreau M (2000) Effects of herbivory on primary production, and plant species replacement. Am Nat 155:734–754
De Ruiter PC, Neutel A, Moore JC (1995) Energetics, patterns of interaction strength, and stability in real ecosystems. Science 269:1257–1260
Díaz S, Cadibo M (2001) Vive la difference: plant functional diversity matters to ecosystem processes. Trends Ecol Evol 16:646–655
Dighton J (1978) Effects of synthetic lime aphid honeydew on populations of soil organisms. Soil Biol Biochem 10:369–376
Duffy JE, Macdonald KS, Rohde JM, Parker JD (2001) Grazer diversity, functional redundancy and productivity in seagrass beds: an experimental test. Ecology 82:2417–2434
Dyer MI, Acra MA, Wang GM, Coleman DC, Freckman DW, McNaughton SJ, Strain BR (1991) Source-sink carbon relations in two *Panicum coloratum* ecotypes in response to herbivory. Ecology 72:1472–1483
Findlay S, Carreiro M, Krishik V, Jones CJ (1996). Effects of damage to living plants on leaf litter quality. Ecol Appl 6:269–275
Grime JP (2001) Plant strategies, vegetation processes and ecosystem properties. Wiley, Chichester
Grime JP, Cornelissen JHC, Thompson K, Hodgson JG (1996) Evidence of a causal connection between anti-herbivore defence and the decomposition rate of leaves. Oikos 77:489–494
Hairston NG, Smith FE, Slobodkin LB (1960) Community structure, population control and competition. Am Nat 94:421–425
Haukioja E, Neuvonen S, Hanhimäki S, Niemelä P (1988) The autumnal moth in Fennoscandia. In: Berryman AA (ed) Dynamics of forest insect populations. Plenum Press, New York, pp 163–178
Holland EA, Parton WJ, Detling JK, Coppock DL (1992) Physiological responses of plant populations to herbivory and their consequences for ecosystem nutrient flow. Am Nat 140:685–706
Holland JN (1995) Effects of above-ground herbivory on soil microbial biomass in conventional and no-tillage agroecosystems. Appl Soil Ecol 2:275–279
Holland JN, Cheng W, Crossley DA (1996) Herbivore-induced changes in plant carbon allocation: assessment of below-ground C fluxes using carbon-14. Oecologia 107:87–94

Loreau M, Naeem S, Inchausti P, Bengtsson J, Grime JP, Hector A, Hooper DU, Huston MA, Raffaelli D, Schmid B, Tilman D, Wardle DA (2001) Biodiversity and ecosystem functioning: current knowledge and future challenges. Science 294:804–808
Lovett GM, Ruesink AE (1995) Carbon and nitrogen mineralization from decomposing gypsy moth frass. Oecologia 104:133–138
Lovett GM, Christenson LM, Groffman PM, Jones CJ, Hart JE, Mitchell MJ (2002) Insect defoliation and nitrogen cycling in forests. BioScience 52:335–341
Masters GJ, Brown VK, Gange AC (1993). Plant mediated interactions between above- and below-ground insect herbivores. Oikos 66:148–151
Mawdsley JL, Bardgett RD (1997) Continuous defoliation of perennial ryegrass (*Lolium perenne*) and white clover (*Trifolium repens*) and associated changes in the microbial population of an upland soil. Biol Fertil Soils 24:52–58
McNaughton SJ (1985) Ecology of a grazing system: the Serengeti. Ecol Monogr 55:259–294
McNaughton SJ, Osterheld M, Frank DA, Williams KJ (1989) Ecosystem level patterns of primary productivity and herbivory in terrestrial habitats. Nature 341:142–144
Mikola J, Yeates GW, Wardle DA, Barker GM, Bonner KI (2001b) Response of soil food web structure to defoliation of different plant species combinations in an experimental grassland community. Soil Biol Biochem 33:205–214
Northup RA, Yu Z, Dahlgren RA, Vogt KA (1995) Polyphenol control of nitrogen release from pine litter. Nature 377:227–229
Oksanen L, Fretwell SD, Arruda J, Niemelä P (1981) Exploitation ecosystems in gradients of primary productivity. Am Nat 118:240–261
Pastor J, Naiman RJ, Dewey B, McInnes P (1988) Moose, microbes and the boreal forest. BioScience 38:770–777
Polis GA, Hurd SD, Jackson CT, Pinero FS (1997) El Niño effects on the dynamics and control of an island ecosystem in the Gulf of California. Ecology 78:1884–1897
Rhoades DF (1985) Offensive-defensive interactions between herbivores and plants: their relevance in herbivore population dynamics and ecological theory. Am Nat 125:205–238
Riedell WE (1989) Western corn rootworm damage in maize: greenhouse technique and plant response. Crop Sci 29:412–415
Root RB (1973) Organisation of a plant-arthropod association in simple and diverse habitats: the fauna of collards (*Brassica oleracea*). Ecol Monogr 43:95–124
Schädler M, Jung G, Auge H, Brandl R (2003) Does the Fretwell-Oksanen model apply to invertebrates? Oikos 100:203–207
Scheu S, Theenhaus A, Jones TH (1999) Links between the detritivore and herbivore system: effects of earthworms and Collembola on plant growth and aphid development. Oecologia 119:541–551
Seastedt TR, Ramundo RA, Hayes DC (1988) Maximization of densities of soil animals by foliage herbivory: empirical evidence, graphical and conceptual models. Oikos 51:243–248
Selas V, Hogstad O, Andersson G, von Proschwitz T (2001) Population cycles of autumnal moth, *Epirrita autumnata*, in relation to birch mast seeding. Oecologia 129:213–219
Siemann E (1998) Experimental tests of effects of plant productivity and diversity on grassland arthropod diversity. Ecology 79:2057–2070
Southwood TRE, Brown VK, Reader PM (1979) The relationships of plant and insect diversities in succession. Biol J Linn Soc 12:327–348
Stadler B, Michalzik B (1998) Linking aphid honeydew, throughfall, and forest floor solution chemistry of Norway spruce. Ecol Lett 1:13–16

Van der Putten WH, Vet LEM, Harvey JA, Wäckers FL (2001) Linking above- and belowground multitrophic interactions of plants, herbivores, pathogens and their antagonists. Trends Ecol Evol 16:547–554
Wardle DA (2002) Communities and ecosystems: linking the aboveground and belowground components. Princeton University Press, Princeton
Wardle DA (2004) Hidden effects: the belowground consequences of introduced browsing mammals in New Zealand forests. In: Allen RB, Lee WG (eds) Biological invasions in New Zealand. Ecological studies, vol 10. Springer, Berlin Heidelberg New York (in press)
Wardle DA, Barker GM (1997) Competition and herbivory in establishing grassland communities: implications for plant biomass, species diversity and soil microbial activity. Oikos 80:470–480
Wardle DA, Zackrisson O, Hörnberg G, Gallet C (1997) The influence of island area on ecosystem properties. Science 277:1296–1299
Wardle DA, Barker GM, Bonner KI, Nicholson KS (1998) Can comparative approaches based on plant ecophysiological traits predict the nature of biotic interactions and plant species effects in ecosystems? J Ecol 86:405–420
Wardle DA, Bonner KI, Barker GM (2000) Stability of ecosystem properties in response to above-ground functional group richness and composition. Oikos 89:11–23
White TRC (1978) The importance of relative shortage of food in animal ecology. Oecologia 33:71–86
Yeates GW, Saggar S, Denton CS, Mercer CF (1998) Impact of clover cyst nematode (*Heterodera trifolii*) infection on soil microbial activity in the rhizosphere of white clover (*Trifolium repens*) – a pulse labelling experiment. Nematologica 44:81–90

4 Biotic Interactions in the Rhizosphere: Effects on Plant Growth and Herbivore Development

M. Bonkowski and S. Scheu

4.1 Summary

Considerable progress has been made in understanding specific interactions of plant roots with rhizosphere microorganisms and interactions with the soil fauna. Due to their function in nutrient mineralization, the role of soil organisms is usually considered important in long-term processes such as decomposition of litter materials. It would be incorrect, however, to assume that effects of decomposer animals on plant performance solely result from improved plant uptake of nutrients. In recent years, our view has profoundly changed, giving soil organisms a much more active role by interacting with living plants, their symbionts and pathogens and thereby shaping ecosystem processes. It has to be appreciated that decomposer animals consist of very different functional groups which differentially affect microbial diversity and function in the rhizosphere, thereby modifying plant physiology, morphology and phenology. These interactions cascade up to herbivores above the ground, ultimately affecting the whole aboveground food web. In addition to changing bottom-up forces on the herbivore community, the decomposer system may strengthen top-down forces on aboveground herbivores by subsidizing generalist predators with prey. The full implications of this integrated view of terrestrial ecosystem function have yet to be explored. In arable systems, intelligent management practices have to be developed employing the decomposer community to help in plant nutrition, to foster plant defence against herbivores and to support the control of herbivore pest populations. Current practices based on soil tillage and inorganic nutrient inputs certainly are inadequate in this respect. In more natural ecosystems the role of the decomposer community as driving agent for plant competition and community composition via modifying the rhizosphere environment needs considerably more attention. Microorganisms have been identified as an important structuring force of natural plant communities in recent years; however, those organisms that regulate the structure and functioning of microbial communities so far have been widely neglected. A comprehensive understanding of

Ecological Studies, Vol. 173
W.W. Weisser and E. Siemann (Eds.)
Insects and Ecosystem Function

regulating forces in arable and natural systems will not be achieved without integrating the animal community below the ground.

4.2 The Rhizosphere – Interface of Intense Microbial and Faunal Interactions

A century ago Hiltner (1904) introduced the term 'rhizosphere' to describe the stimulation of biomass and activity of microorganism in soil around plant roots. However, even today it is often not fully acknowledged that all nutrients that plants absorb from soil pass through a region of intense microbial and faunal activity (cf. Hartley and Jones, Chap. 2, this Vol.; Wardle and Bardgett, Chap. 3, this Vol.).

Soil is densely packed with organisms and plant roots are continuously confronted by a vast array of organisms, including a multitude of saprophytic, beneficial and deleterious microorganisms, soil fauna (including herbivores and decomposers), and the roots of the members of the plant community that is present in their vicinity.

Plants are not passive in these interactions; instead plants integrate information from the environment into their decisions on belowground investments like root proliferation (Huber-Sannwald et al. 1997; Hodge et al. 1998, 1999), formation of symbiotic relationships with infecting microorganisms (e.g. mycorrhizal fungi, Fitter and Merryweather 1992; Smith and Read 1997; or N_2-fixing bacteria, Ryle et al. 1979), alteration in exudation rates (Kraffczyk et al. 1984; Jones and Darrah 1995; Bonkowski et al. 2001b; Wamberg et al. 2003), interactions with free-living bacteria (Mathesius et al. 2003) or production of secondary compounds to defend herbivores (Baldwin and Hamilton 2000; Cipollini et al. 2003; Voelckel and Baldwin, Chap. 17, this Vol.). To cope with these interactions a dynamic strategy needs to be adopted by a non-motile organism. However, since plants are in the dilemma of growing and outcompeting their neighbours whilst at the same time defending themselves against herbivores (Herms and Mattson 1992), any potential trade-off may reduce plant fitness and open up opportunities for insect herbivore attack. The balance between carbon (and other) costs associated with such morphological and physiological changes and nutrient gain below ground will determine aboveground investments in structural compounds and ultimately feeds back to defence mechanisms against insect herbivores.

We will first consider plant C investments below ground and the feedback in nutrient gain from food web interactions and then provide examples of how outcomes of these interactions cascade up above ground to influence insect herbivore performance.

4.2.1 Plants as Drivers of Rhizosphere Interactions

Most nutrients in soil are bound to organic matter, absorbed by mineral surfaces, or locked up in microbial biomass and thus not easily available. Nutrient acquisition by plants therefore includes a number of mechanisms, often involving the aid of different microorganisms simultaneously. Common to all these strategies is that they are highly energy demanding and may cause a significant trade-off in plant C allocation.

Most land plants live in close symbiosis with mycorrhizal fungi. Most widespread are infections with arbuscular- and/or ecto-mycorrhizae (Smith and Read 1997; Cornelissen et al. 2001). Mycorrhizal fungi increase the accessibility and uptake of water and nutrients (particularly phosphorus, and often nitrogen), and may help in protecting roots against pathogens (Smith and Read 1997). Subsidizing the production of hyphae may be less energy demanding than the production of roots (Fitter 1994). Nevertheless, the maintenance of mycorrhizae is costly, with an estimated 10–20 % of total photosynthetically fixed C that plants translocate to the fungus (Christensen 1989; Marschner 1992; Söderström 1992). A trade-off may arise between mycorrhizal colonization and other symbionts (Bonkowski et al. 2001b; Wamberg et al. 2003) or a carbon shunt to neighbouring plants (Newman 1988; Graves et al. 1997) and not all plants benefit from mycorrhizal infections, since mycorrhizal species differ greatly in their degree of infection and host exploitation (Johnson et al. 1997; Gange and Ayres 1999).

The mutualistic relationship between plants and N_2-fixing bacteria or actinomycetes (e.g. *Rhizobium*, *Frankia*) ensures independence from soil nitrogen availability, but carries additional costs of an estimated 10–30 % of the total photosynthetically fixed carbon (Bezdicek and Kennedy 1979; Ryle et al. 1979). The benefit, i.e. the effectiveness, of rhizobial strains in N_2 fixation varies widely among plant species (Bala and Giller 2001). However, the few plant species that evolved this trait seem to be in control of the symbiosis. White clover, for example, has the ability to rapidly downregulate N_2 fixation and compete with other plant species for soil N when mineral N levels in soil increase (Griffith et al. 2000).

Only part of the plant's C release in the rhizosphere is channeled to specific infecting microbial partners. The stimulation of microbial biomass and activity around roots results from the fact that plants secrete an array of low- and high-molecular-weight molecules into the soil as exudates which account for 10–20 % and up to 40 % of total photosynthetically fixed C (Lynch and Whipps 1991; Rovira 1991). Free-living soil microorganisms are strongly carbon limited (Wardle 1992) and triggered into activity by the carbon pulses from exudates (Hawes 1991), leading to a short-term increase in decomposition and mineralization of nutrients, known as the 'priming effect' (Kuzyakov 2002). These interactions are particularly complex because plant-available nutrients will be strongly sequestered during microbial growth (Kaye and

Hart 1997; Wang and Bakken 1997) and would remain locked up in microbial biomass if consumption by protozoa and nematodes did not constantly remobilize essential nutrients for plant uptake (Christensen et al. 1992; Griffiths and Caul 1993; Griffiths et al. 1993; Bonkowski et al. 2000b); furthermore, by fostering specific types of bacteria and fungi, they directly influence plant root growth (Bonkowski and Brandt 2002).

Thus, plant roots are continuously exposed to strong interactions with soil organisms and, whatever strategy a plant subscribes to in order to obtain nutrients in the rhizosphere, the amount of carbon translocated below ground is substantial. The costs of these interactions and the gain in the form of nutrients may have important implications for the efficiency with which plants produce biomass, consequently affecting litter quality for insect herbivores; but also for plant fitness and defence mechanisms against herbivores if energy diverted to growth is lacking for defence (Lorio 1986; Herms and Mattson 1992; Gershenzon 1994; Halitschke et al. 2000; Mutikainen et al. 2002; Voelckel and Baldwin, Chap. 17, this Vol.).

4.3 Belowground Interactions and the Herbivore System

To develop a framework of how food web interactions below ground may feed back to plant community composition and affect insect herbivores above ground, we will distinguish three fundamental functional groups of soil organisms (modified after Brussaard 1998) and their connections to aboveground foliar herbivores which are mediated either by influencing host plant quality or by subsidizing generalist predators as antagonists to insect herbivores (Fig. 4.1).

Root biota are either beneficial (symbionts: e.g. mycorrhizal fungi, rhizobia) or detrimental (e.g. root herbivores, root pathogens) to plant growth. Root biota strongly depend on plant nutrient and carbon allocation below ground and their effects instantaneously feed back to plant fitness.

Micro-decomposers include both free-living microorganisms responsible for the mineralization processes (e.g. bacteria, saprophytic fungi) and their respective faunal grazers (microfauna, e.g. protozoa and nematodes, and mesofauna, e.g. Collembola, mites and enchytraeid worms) which not only regulate microbial biomass, turnover and activity, but also often affect microbial diversity and function. The interplay between microorganisms and microbivores determines the rates of nutrient cycling and strongly influences the availability of mineral nutrients to plants. These micro-food webs reach their highest activity in the presence of easily available carbon sources (i.e. litter patches and rhizosphere).

Members of the macrofauna (diplopods, earthworms and others) function as litter transformers and/or ecosystem engineers (sensu Jones et al. 1994;

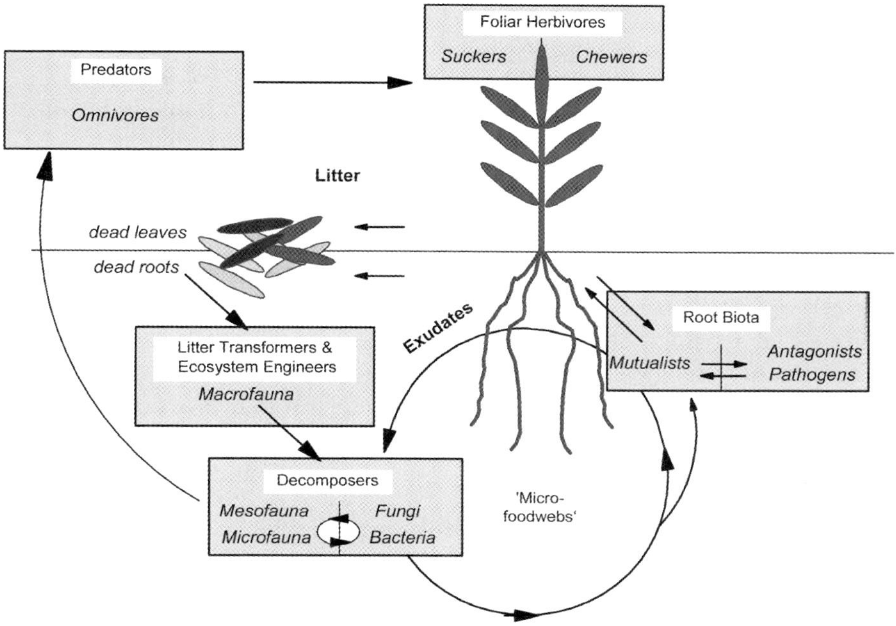

Fig. 4.1. Conceptual model depicting effects on foliar insect herbivores through biotic interactions in the rhizosphere. We distinguished three fundamental functional groups of soil organisms (litter transformers and ecosystem engineers, decomposers, root biota) that affect insect herbivores (1) by affecting nutrient contents and defence mechanisms (e.g. secondary compounds) in plants or (2) by subsidizing generalist predators antagonistic to foliar herbivores; for details see text

Lavelle et al. 1997) through fragmentation of litter, mixing of organic matter with soil, release of physically protected nitrogen from soil organic matter (Scheu 1993; Schulman and Tiunov 1999), and creation of microhabitats for other soil biota, including plant roots (Wurst et al. 2003). Therefore, macrofauna organisms strongly influence the physicochemical environment in which plant roots interact with belowground communities.

Foliar insect herbivores may be differently affected by these interactions depending on whether they live on phloem sap (suckers) or have to ingest whole parts of the plant foliage (chewers; Gange and West 1994).

Finally, generalist predators (spiders, staphylinid and carabid beetles) antagonistic to foliar herbivores are subsidized by the decomposer food web. Particularly Collembola, an abundant group of tiny insects (mesofauna), are an important prey of generalist predators, especially of juveniles and at times when foliar herbivores are scarce and therefore unable to keep predator levels up.

The functional significance of insects in soils falls into two broad categories, root herbivores and decomposers. In fact, most insect species we see above ground spend the greatest part of their life cycle in litters and soil. These insects influence many ecosystem processes, either directly by affecting plant primary production as root herbivores (e.g. beetle larvae, such as Curculionidae and Elateridae; Masters, Chap. 5, this Vol.) or indirectly by affecting fungal growth rate and decomposition (many dipteran larvae, such as Sciaridae, Mycetophilidae and Bibionidae among others) and subsequent effects on recycling of nutrients to plants. The most numerous insects in soil, however, are Collembola, tiny micro-decomposers well adapted to a permanent life in the porous soil environment. Collembola serve as important prey for generalist predatory insects and spiders, and as such are critical link species that subsidize aboveground food webs and ultimately determine the efficiency of biological control of insect herbivores above ground. Thus insects play a multitude of functional roles in below- and aboveground interactions.

4.3.1 Effects of Mycorrhiza and Rhizobacteria on Aboveground Herbivores

The most widespread interactions between plant roots and microorganisms are those with mycorrhizal fungi and interactions with free-living rhizosphere bacteria. Infectious rhizosphere microorganisms, such as mycorrhizae, form permanent partnerships with their host plants and strongly influence phytophagous insects by both changes in the nutritional quality and the production of secondary compounds by the plant host (Gange and West 1994; Gange and Nice 1997). Mycorrhizal fungi may have played a significant role in shaping the co-evolution of insect herbivores with their host plants (Gange et al. 2002). The effects of mycorrhizae on insect herbivores have recently been reviewed by Brown and Gange (2002) and Gehring and Whitham (2002) and therefore will only be discussed briefly here. Among the major patterns that emerge, mycorrhizal colonization and mycorrhizal fungal community composition are stronger influenced by herbivores, particularly with increasing herbivore pressure, than vice versa. However, mycorrhization also feeds back to herbivores. Not only the presence of mycorrhiza but also the identity of the mycorrhizal species have been shown to influence herbivore performance (Goverde et al. 2000). Because different mycorrhizal species differentially affect plants, the effects on herbivores showed a wide range from positive to negative (Brown and Gange 2002; Gehring and Whitham 2002). Nevertheless, general patterns appear: generalist herbivores seem to be more responsive than specialists, and leaf chewers are more often detrimentally affected, while sap feeders more often benefit from mycorrhiza (Gange and West 1994; Borowicz 1997; Gange et al. 1999).

Many of the free-living bacteria in the soil–plant interface can enhance plant performance by increasing the solubilization of minerals (Kloepper et al. 1980; Antoun et al. 1998), fixing of nitrogen (Obreht et al 1993; Reinhold-Hurek and Hurek 1997; Steenhoudt and Vanderleyden 2000), competitively suppressing plant pathogens (Chet et al. 1991) or by producing hormones (Brown 1972; Costacurta and Vanderleyden 1995; Arshad and Frankenberger 1998; Lambrecht et al. 2000). Up to 80 % of the bacteria isolated from plant rhizospheres are considered to produce auxins (Barea et al. 1976; Patten and Glick 1996). This widespread ability of both beneficial and deleterious rhizosphere microorganisms to produce plant hormones suggests that free-living rhizosphere bacteria play an important role in manipulating root and plant growth (Shishido et al. 1996; Holland 1997; Rolfe et al. 1997; Bonkowski and Brandt 2002, Phillips et al. 2003).

Recently, Mathesius et al. (2003) demonstrated that plant auxin responses and investment in defence are directly affected by free-living rhizosphere bacteria. The legume *Medicago truncatula* activated its defence genes in response to nanomolar concentrations of the bacterial signal molecule N-acyl homoserine lactone from both symbiotic (*Sinorhizobium meliloti*) and pathogenic (*Pseudomonas aeruginosa*) bacteria (Mathesius et al. 2003). Their results show that plants alerted by certain densities of specific bacterial species can activate their systemic resistance mechanisms to defend pathogens – a defence reaction ultimately feeding back to insect herbivore performance above ground.

Direct microbial interactions with plant roots thus may reduce insect herbivore load and plant damage by increasing plant vigour or activation of resistance mechanisms. There is strong evidence that plants possess an extensive range of complex responses to rhizosphere microorganisms that may play important roles in the outcomes of plant interactions with insect herbivores. We expect significant progress in this rapidly developing field of research by integrating molecular techniques into ecological research (Rolfe et al. 1997; Phillips and Strong 2003).

4.3.2 Interactions with the Micro-Decomposer Food Web

Recent studies provide convincing evidence that decomposers can induce subtle host-mediated changes that determine the disposition of plants to herbivore attack (Scheu et al. 1999; Bonkowski et al. 2001a; Wurst and Jones 2003; Wurst et al. 2003, 2004). The energy provided by primary producers is channeled through two distinct compartments, the bacterial and fungal energy channel within the detritus food web (Moore and Hunt 1988). Ecosystem engineers modulate the interactions in both microbial energy channels and accelerate decomposition by physicochemical processes (Scheu and Setälä 2002). We will therefore consider in turn how the bacterial and fungal food chain and ecosystem engineers affect plant growth and how this feeds back to the herbivore system.

4.3.2.1 The Bacterial Loop and Herbivore Performance

Since microorganisms in the rhizosphere are much stronger competitors for nutrients than plant roots (Jingguo and Bakken 1997; Puri and Ashman 1999; Lipson et al. 1999a, 1999b), the stimulation of bacterial activity and growth by exudates is counterintuitive. However, the bacterial biomass in soil is strongly regulated by the feeding of protozoa and nematodes (Wardle and Yeates 1993; Verhagen et al. 1994; Bonkowski et al. 2000b). Their grazing ensures that essential nutrients, particularly nitrogen, locked up in bacterial biomass are constantly released for root uptake, a mechanism known as 'microbial loop in soil' (Clarholm 1985; Griffiths 1994). Bacterial-grazing protozoa are known to strongly increase plant growth and plant nitrogen acquisition (Clarholm 1984; Ritz and Griffiths 1987; Kuikman et al. 1990; Jentschke et al. 1995; Alphei et al. 1996; Bonkowski et al. 2000b, 2001a). Since plant growth and plant tissue nitrogen concentrations are among the most important drivers of herbivore performance (Price 1991; White 1993), one would expect significant effects on aboveground herbivores from bacterial grazers in the rhizosphere.

Bonkowski et al. (2001a) investigated the effects of bacterial-feeding protozoa and earthworms on aphid performance on barley. The biomass of barley increased by ca. 40 % in the presence of protozoa. Concomitantly, aphid numbers and biomass more than doubled on plants grown in the presence of protozoa (Fig. 4.2). However, protozoa also increased plant reproduction (biomass of ears, number of seeds and individual seed weight). Apparently, the plants in protozoan treatments tolerated higher levels of herbivory and even increased their fitness. Effects of protozoa on plant biomass and nutrient turnover considerably exceeded the effects of earthworms on most parameters measured. This indicates that indirect effects due to grazing of protozoa on bacteria were more important for plant growth and aphid performance than the direct physicochemical effects on nutrient mobilization by earthworms.

Protozoa, however, can also enhance plant growth without increasing nutrient contents in plant tissue (Alphei et al. 1996). An extensive and highly branched root system is formed under the influence of protozoa (Jentschke et al. 1995), resembling hormonal effects on root growth by beneficial rhizobacteria (Chanway et al. 1988; Petersen et al. 1996; Rolfe et al. 1997). Bonkowski and Brandt (2002) introduced a novel mechanism by which protozoan grazing stimulates auxin-producing rhizobacteria. Accordingly, growth of the root system is stimulated and allows more nutrients to be absorbed, but will also increase exudation rates, thereby further stimulating bacterial–protozoan interactions. These interactions may result in a dilution of nitrogen in plant tissues (Alphei et al. 1996) and lead to strong negative effects on plant herbivores, as recently confirmed by Bonkowski, Omacini and Jones (unpublished) in a study on aphid herbivore development on the grass *Lolium multiflorum*.

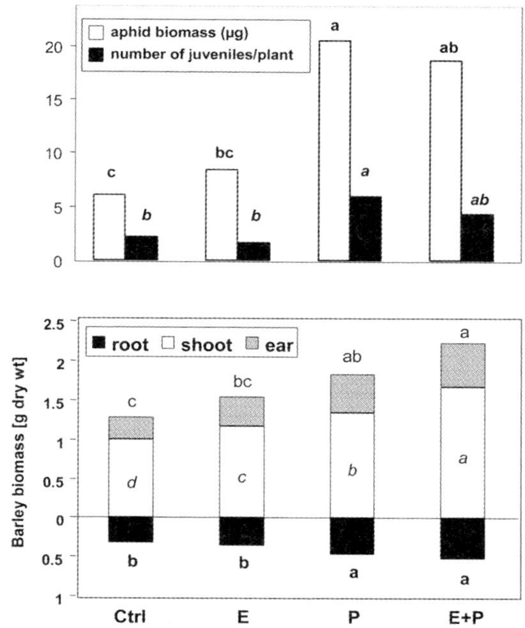

Fig. 4.2. Effects of protozoa and earthworms on aphid (*Sitobion avenae*) biomass (µg) and numbers of juveniles per plant, and biomass of barley plants (roots, shoots, ears) after 9 weeks. *Ctrl* animal-free control; *E* earthworms; *P* protozoa; *E+P* earthworms and protozoa. *Different letters* indicate a significant difference between means ($P < 0.05$, Tukey-test). (Redrawn from Bonkowski et al. 2001a)

4.3.2.2 The Fungal Food Chain and Herbivore Performance

Scheu et al. (1999) demonstrated that indirect effects of fungal-feeding Collembola and endogeic earthworms significantly affected the development of aphids on a grass (*Poa annua*) and a legume (*Trifolium repens*); however, the effects differed among animal and plant species. Collembola (*Heteromurus nitidus* and *Onychiurus scotarius*) caused a reduction in plant biomass, particularly of grass roots. The presence of Collembola enhanced the reproduction of aphids (*Myzus persicae*) nearly threefold on the grass, but decreased aphid reproduction on legumes on average by half, although nitrogen concentrations in plant tissue of clover even slightly increased. Thus decomposer insects below ground may strongly influence insect herbivores above ground (Fig. 4.3).

Scheu et al. (1999) concluded that Collembola decrease aphid reproduction on more palatable host plants, such as *T. repens*, but increase that on less palatable ones, such as *P. annua*. The mechanisms, however, remained unclear.

Assuming that decomposers affect herbivores mainly through nutrient effects on plant growth, the impact of Collembola on nutrient mineralization, plant growth and herbivore performance should be highest under low nutrient, particularly nitrogen, availability because nitrogen is one of the main factors limiting herbivore development (Lawton and McNeill 1979; Mattson

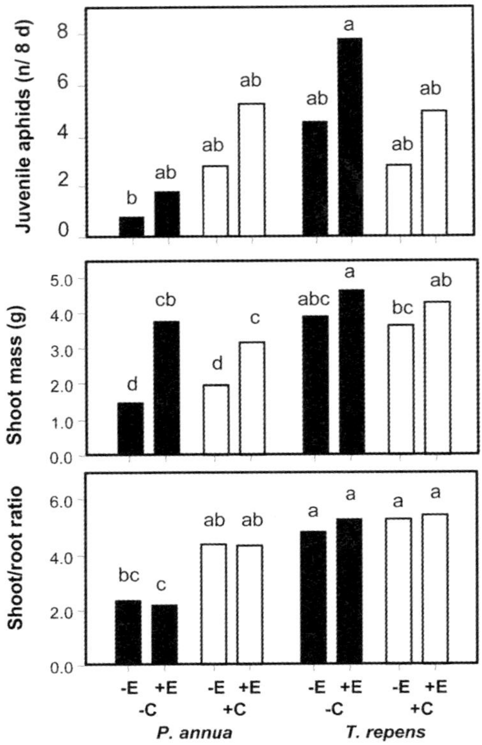

Fig. 4.3. Effects of earthworms (*E*) and Collembola (*C*) on (*above*) number of juvenile produced by the aphid *Mycus persicae* clamped on freshly grown leaves of the grass *P. annua* and clover *T. repens*. *Middle* Shoot biomass and *below* shoot/root ratio of the respective plant species 15 weeks after the experiment started. *Different letters* indicate a significant difference between means ($P < 0.05$, Tukey-test). (Redrawn from Scheu et al. 1999)

1980; White 1993). In order to separate non-nutrient and nutrient effects of Collembola (*Onychiurus fimatus*), Schuetz, Bonkowski and Scheu (unpublished) investigated how the performance of an aphid (*Rhopalosiphum padi*) on wheat (*Triticum aestivum* L. var. Naxos) in differently fertilized soil was influenced by *O. fimatus*. The soil was either unfertilized, fertilized with nitrogen (+N) or fertilized with nitrogen, phosphorus and potassium (+NPK).

The biomass of wheat increased threefold when plants were fertilized. Correspondingly, the average number of offspring produced by an adult aphid increased by factors of 2 and 2.6 in N- and NPK-fertilized plants, respectively (Fig. 4.4); reflecting the increased nutrient availability to herbivores. In contrast to our expectations, Collembola did not affect aphid reproduction on control plants, but reduced aphid performance by 22 and 10% on N- and NPK-fertilized plants, respectively (Fig. 4.4). The results of Schuetz, Bonkowski and Scheu (unpublished) demonstrate that Collembola affect herbivores by more complex interactions than by simply improving plant nutrition. The results also suggest that decomposer organisms may actually help in reducing the susceptibility of crop plants to herbivore attack.

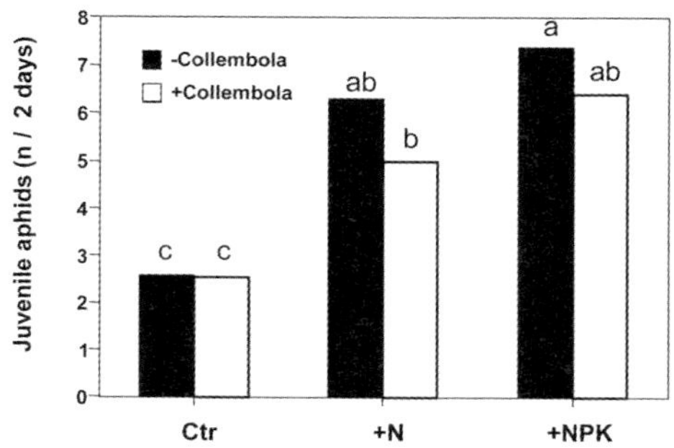

Fig. 4.4. Effects of fertilizer addition (*Ctr* control, +N fertilizer, +NPK fertilizer) in treatments without (-) and with (+) Collembola on the total number of offspring per plant produced by the aphid *Rhopalosiphum padi* on wheat. *Different letters* indicate a significant difference between means ($P < 0.05$, Tukey-test). (Schuetz, Bonkowski and Scheu, unpublished)

4.3.2.3 Ecosystem Engineers and Herbivore Performance

The study of Scheu et al. (1999) demonstrated that earthworms (*Aporrectodea caliginosa* and *Octolasion tyrtaeum*) enhanced plant growth by increasing the availability of nutrients. The more than twofold increase in shoot and root biomass of the grass considerably exceeded that of the legume which is more independent of soil nitrogen supply. Although earthworms strongly increased aphid reproduction in one experimental period, they had no effect in another, making earthworm effects on aphid herbivores less predictable than those of Collembola (Fig. 4.3).

Wurst et al. (2003) investigated the effects of *A. caliginosa* and *O. tyrtaeum* on aphid performance in more detail by including treatments with homogeneous and patchy distribution of ^{15}N-labelled grass litter. Earthworms were expected to further affect litter distribution and mineralization by their burrowing and casting activity, thereby influencing plant nutrient status and their susceptibility to herbivores. Because the exploitation of organic patches depends on both soil animals and root-foraging strategies of plants (Robinson 1994; Bonkowski et al. 2000b), Wurst et al. (2003) included plants of different functional groups (the grass *Lolium perenne*, the forb *Plantago lanceolata*, the legume *Trifolium repens*) which differed in root morphology and N allocation strategies.

Earthworm activity generally enhanced nitrogen mobilization from litter and from soil. However, the earthworm-mediated increase in plant nitrogen uptake differed between plant species. Earthworms enhanced N uptake from litter and soil in all plant species but enhanced shoot and root growth only in *L. perenne* and *P. lanceolata*. Clover exploited more of the ^{15}N from added litter than the other plants, but, similar to the study of Scheu et al. (1999), the growth of the legume was more independent of biotic (earthworms) and abiotic (litter distribution) soil conditions, presumably due to its symbiosis with N_2-fixing bacteria. Earthworms increased aboveground biomass and contents of total nitrogen and ^{15}N in both *L. perenne* and *P. lanceolata* and enhanced root growth in the grass. Similarly, root biomass of *P. lanceolata* doubled in litter patches. However, this effect occurred exclusively in treatments with earthworms, presumably due to a local increase in nutrient concentrations resulting from earthworm activity. Although root proliferation of *P. lanceolata* was strongly stimulated in the presence of both a litter patch and earthworms, the increased investment in roots apparently was not repaid since shoot biomass remained unchanged.

Due to the increase in plant nitrogen content one would expect positive effects of earthworms on aphid reproduction (Dixon 1985). However, despite total nitrogen content in aboveground plant tissue increasing, reproduction of *M. persicae* was reduced on *P. lanceolata* in the presence of earthworms (Fig. 4.5; Wurst et al. 2003).

Most likely, earthworm effects on plant defence compounds were responsible for the reduced aphid reproduction. Wurst et al. (2004) subsequently confirmed that earthworms and organic matter distribution strongly affected the contents of phytosterols in *P. lanceolata*. Phytosterols serve as precursors of moulting hormones in the diet of herbivorous insects, including aphids (Campell and Nes 1983). Phytosterols and iridoid glycosides were positively correlated with plant nitrogen content, strongly suggesting that the production of defence compounds might be indirectly driven by increased N availability as a result of earthworm activity. Unfortunately, it is difficult to predict which plant metabolites might be responsible for effects on herbivore performance. Although phytosterol content and aphid performance were not directly correlated in the study of Wurst et al. (2004), the results clearly demonstrate that decomposers affected plant compounds involved in herbivore defence.

Earthworms are known to mobilize nutrients protected in soil organic matter during gut passage (Scheu 1993; Brussaard 1998; Schulman and Tiunov 1999). The examples provided here confirm that earthworms affect plant growth by increasing nutrient availability (Scheu et al. 1999; Bonkowski

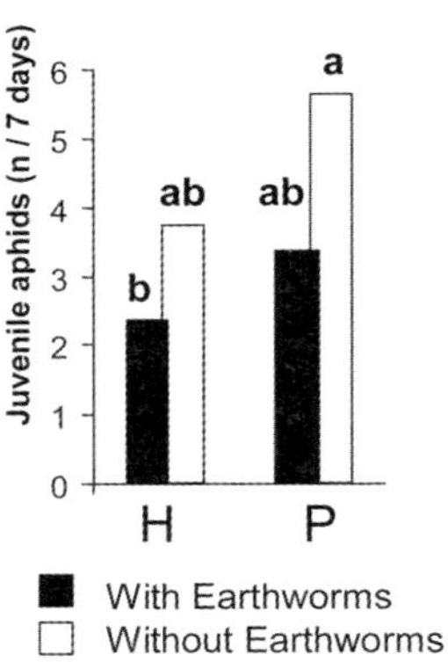

Fig. 4.5. Effect of litter distribution (*H* homogeneous; *P* patchy) in soil and earthworms on number of offspring of *M. persicae* on *P. lanceolata* 7 days after inoculation. *Different letters* indicate a significant difference between means ($P < 0.05$, Tukey-test). (Data from Wurst et al. 2003)

et al. 2001a; Wurst et al. 2003) and this may affect plant defence chemistry (Bryant et al 1983; Lerdau and Coley 2002, but see Hamilton et al. 2001; Nitao et al. 2002). However, other indirect effects include those mediated by changing soil structure and the foraging strategy of plant roots (Wurst et al. 2003), modulating the infection of roots by fungal pathogens (Stephens and Davoren 1997) or modifying the activity of protozoa (Alphei et al. 1996; Bonkowski et al. 2001a), and Collembola in the rhizosphere (Scheu et al. 1999). These results demonstrate that plant vigour and susceptibility to insect herbivores are driven by a complexity of interactions with soil organisms reaching far beyond simple nutrient effects (Bonkowski et al. 2000a; Scheu and Setälä 2002).

4.4 Top-Down Effects by Subsidizing Generalist Predators

Generalist predators are considered to be among the most important predators in terrestrial ecosystems (Hawkins et al. 1999; van der Meijden and Klinkhamer 2000). In their review of manipulative field studies, Symondson et al. (2002) showed that generalist predators such as spiders, or staphylinid and carabid beetles in about 75 % of cases, whether single species or species assemblages, significantly reduced herbivore numbers, leading to a reduction of yield losses and even preventing crop damage (Symondson et al. 2002). However, the link between generalist predators and insect herbivores is not straightforward. Generalist predators may not exclusively feed on herbivores but also feed on prey from the decomposer system (Scheu 2001). Furthermore, generalist predators may prey upon each other (cannibalism, intraguild predation), thereby diminishing their effect on herbivore pest populations. However, additional prey may in fact strengthen rather than weaken the potential of predators to effectively control pest populations (Holt and Lawton 1994; Beggs and Rees 1999). Polis (1991) was among the first documenting that generalist predators are intimately linked to prey from the decomposer community. Polis (1994) and Polis and Strong (1996) stressed that allochthonous resource inputs, i.e. additional prey originating from other systems, strengthen top-down effects of generalist predators on insect herbivores.

In their detailed analysis, Settle et al. (1996) describe how generalist predator populations in rice fields of central Java were fostered by prey from the decomposer community before planting. Later in the season when the plants had grown up, predators switched to aboveground prey and reduced insect herbivores on rice. Application of pesticides detrimentally affected generalist predators by reducing their decomposer prey, leading to reduced herbivore control later in the year (Settle et al. 1996). In a series of experiments, Wise and coworkers convincingly documented that generalist predator species, such as lycosid spiders and carabid beetles, benefit from increased resource supply to decomposer invertebrates and that increased predator populations help to control herbivore pest populations above ground, resulting in improved fruit yield in squash and cucumber gardens (Chen and Wise 1999; Snyder and Wise 1999; Wise et al. 1999; Halaj and Wise 2001, 2002) Since the input of plant litter materials to the decomposer system considerably exceeds the amount of living plant tissue consumed by herbivores above ground (Reichle et al. 1975; McNaughton et al. 1989), decomposer organisms may in fact be more important for generalist predators than prey from the herbivore system. Rather than part of the aboveground system, generalist predators therefore may better be viewed as members of the belowground system which are subsidized by resources (prey) from the herbivore community (Scheu 2001). The alternative view of decomposer animal species functioning as subsidy for generalist predators above the ground resembles the model of Oksanen et al. (1997) for taiga systems where aboveground predators essentially rely on energy shunts from the decomposer system.

An essential component of the view of generalist predators as potential biocontrol agents which primarily live on prey from the decomposer community is that predators switch from decomposer to herbivore prey. This switch may necessarily be triggered by the onset of plant growth since plants expose the soil water pool to the atmosphere, thereby reducing soil humidity. Since the activity of most decomposer animal species heavily relies on high soil humidity, plant growth diminishes the availability of prey from the decomposer community to generalist predators which then are forced to switch to other prey. Switching back and forth between prey from the decomposer to the aboveground herbivore system may be essential for stabilizing generalist predator populations in the long term (Scheu 2001).

The view outlined above suggests that knowledge of the relative contribution of prey from the decomposer community to generalist predator nutrition is a prerequisite for the understanding of aboveground predator–prey interactions in more natural systems and for developing strategies for biocontrol of herbivore pest species in arable systems. In both systems, generalist predators, such as spiders, staphylinid or carabid beetles, are among the most important antagonists of herbivores. A more complete understanding of the functioning of terrestrial ecosystems can only be achieved by integrating both the above- and the belowground system (Fig. 4.1).

Our view on plant–insect herbivore interactions is traditionally skewed by an aboveground-centered focus. We have to acknowledge that the hidden half of plants in the soil is simultaneously confronted by interactions with a multitude of organisms and the outcome of these interactions determines the availability of nutrients to plants, influences the productivity and fitness of plants in a community and consequently the responsiveness of a plant to insect herbivore attack. Decomposers, among them many insects, serve as critical link species that regulate important belowground ecosystem processes, such as microbial growth rate, decomposition and rates of recycling of nutrients to plant growth. However, interactions between microorganisms, decomposers and plants in the rhizosphere are not simply limited to the mineralizing activities of the fauna. Rather, complex indirect interactions exist, with symbiotic microorganisms and soil fauna affecting insect herbivore load by influencing plant metabolic pathways, or through soil insects subsidizing generalist (insect) predators and thus strengthening top-down biocontrol of aboveground insect herbivores. The examples given above demonstrate that investigations of multitrophic interactions between below- and aboveground systems are a crucial step in furthering our understanding of plant–insect herbivore interactions.

Acknowledgements. This study was supported by the German Ministry for Education and Research (BMBF; BIOLOG program 'Microbial terrestrial biodiversity', project 7) (M.B.) and the German Science Foundation (DFG; 'Jena experiment', project 5) (S.S.).

References

Alphei J, Bonkowski M, Scheu S (1996) Protozoa, Nematoda and Lumbricidae in the rhizosphere of *Hordelymus europaeus* (Poaceae): faunal interactions, response of microorganisms and effects on plant growth. Oecologia 106:111–126

Antoun H, Beauchamp CJ, Goussard N, Chabot R, Lalande R (1998) Potential of *Rhizobium* and *Bradorhizobium* species as plant growth promoting rhizobacteria on non-legumes: effect on radishes (*Raphanus sativus* L.). Plant Soil 204:57–67

Arshad M, Frankenberger WT (1998) Plant growth-regulating substances in the rhizosphere: microbial production and functions. Adv Agron 62:45–151

Bala A, Giller KE (2001) Symbiotic specificity of tropical tree rhizobia for host legumes. New Phytol 149:495–507

Baldwin IT, Hamilton W (2000) Jasmonate-induced responses of *Nicotiana sylvestris* results in fitness costs due to impaired competitive ability for nitrogen. J Chem Ecol 26:915–952

Barea JM, Navarro E, Montoya E (1976) Production of plant growth regulators by rhizosphere phosphate-solubilizing bacteria. J Appl Bacteriol 40:129–134

Beggs JR, Rees JS (1999) Restructuring of Lepidoptera communities by introduced *Vespula* wasps in a New Zealand beech forest. Oecologia 119:565–571

Bezdicek DF, Kennedy AC (1979) Economic microbial ecology: symbiontic nitrogen fixation and nitrogen cycling in terrestrial environments. In: Lynch JM, Hobbie JE (eds) Micro-organisms in action: concepts and applications in microbial ecology. Blackwell, Oxford, pp 241–260

Bonkowski M, Brandt F (2002) Do soil protozoa enhance plant growth by hormonal effects? Soil Biol Biochem 34:1709–1715

Bonkowski M, Cheng W, Griffiths BS, Alphei J, Scheu S (2000a) Microbial–faunal interactions in the rhizosphere and effects on plant growth. Eur J Soil Biol 36:135–147

Bonkowski M, Griffiths BS, Scrimgeour C (2000b) Substrate heterogeneity and microfauna in soil organic 'hotspots' as determinants of nitrogen capture and growth of rye-grass. Appl Soil Ecol 14:37–53

Bonkowski M, Geoghegan IE, Birch ANE, Griffiths BS (2001a) Effects of soil decomposer invertebrates (protozoa and earthworms) on an above-ground phytophagous insect (cereal aphid), mediated through changes in the host plant. Oikos 95:441–450

Bonkowski M, Jentschke G, Scheu S (2001b) Contrasting effects of microbes in the rhizosphere: interactions of mycorrhiza (*Paxillus involutus* (Batsch) Fr.), naked amoebae (Protozoa) and Norway spruce seedlings (*Picea abies* Karst.). Appl Soil Ecol 18:193–204

Borowicz VA (1997) A fungal root symbiont modifies plant resistance to an insect herbivore. Oecologia 112:534–542

Brown ME (1972) Plant growth substances produced by micro-organisms of soil and rhizosphere . J Appl Bacteriol 35:443–451

Brown VK, Gange AC (2002) Tritrophic below- and above-ground interactions in succession. In: Tscharntke T, Hawkins BA (eds) Multitrophic level interactions. Cambridge University Press, Cambridge, pp 197–222

Brussaard L (1998) Soil fauna, guilds, functional groups and ecosystem processes. Appl Soil Ecol 9:123–135

Bryant JP, Chapin FS, Klein DR (1983) Carbon/nutrient balance of boreal plants in relation to vertebrate herbivory. Oikos 40:357–368

Campell BC, Nes WD (1983) A reappraisal of sterol biosynthesis and metabolism in aphids. J Insect Physiol 29:149–156

Chanway CP, Nelson LM, Holl FB (1988) Cultivar specific growth promotion of spring wheat (*Triticum aestivum* L.) by co-existent *Bacillus* species. Can J Microbiol 34:925–929

Chen BR, Wise DH (1999) Bottom-up limitation of predaceous arthropods in a detritus-based terrestrial food web. Ecology 80:761–772

Chet I, Ordentlich A, Shapira R, Oppenheim A (1991) Mechanisms of biocontrol of soil-borne plant pathogens by rhizobacteria. In: Kleister DL, Cregan PB (eds) The rhizosphere and plant growth. Kluwer, Dordrecht, pp 229–236

Christensen M (1989) A view of fungal ecology. Mycologia 81:1–19

Christensen S, Griffiths BS, Ekelund F, Rønn R (1992) Huge increase in bacterivores on freshly killed barley roots. FEMS Microbiol Ecol 86:303–310

Cipollini D, Purrington CB, Bergelson J (2003) Costs of induced responses in plants. Basic Appl Ecol 4:79–85

Clarholm M (1984) Microbes as predators or prey – heterotrophic, free-living protozoa: neglected microorganisms with an important task in regulating bacterial populations. In: Klug MJ, Reddy CA (eds) Current perspectives on microbial ecology. American Society of Microbiology, Washington, pp 321–326

Clarholm M (1985) Interactions of bacteria, protozoa and plants leading to mineralization of soil nitrogen. Soil Biol Biochem 17:181–187

Cornelissen JHC, Aerts R, Cerabolini B, Werger MJA, van der Heijden MGA (2001) Carbon cycling traits of plant species are linked with mycorrhizal strategy. Oecologia 129:611–619

Costacurta A, Vanderleyden J (1995) Synthesis of phytohormones by plant-associated bacteria. Crit Rev Microbiol 21:1–18

Dixon AFG (1985) Aphid ecology. Blackie, Glasgow

Fitter AH (1994) Architecture and biomass allocation as components of the plastic response of root systems to soil heterogeneity. In: Caldwell MM, Pearcey RW (eds) Exploitation of environmental heterogeneity by plants: ecophysiological processes above- and belowground. Academic Press, San Diego, pp 305–323

Fitter AH, Merryweather JW (1992) Why are some plants more mycorrhizal than others? An ecological enquiry. In: Read DJ, Lewis DH, Fitter AH, Alexander IJ (eds) Mycorrhizas in ecosystems. CAB International, Wallingford, pp 26–36

Gange AC, Ayres RL (1999) On the relation between mycorrhizal colonization and plant "benefit". Oikos 87:615–621

Gange AC, Nice HE (1997) Performance of the thistle gall fly, *Urophora cardui*, in relation to host plant nitrogen and mycorrhizal colonization. New Phytol 137:335–343

Gange AC, West HM (1994) Interactions between arbuscular mycorrhizal fungi and foliar-feeding insects in *Plantago lanceolata* L. New Phytol 128:79–87
Gange AC, Bower E, Brown VK (1999) Positive effects of an arbuscular mycorrhizal fungus on aphid life history traits. Oecologia 120:123–131
Gange AC, Stagg PG, Ward LK (2002) Arbuscular mycorrhizal fungi affect phytophagous insect specialism. Ecol Lett 5:11–15
Gehring GA, Whitham TG (2002) Mycorrhizae–herbivore interactions: population and community consequences. In: van der Heijden MGA, Sanders IR (eds) Mycorrhizal ecology. Springer ecological studies analysis and synthesis, vol 157. Springer, Berlin Heidelberg New York, pp 295–320
Gershenzon J (1994) The cost of plant chemical defense against herbivory: a biochemical perspective. In: Bernays EA (ed) Insect–plant interactions. CRC Press, Boca Raton, pp 105–173
Goverde M, van der Heijden MGA, Wiemken A, Sanders IR, Erhardt A (2000) Arbuscular mycorrhizal fungi influence life history traits of a lepidopteran herbivore. Oecologia 125:362–369
Graves JD, Watkins NK, Fitter AH, Robinson D, Scrimgeour C (1997) Intraspecific transfer of carbon between plants linked by a common mycorrhizal network. Plant Soil 192:153–159
Griffiths BS (1994) Soil nutrient flow. In: Darbyshire JF (ed) Soil protozoa. CAB International, Wallingford, pp 65–91
Griffiths BS, Caul S (1993) Migration of bacterial-feeding nematodes, but not protozoa, to decomposing grass residues. Biol Fert Soils 15:201–207
Griffiths BS, Ekelund F, Rønn R, Christensen S (1993) Protozoa and nematodes on decomposing barley roots. Soil Biol Biochem 25:1293–1295
Griffith GS, Cresswell A, Jones S, Allen DK (2000) The nitrogen handling characteristics of white clover (*Trifolium repens* L.) cultivars and a perennial ryegrass (*Lolium perenne* L.) cultivar. J Exp Bot 51:1879–1892
Halaj J, Wise DH (2001) Terrestrial trophic cascades: how much do they trickle? Am Nat 157:262–281
Halaj J, Wise DH (2002) Impact of a detrital subsidy on trophic cascades in a terrestrial grazing food web. Ecology 83:3141–3151
Halitschke R, Keßler A, Kahl J, Lorenz A, Baldwin IT (2000) Ecophysiological comparison of direct and indirect defenses in *Nicotiana attenuata*. Oecologia 124:408–417
Hamilton JG, Zangerl A, DeLucia EH, Berenbaum MR (2001) The carbon–nutrient balance hypothesis: its rise and fall. Ecol Lett 4:86–95
Hawes MC (1991) Living plant cells released from the root cap: a regulator of microbial populations in the rhizosphere? In: Kleister DL, Cregan PB (eds) The rhizosphere and plant growth. Kluwer, Dordrecht, pp 51–59
Hawkins BA, Mills NJ, Jervis MA, Price PW (1999) Is the biological control of insects a natural phenomenon? Oikos 86:493–506
Herms DA, Mattson WJ (1992) The dilemma of plants: to grow or to defend. Q Rev Biol 67:283–335
Hiltner L (1904) Über neue Erfahrungen und Probleme auf dem Gebiet der Bodenbakteriologie unter besonderer Berücksichtigung der Gründüngung und Brache. Arb Dtsch Landw Ges 98:59–78
Hodge A, Stewart J, Robinson D, Griffiths BS, Fitter AH (1998) Root proliferation, soil fauna and plant nitrogen capture from nutrient-rich patches in soil. New Phytol 139:479–494
Hodge A, Stewart J, Robinson D, Griffiths BS, Fitter AH (1999) Plant, soil fauna and microbial responses to N-rich organic patches of contrasting temporal availability. Soil Biol Biochem 31:1517–1530

Holland MA (1997) Occam's razor applied to hormonology: are cytokinins produced by plants? Plant Physiol 115:865–868

Holt RD, Lawton JH (1994) The ecological consequences of shared natural enemies. Annu Rev Ecol Syst 25:495–520

Huber-Sannwald E, Pyke DA, Caldwell MM (1997) Perception of neighbouring plants by rhizomes and roots: morphological manifestations of a clonal plant. Can J Bot 75:2146–2157

Jentschke G, Bonkowski M, Godbold DL, Scheu S (1995) Soil protozoa and forest tree growth: non-nutritional effects and interaction with mycorrhizas. Biol Fertil Soils 20:263–269

Jingguo W, Bakken LR (1997) Competition for nitrogen during mineralization of plant residues in soil: microbial response to C and N availability. Soil Biol Biochem 29:163–170

Johnson NC, Graham JH, Smith FA (1997) Functioning of mycorrhizal associations along the mutualism–parasitism continuum. New Phytol 135:575–585

Jones CG, Lawton JH, Shachak M (1994) Organisms as ecosystem engineers. Oikos 69:373–386

Jones DL, Darrah PR (1995) Influx and efflux of organic acids across the soil–root interface of *Zea mays* L. and its implications in rhizosphere C flow. Plant Soil 173:103–109

Kaye JP, Hart SC (1997) Competition for nitrogen between plants and soil microorganisms. Trends Ecol Evol 12:139–143

Kloepper JW, Leong J, Teitze M, Schroth MN (1980) Enhanced plant growth by siderophores produced by plant growth-promoting rhizobacteria. Nature 286:885–886

Kraffczyk I, Trolldenier G, Beringer H (1984) Soluble root exudates of maize: influence of potassium supply and rhizosphere microorganisms. Soil Biol Biochem 16:315–322

Kuikman PJ, Jansen AG, van Veen JA, Zehnder AJB (1990) Protozoan predation and the turnover of soil organic carbon and nitrogen in the presence of plants. Biol Fertil Soils 10:22–28

Kuzyakov Y (2002) Review: factors affecting rhizosphere priming effects. J Plant Nutr Soil Sci 165:382–396

Lambrecht M, Okon Y, Vande Broek A, Vanderleyden J (2000) Indole-3-acetic acid: a reciprocal signalling molecule in bacteria–plant interactions. Trends Microbiol 8:298–300

Lavelle P, Bignell D, Lepage M, Wolters V, Roger P, Ineson P, Heal OW, Dhillion S (1997) Soil function in a changing world: the role of invertebrate ecosystem engineers. Eur J Soil Sci 33:159–193

Lawton JH, McNeill S (1979) Between the devil and the deep blue sea: on the problems of being a herbivore. In: Anderson RM, Turner BD, Taylor LR (eds) Population dynamics. Blackwell, Oxford, pp 223–244

Lerdau M, Coley PD (2002) Benefits of the carbon–nutrient balance hypothesis. Oikos 98:533–535

Lipson DA, Raab TK, Schmidt SK, Monson RK (1999a) Variation in competitive abilities of plants and microbes for specific amino acids. Biol Fertil Soils 29:257–261

Lipson DA, Schmidt SK, Monson RK (1999b) Links between microbial population dynamics and nitrogen availability in an alpine ecosystem. Ecology 80:162–163

Lorio P (1986) Growth-differentiation balance: a basis for understanding southern pine beetle–tree interactions. For Ecol Manage 14:259–273

Lynch JM, Whipps JM (1990) Substrate flow in the rhizosphere. Plant Soil 129:1–10

Marschner H (1992) Nutrient dynamics at the soil–root interface (rhizosphere). In: Read DJ, Lewis DH, Fitter AH, Alexander IJ (eds) Mycorrhizas in ecosystems. CAB International, Wallingford, pp 3–12

Mathesius U, Mulders S, Gao M, Teplitski M, Caetano-Anollés G, Rolfe BG, Bauer WD (2003) Extensive and specific responses of a eukaryote to bacterial quorum-sensing signals. Proc Natl Acad Sci USA 100:1444–1449
Mattson WJ (1980) Herbivory in relation to plant nitrogen content. Annu Rev Ecol Syst 11:119–161
McNaughton SJ, Oesterheld M, Frank DA, Williams KJ (1989) Ecosystem-level patterns of primary productivity and herbivory in terrestrial habitats. Nature 341:142–144
Moore JC, Hunt HW (1988) Resource compartmentation and the stability of real ecosystems. Nature 333:261–263
Mutikainen P, Walls M, Ovaska J, Keinänen M, Julkunen-Tiitto R, Vapaavuori E (2002) Costs of herbivore resistance in clonal saplings of *Betula pendula*. Oecologia 133:364–371
Newman EI (1988) Mycorrhizal links between plants: their functioning and ecological significance. Adv Ecol Res 18:243–271
Nitao JK, Zangerl AR, Berenbaum MR (2002) CNB: requiescat in pace? Oikos 98:540–546
Obreht Z, Kerby NW, Gantar M, Rowell P (1993) Effects of root-associated N_2-fixing cyanobacteria on the growth and nitrogen content of wheat (*Triticum vulgare* L.) seedlings. Biol Fertil Soils 15:68–72
Oksanen L, Aunapuu M, Oksanen T, Schneider M, Ekerholm P, Lundberg PA, Armulik T, Aruoja V, Bondestad L (1997) Outlines of food webs in a low arctic tundra landscape in relation to three theories on trophic dynamics. In: Gange AC, Brown VK (eds) Multitrophic interactions in terrestrial systems. Blackwell, Oxford, pp 351–373
Patten CL, Glick BR (1996) Bacterial biosynthesis of indole-3-acetic acid. Can J Microbiol 42:207–220
Petersen DJ, Srinivasan M, Chanway CP (1996) *Bacillus polymyxa* stimulates increased *Rhizobium etli* populations and nodulation when co-resident in the rhizosphere of *Phaseolus vulgaris*. FEMS Microbiol Lett 142:271–276
Phillips DA, Strong DR (2003) Rhizosphere control points: molecules to food webs. Ecology 84:815
Phillips DA, Ferris H, Cook DR, Strong DR (2003) Molecular control points in rhizosphere food webs. Ecology 84:816–826
Polis GA (1991) Complex trophic interactions in deserts: an empirical critique of food-web theory. Am Nat 138:123–155
Polis GA (1994) Food webs, trophic cascades and community structure. Aust J Ecol 19:121–136
Polis GA, Strong DR (1996) Food web complexity and community dynamics. Am Nat 147:813–846
Price PW (1991) The plant vigor hypotheses and herbivore attack. Oikos 62:244–251
Puri G, Ashman MR (1999) Microbial immobilization of ^{15}N-labelled ammonium and nitrate in a temperate woodland soil. Soil Biol Biochem 31:929–931
Reichle DE, O'Neill RV, Harris WF (1975) Principles of energy and material exchange in ecosystems. In: Van Dobben WH, Lowe McConnell RH (eds) Unifying concepts in ecology. Junk, The Hague, pp 27–43
Reinhold-Hurek B, Hurek T (1997) *Azoarcus* spp. and their interactions with grass roots. Plant Soil 194:57–64
Ritz K, Griffiths BS (1987) Effects of carbon and nitrate additions to soil upon leaching of nitrate, microbial predators and nitrogen uptake by plants. Plant Soil 102:289–237
Robinson D (1994) The response of plants to non-uniform supplies of nutrients. New Phytol 127:635–674
Rolfe BG, Djordjevic MA, Weinman JJ, Mathesius U, Pittock C, Gärtner E, Ride EM, Dong Z, McCully M, McIver J (1997) Root morphogenesis in legumes and cereals and the effect of bacterial inoculation on root development. Plant Soil 194:131–144

Rovira AD (1991) Rhizosphere research – 85 years of progress and frustration. In: Kleister DL, Cregan PB (eds) The rhizosphere and plant growth. Kluwer, Dordrecht, pp 3–13

Ryle GJA, Powell CE, Gordon AJ (1979) The respiratory costs of nitrogen fixation in soybean, cowpea and white clover. J Exp Bot 30:145–153

Scheu S (1993) There is an earthworm mobilizable nitrogen pool in soil. Pedobiologia 37:1–7

Scheu S (2001) Plants and generalist predators as links between the below-ground and above-ground system. Basic Appl Ecol 2:3–13

Scheu S, Setälä H (2002) Multitrophic interactions in decomposer food webs. In: Tscharntke T, Hawkins BA (eds) Multitrophic level interactions. Cambridge University Press, Cambridge, pp 223–264

Scheu S, Theenhaus A, Jones H (1999) Links between the detritivore and the herbivore system: effects of earthworms and Collembola on plant growth and aphid development. Oecologia 119:541–551

Schulman OP, Tiunov AV (1999) Leaf litter fragmentation by the earthworm *Lumbricus terrestris* L. Pedobiologia 43:453–458

Settle WH, Ariawan H, Tri Astuti E, Cahyana W, Hakim AL, Hindayana D, Sri Lestari A, Sartano P (1996) Managing tropical rice pests through conservation of generalist natural enemies and alternative prey. Ecology 77:1975–1988

Shishido M, Massicotte HB, Chanway CP (1996) Effect of plant growth promoting *Bacillus* strains on pine and spruce seedling growth and mycorrhizal infection. Ann Bot 77:433–441

Smith SE, Read DJ (1997) Mycorrhizal symbiosis. Academic Press, London

Snyder WE, Wise DH (1999) Predator interference and the establishment of generalist predator populations for biocontrol. Biol Control 15:283–292

Söderström B (1992) The ecological potential of the ectomycorrhizal mycelium. In: Read DJ, Lewis DH, Fitter AH, Alexander IJ (eds) Mycorrhizas in ecosystems. CAB International, Wallingford, pp 77–83

Steenhoudt O, Vanderleyden J (2000) *Azospirillum*, a free-living nitrogen-fixing bacterium closely associated with grasses: genetic, biochemical and ecological aspects. FEMS Microbiol Rev 24:487–506

Stephens PM, Davoren CW (1997) Influence of the earthworms *Aporrectodea trapezoides* and *A. rosea* on the disease severity of *Rhizoctonia solani* on subterranean clover and ryegrass. Soil Biol Biochem 29:511–516

Symondson WOC, Sunderland KD, Greenstone MH (2002) Can generalist predators be effective biocontrol agents? Annu Rev Entomol 47:561–594

van der Meijden E, Klinkhamer PGL (2000) Conflicting interests of plants and the natural enemies of herbivores. Oikos 89:202–208

Verhagen FJM, Hagemann PEJ, Woldendorp JW, Laanbroek HJ (1994) Competition for ammonium between nitrifying bacteria and plant roots in soil in pots; effects of grazing by flagellates and fertilization. Soil Biol Biochem 26:89–96

Wamberg C, Christensen S, Jakobsen I (2003) Interaction between foliar-feeding insects, mycorrhizal fungi, and rhizosphere protozoa on pea plants. Pedobiologia 47:281–287

Wang JG, Bakken LR (1997) Competition for nitrogen during mineralization of plant residues in soil: microbial response to C and N. Soil Biol Biochem 29:163–170

Wardle DA (1992) A comparative assessment of factors which influence microbial biomass carbon and nitrogen levels in soil. Biol Rev 67:321–358

Wardle DA, Yeates GW (1993) The dual importance of competition and predation as regulatory forces in terrestrial ecosystems: evidence from decomposer food webs. Oecologia 93:303–306

White TCR (1993) The inadequate environment: nitrogen and the abundance of animals. Springer, Berlin Heidelberg New York

Wise DH, Snyder WE, Tuntibunpakul P, Halaj J (1999) Spiders in decomposition food webs of agroecosystems: theory and evidence. J Arachnol 27:363–370

Wurst S, Jones TH (2003) Indirect effects of earthworms (*Aporrectodea caliginosa*) on an above-ground tritrophic interaction. Pedobiologia 47:91–97

Wurst S, Langel R, Reineking A, Bonkowski M, Scheu S (2003) Effects of earthworms and organic litter distribution on plant performance and aphid reproduction. Oecologia 137:90–96

Wurst S, Dugassa-Gobena D, Scheu S (2004) Earthworms and litter distribution affect plant defensive chemistry. J Chem Ecol 30(4)

5 Belowground Herbivores and Ecosystem Processes

G.J. Masters

5.1 Summary

There is increasing awareness of the importance of belowground herbivores in the structure and function of ecological communities. However, experimenting with root feeders, especially to elucidate ecosystem effects, is difficult and involves techniques that may have indirect effects on the parameters of interest. Belowground insect herbivory affects many aspects of ecosystem function, particularly productivity, allocation patterns, nutrient cycling and interactions between component species. However, there is a likely general continuum of response to belowground herbivory. Low levels of root feeding possibly result in a root system that is more efficient and so may benefit the host plant; high levels of root feeding have dramatic negative impacts on plant growth, often causing plant death. At moderate levels of root herbivory, whether there is a negative, positive or no effect on plant growth is likely to be determined by other factors such as soil nutrient and water content and competition.

Root feeding changes the nutrient content of host plants, leading to increased aboveground insect herbivore growth and fecundity, resulting in changed population sizes and shifts in community structure. Recent evidence suggests that these plant-mediated interactions can also affect higher trophic levels: tephritid seed predators and their parasitoids were recorded in greater abundance on thistles subjected to root herbivory. Changes in the quality, quantity and composition of the foliage, induced by root feeding, can lead to increased litter quality, quantity and composition. This is known to impact on the size and structure of comminutor assemblages. Consequently, underground herbivory may affect decomposition, mineralization and subsequent nutrient availability.

As quoted by MoronRios et al. (1997), "Below-ground herbivores have been poorly studied regardless of their importance for the establishment and composition of plant communities".

Ecological Studies, Vol. 173
W.W. Weisser and E. Siemann (Eds.)
Insects and Ecosystem Function

5.2 Introduction

Until relatively recently, belowground herbivory by insects has been largely ignored by ecologists, however, there is increasing awareness of the importance of belowground herbivores in natural communities (e.g. Brown and Gange 1990; Masters and Brown 1997; van Dam et al. 2003). In comparison with aboveground insect herbivores, only a few families have evolved the belowground habit. Of these families though, belowground herbivores have become, generally, major pests of crops, horticulture and pasture on a global scale (e.g. Potter and Held 2002); for example, feeding by members of the Tipulidae, Scarabaeidae and Noctuidae led to a 32 % reduction of permanent pasture yield loss (Clements and Bentley 1983; Clements 1984). Root-feeding insects are damaging due to the location of their feeding. Belowground plant structures provide stability, anchorage, the ability to forage and absorb water and nutrients, storage for metabolites and photosynthates and are the site for many processes such as the initiation of vegetative reproduction or the synthesis of compounds for shoot growth and development (Jeffrey 1987). It has been estimated that 50–90 % of net primary production is allocated belowground in habitats dominated by perennial plant species (e.g. Coleman 1976; Fogel 1985). Thus, root feeding can have severe consequences for plant processes, particularly nutrient/water uptake and C allocation.

It is not surprising that belowground insect herbivores have the potential to impact significantly ecosystem function, dynamics and structure. A problem is designing experiments that demonstrate conclusively such impacts considering that the test organism is not readily observable – hence assumptions have to be made regarding its activities. In this chapter, the manipulation of insect herbivores below ground is discussed before considering the results of such experiments through examining the consequences of belowground herbivory for plant productivity, community structure, biodiversity and implications for nutrient cycling. Further impacts on ecosystem structure and function will be considered by detailing the indirect effects of root feeding for other components of the community, particularly aboveground insect herbivores.

5.3 Experimenting with Belowground Insect Herbivores

There are two approaches for testing the implications of root herbivory for ecosystem structure and function: adding or eliminating root feeders from the system and recording the response (cf. Chaps. 12–17, this Vol.). Field observations provide further insight, for example comparing plant community structure between root-herbivore-infested and non-infested patches of semi-natural grassland. Due to the nature and potential consequences of belowground insect herbivory and the patchy distribution of many root feeders, it is important to examine effects over a range of scales: from developing hypotheses from conceptual models and field observations, to tightly controlled laboratory, controlled environment or greenhouse experiments, through isolated garden trials (where natural variability is introduced), to large-scale manipulative field experiments with communities conducted over longer time periods.

Examining, for example, the impact of root herbivory on plant growth, or productivity, may involve adding root feeders at different densities in addition to exclusion (e.g. Masters 1995a, 1995b). Additions imply a ready supply of root-feeding insects, either from a field collection, in which the insects may be held in a dormant state in a low-temperature-controlled environment, or through the development of some culturing technique. Field collections do not guarantee that the root feeders are all the same species (given the difficulty of identifying root-feeding beetle larvae, e.g. chafer grubs, for example) or are all at the same stage of development, although this could be assumed/standardized by using root feeders of the same size and mass. Culturing ensures that the herbivores are of the same species, are similar if not identical genetically and are at the same stage of development. However, establishing a culture can be difficult, with only a few species seeming to be amenable to mass rearing, and owing to the problem of having access to suitable facilities to enable the production of possibly large numbers of insects. With a ready supply of insects, conducting laboratory, greenhouse or microcosm experiments is relatively straightforward. However, as Masters (1992) illustrated, the choice of substrate is important to ensure normal feeding behaviour.

Exclusion in the field often involves the use of soil insecticide. Limitations with insecticide use have been well known since early experimental work by Shure (1971), and in the context of ecosystem function there may well be interference with measurements of C and N fluxes, although this remains to be fully tested (Siemann et al., Chap. 15, this Vol.) It is essential to use a tried and tested soil insecticide to have any confidence in its effectiveness and in the validity/power of the results (see Evans 1991). One of the main problems with a soil insecticide is that its application to the treated area invariably involves some form of contact with the aboveground biota and so can potentially bias the results. An effective way around this is to use granules as a carrier for the active ingredient. Granules can be sprinkled over the vegetation when dry, so the grains do not stick to the vegetation and fall through the sward to the soil surface, and then are irrigated straight into the soil. This technique has been used successfully many times (e.g. Brown and Gange 1989a, b; Masters 1995a; Masters et al. 2001). A compound commonly used is the granular formulation of chlorpyrifos (5 % w/w), Dursban 5G (DowElanco Chemical Company), applied at the recommended agricultural rate (2 g m^{-2}).

Dursban 5G has little effect on micro-organisms and nematodes (Anonymous 1985), no toxic effects on nitrogen-fixing bacteria, nitrifying and denitrifying bacteria and fungal populations (Pozo et al. 1995), small mammals and birds (Clements and Bale 1988) or earthworms (Clements et al. 1986), but is known to be highly effective in reducing the total number of insect root herbivores (Hector 1996). However, chlorpyrifos is likely to have side effects on Collembola, which perform an important function as comminutors [generally arthropods, particularly insects with hard biting jaws, which break down material, e.g. in the soil or litter, into smaller pieces (comminution), thus allowing, for example, decomposition] and fungal feeders, hence being important for ecosystem function (Bonkowski and Scheu, Chap. 4, this Vol.; see also below). Additionally, Dursban 5G is reported to have no phytotoxic or phytostimulant effects on a range of plant species (Brown and Gange 1989a; Whitehouse 1993; Maron 1998). Indeed, Masters (1992) showed that Dursban 5G had little direct contact effect on foliar-feeding aphid abundance when sprinkled over plants in a pot trial simulating field application. Unfortunately, this granular formulation of chlorpyrifos appears to be no longer readily available in some countries.

Insecticides used on turf-grass systems can be applied to areas with flowering weeds that attract native pollinators. Gels et al. (2002) found that granular insecticide for control of root-feeding grubs had no effect on pollinator colony vitality or bumble bee worker behaviour (*Bombus impatiens*; Hymenoptera: Aphidae) when applied with post-treatment irrigation to wash it straight into the soil. However, dry non-irrigated residues were found to have a detrimental effect on colony vitality.

A means of delivering insecticide to targeted areas of the plant on a small scale was demonstrated by Zhou and Smith (1996). They used stem infusion methods that delivered pressurized solutions through syringe needles sealed to the stem with latex. This technique can be used in the field or greenhouse but is limited by the number of plants that can be injected; hence for ecologists working at the population or community level this technique may not be appropriate.

Alternative exclusion techniques involve the use of barriers to prevent root-feeding insects entering the experimental arena. However, such methods rely on the fact that there were no root herbivores around the excluded area to start with. Selective exclusion could be achieved using material of different pore/mesh size as the barrier. This enables organisms of appropriate size to enter the exclusion zone and the relative importance of the macro, meso- and micro-soil fauna to be tested.

5.4 Belowground Herbivory and Plant Productivity: Allocation and Biomass

Generally, herbivory, by its nature, imposes a cost on host plants, whether it is reduced biomass, altered allocation patterns, reallocation of resources to compensatory or defensive mechanisms. Root feeding by insects is no exception. Typical consequences of root herbivory for productivity were demonstrated in a study by Preus and Morrow (1999): herbivory by the rhizome galler *Eurosta comma* (Diptera: Tephritidae) reduced the number of new rhizomes of goldenrod *Solidago missouriensis* (Asteraceae) by up to 20 % and the plant allocated less biomass to the leaves and stems but increased biomass allocation to the roots. An earlier study by Notzold et al. (1998) also demonstrated a range of negative effects. They examined the impact of *Hylobius transversovittatus* (Coleoptera: Curculionidae) larval feeding on the roots of *Lythrum salicaria* (Lythraceae) and demonstrated that belowground herbivory resulted in decreased plant height (stature), decreased biomass of all plant parts, delayed and shortened flowering period and changed biomass allocation patterns. Both of these studies suggest that herbivory by belowground insects resulted in reduced plant growth and changed C allocation patterns.

Root herbivory by either chewing or sucking insects can induce a stress response within the host plant (e.g. Ridsdill Smith 1977; Gange and Brown 1989; Masters et al. 1993; van Dam et al. 2003; Wäckers and Bezemer 2003). Root chewers remove root material, often the finer roots that are important in the foraging and uptake of water and nutrients. Although root-sucking insects do not physically *remove* roots, they do generate a sink within the root system, thus preventing normal root function and disrupting foraging/uptake of water and nutrients. In either case, the plant is limited in its ability to obtain sufficient nutrients/water for sustained growth and a stress response is induced (Masters 1992; Masters et al. 1993). During such a response the plant mobilizes or translocates soluble carbohydrate (sugars) and soluble N (amino acids) to the foliage (e.g. Hsiao 1973; Brodbeck and Strong 1987). Hence, root herbivory leads to a shift in allocation with more soluble C and soluble N being found in the foliage and a reduction in the amount of water in the foliage as an indication of water stress (Fig. 5.1).

Contrary to the increased levels of foliar soluble N and soluble carbohydrate shown in Fig. 5.1, Dawson et al. (2002) and Murray et al. (1996) recorded decreased levels of plant N and total N and C respectively in *Trifolium repens* (Fabaceae) subjected to root herbivory by tipulid larvae. Such a discrepancy could be due to *T. repens* being an N-fixer and so potentially having a different physiological response to root herbivory. Another possibility though is the possible disparity between recording the total and soluble fractions of N and C. Soluble N is a small component of total N within a plant, so a change in the

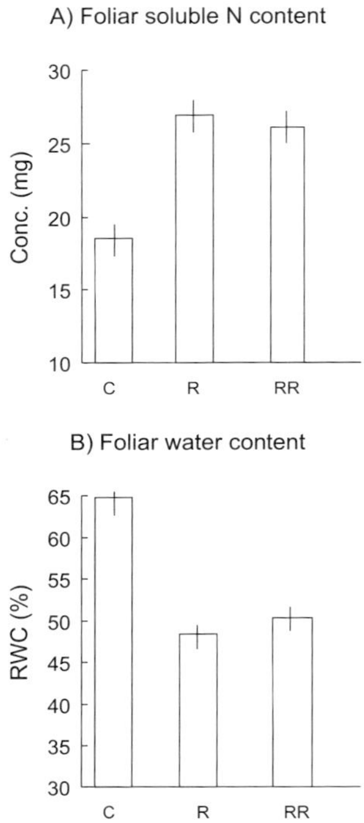

Fig. 5.1. Effects of root herbivory by the chafer *Phyllopertha horticola* (Coleoptera: Scarabaeidae) at two densities (*R* 70 individuals m^{-2} and *RR* 210 individuals m^{-2}) on foliar physiology, soluble N (**A**) and relative water content (*RWC*) (**B**) of the annual *Sonchus oleraceus* (Asteraceae). Root feeding leads to a significant increase in soluble N (ANOVA $F_{5,17}=3.11$, $P<0.05$; see Masters 1995b for full details) and reduction in RWC (ANOVA $F_{5,59}=9.96$, $P<0.001$; see Masters 1995b for full details), consistent with the stress hypothesis (see Masters et al. 1993). *Error bars* are ±1 SEM. (Modified from Masters 1995b)

levels of soluble N may not be recorded if only total N is measured. Soluble N is important for vascular bundle feeders ('sucking insects') and more specialist guilds such as cell feeders, miners and gallers. This is a continuing problem in ecology, with reports of the N concentration but no clarification of what form of N was measured becoming common, and probably inhibits our ability to elucidate general mechanisms of the interaction between herbivores and plants.

Changes in allocation and physiology would be ultimately expressed as changes in biomass and the root:shoot ratio (cf. Voelckel and Baldwin, Chap. 17, this Vol.). *Popillia japonica* and *Cyclocephala lurida* (both species are Coleoptera: Scarabaeidae) (Crutchfield and Potter 1995) stimulated foliage growth of a range of turf grasses through root feeding. However, decreased plant growth has also been recorded as a consequence of root feeding, although the thistle, *Carthamus lanatus* (Asteraceae), was later found to compensate for belowground insect herbivory (Sheppard and Vitou 2000). Recent growth chamber experiments showed that root-nodule feeding by *Sitona lepidus* (Coleoptera: Curculionidae) larvae on *Trifolium repens* reduced plant biomass and led to a reduction in the root:shoot ratio. This reduction in biomass could be a consequence of a reduction in C accumulation (and hence assimilation) within the plant (Murray et al. 2002). When root herbivores are given a choice between two host plants then preferences can be often demonstrated. For example, Grayston et al. (2001) conducted a pot trial with *Agrostis capillaris* (Poaceae) and *T. repens* as pure stands and a mixture. The root-chewing *Tipula paludosa* (Diptera: Tipulidae) larvae were introduced at field density into pots and root herbivory led to a general reduction in the shoot growth of both plant species. However, a reduction in root biomass was observed for *Agrostis* alone. In the mixed swards, the larvae had a distinct preference for *T. repens*. Root aphids can have similarly dramatic effects on productivity as the root chewers. Root feeding by the woolly apple aphid led to a decrease in apple tree growth and production (yield); again this could be due to a decrease in the amount of stored carbohydrates in galled roots (Brown et al. 1995).

Masters and Brown (1992) found that the aboveground biomass of *Sonchus oleraceus* (Asteraceae) was not reduced by root herbivory by *Phyllopertha horticola* (Coleoptera: Scarabaeidae), even though there were fewer leaves. A possible explanation for a reduction in aboveground architecture but with no expected decrease in foliar biomass is compensatory photosynthesis (e.g. Nowak and Caldwell 1984). In a series of controlled environment experiments, all involving *P. horticola* and *S. oleraceus*, root feeding never had a significant effect on foliar biomass except when the root herbivore was at three times natural field density (210 individuals m^{-2}), when there was a reduction (Masters 1992, 1995a, 1995b). Also, there was a density-dependent response of the root:shoot ratio to root herbivory, with lower densities of chafer larvae increasing the root biomass and higher densities (similar to a pest infestation) decreasing root biomass (Masters 1992, 1995a, b). It is interesting to note that the increase in soluble N and soluble C shown in Fig. 5.1 was a consistent significant effect across all of these experiments, as was a subsequent increase in foliar-feeder performance (see below).

The observation that root herbivory had little impact on *S. oleraceus* aboveground biomass may be a feature of its annual life history. In a garden experiment with plant competition controlled for, Masters (1992) found no consistent effect of root herbivory (tested by the application of soil insecticide, Dursban 5G) on the annual *Arabidopsis thaliana* (Brassicaceae), while there was a reduction in the aboveground biomass of the perennial *Plantago major* (Plantaginaceae) by belowground insect herbivory. The annual has a rapid growth pattern with resources being allocated to meet this demand. Such rapid allocation to maintain growth may mask any effects of root herbivory. The perennial, however, has a comparatively slower allocation to growth and a greater investment in storage and defence. Hence, a root-feeding insect may have a greater potential to disrupt allocation above ground and so reduce vegetative biomass.

Similar shifts in plant biomass have been reported within natural communities. A common method of non-destructive recording is to estimate plant cover (e.g. Goldsmith et al. 1986). A modification of this, assessing cover abundance through recording the number of contacts of tissue from each plant species found to be touching a point quadrat pin (Goldsmith et al. 1986), was extensively used by workers such as Southwood, Gibson and Brown and has been directly related to aboveground plant biomass (Goodall 1952 cited in Grime et al. 2000). A series of manipulative field experiments designed to examine the impacts of insect herbivory [and latterly arbuscular mycorrhizal (AM) fungi as well] on plant community development were conducted by Brown, Gange and colleagues. This work demonstrated that root herbivory reduced the cover abundance of the plant community – particularly the grasses (e.g. Brown and Gange 1989a; Masters 1992), decreased plant species richness – through limiting forb establishment (e.g. Brown and Gange 1989b; Masters 1992), and thus accelerated the rate of succession (e.g. Gange and Brown 2002). In a related study, Müller-Schärer and Brown (1995) estimated the density of *Tripleurospermum perforatum* (Asteraceae) plants in an early plant succession by recording its cover abundance (reflection of the number of stems), and found that density was increased by a reduction in belowground herbivory through the application of soil insecticide. However, Clark et al. (2001) found no effect of two root-feeding insects [*Agapeta zoegana* and *Cyphocleonus achates* (Lepidoptera: Cochylidae and Coleoptera: Curculionidae respectively] on spotted knapweed [*Centaurea maculosa* (Asteraceae)] stem density.

Plant responses to herbivory are variable and where there are differences in the effects of root feeding, such as those detailed above, then one may turn to ecosystem quality/function as a possible explanation. If the host plant is growing in non-stressful or plentiful conditions (e.g. soil fertility or water status), then the host plant may be able to either resist or compensate more readily for root herbivory. For example, the effects of root herbivory by *Phyllopertha horticola* on *Capsella bursa-pastoris* (Brassicaceae) were negated by over-watering the host plant (Gange and Brown 1989). Saner and Müller-Schärer (1994) detailed an architectural response to root herbivory: the number of *Linaria vulgaris* (Scrophulariaceae) stems doubled in spring when there had been root mining by *Eteobalea* (syn. *Stagmatophora*) spp. (Lepidoptera: Cosmopterygidae) throughout the season before. This could possibly be due to root herbivory disrupting apical dominance; however, there was no effect on plant biomass.

Herbivory, by its nature, will have a negative impact on plant growth and productivity at some level if all other factors are constant. However, root herbivory by insects seems to have negative, null or stimulatory (compensatory) effects on aboveground biomass at all scales (individual, population and community). It is likely that there is a continuum of response to belowground herbivory. Low levels of root feeding possibly result in a root system that is more efficient through the removal of old roots which are replaced by more efficient young roots/rootlets. High levels of root feeding have dramatic negative impacts on plant growth (sometimes with the entire root system removed, causing plant death). At moderate levels of root herbivory, whether there is a negative, positive or no effect on plant growth is likely to be determined by other factors such as soil nutrient content, competition, soil water content, plant growth stage and life history. The stress or nutrient hypothesis first proposed by Masters (1992) and Masters et al. (1993) and developed in Masters and Brown (1997) suggests that if the stress to the host plant caused by root herbivory can be alleviated then there should be little or no effect on plant biomass. This hypothesis can be extrapolated to include biotic stress (e.g. competition) where stressed plants will be more susceptible to belowground insect herbivory.

5.5 Implications of Belowground Herbivory for Nutrient Cycling

There is little doubt of the importance of physiological responses of plants to (aboveground) herbivory for cornerstone ecosystem processes such as nutrient mineralization and decomposition through affecting soil organism populations and communities (Bardgett et al. 1998; Wardle and Bargett, Chap. 3, this Vol.). Indeed, these authors identified two broad pathways by which foliar herbivory indirectly affects the soil biota and associated ecosystem processes:

1. Changes in root exudation and C allocation in the short term, or as longer-term changes in root biomass and architecture, thus affecting the biomass and activity of the soil microbial community, hence impacting nutrient supply; and
2. Soil organism responses to shifts in plant litter quality either directly through the nutrient and secondary chemical composition of the litter, or indirectly through a foliar-herbivore-induced change in plant community composition, thus affecting the nature of the litter input.

Root-feeding insects, to a lesser or greater extent, have very similar effects, affecting the allocation and nutrient/secondary chemical composition of foliage (future litter) (Fig. 5.1) and above- and belowground biomass, thus affecting the quantity of litter input and plant community composition. It is likely that belowground insect herbivory can affect nutrient cycling, particularly the linked ecosystem processes of decomposition and nutrient mineralization, through the pathways described by Bardgett et al. (1998). To my knowledge, however, there has been no direct examination of such links for belowground insects. A simple conceptual model of possible host-plant-mediated implications of root herbivory by insects for soil processes is presented in Fig. 5.2. This model illustrates pathways by which root herbivory may lead, in certain circumstances which will probably be system/species specific, to a greater ability of the plant to take up nutrients and thus tolerate or even compensate for herbivory.

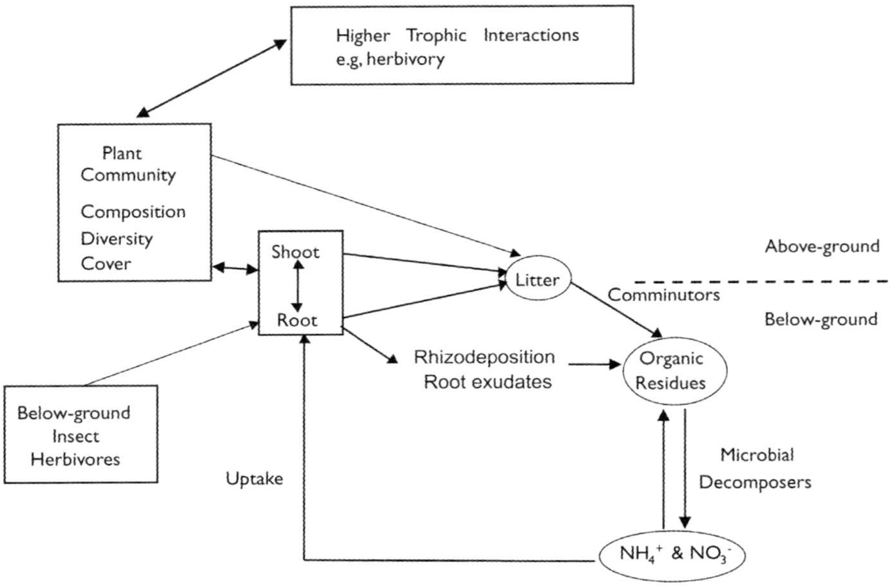

Fig. 5.2. A simple conceptual model of the role of belowground insect herbivory in nutrient cycling. Only direct interactions and processes are shown; indirect or host-plant-mediated interactions, such as effect of root feeding on plant communities and on higher trophic levels, are implied through *arrows* denoting direct interactions but are not explicitly illustrated. Changes in quantity, quality and composition of the litter input all have knock-on effects on rate of decomposition and the mineralization process by affecting soil biota. Root herbivory can affect the nutrient and water content of foliage (future litter) (Fig. 5.1) and cause shifts in plant community structure and thus the nature of litter input to the soil. If litter of a higher quality enters the system, for example, then there probably will be more available N after decomposition and mineralization for the plant. Although not tested, if this mechanism occurred then plants might be in a compensatory loop, enabling them to tolerate low/medium levels of root herbivory. While these plant-mediated effects are important, their direction cannot be predicted because of the inherent complexity of soil food web interactions, the number of mechanisms involved, and our limited understanding of the role of soil organisms in ecosystem processes

Some evidence for the model presented in Fig. 5.2 comes from combining the effects of root-feeding insects on plants (as described previously) with the additional interaction of insects with other members of the soil biota, notably AM fungi, and from studies of nematodes and ecosystem processes. Filser (2002) identified root feeding as a key mechanism, along with fungal feeding, distribution of fungal propagules and predation on nematodes, in the role Collembola play in C and N cycling. In particular, manipulative field experiments illustrated positive effects of Collembola on a range of ecosystem processes, particularly N mineralization, soil respiration, dissolved organic C leaching and plant growth. More direct evidence of insect root herbivores affecting the soil biota was provided by Grayston et al. (2001). They conducted pot trials with the grass *Agrostis capillaris* and the legume *Trifolium repens*, as pure stands and as a mixture. Tipulid larvae (*Tipula paludosa*) were introduced into pots at field density. Where the root herbivore was present, the soil microbial community structure was altered, with populations of pseudomonads increasing greatly. This shift in the soil biota was suggested as a possible factor accounting for the increased quantity and qualitative changes in C flux to the soil as a result of root herbivory, particularly increased utilization of some sugars, carboxylic and amino acids. Belowground insect herbivores have also been suggested to interact with AM fungi. Gange and Brown (2002) demonstrated that within a ruderal plant community, AM fungi had the greatest impact on plant community diversity when soil insecticide was applied to exclude insect root feeders. They suggested that this result may indicate that insect root herbivores might have some disruptive influence on the AM fungi–plant association. Indeed, Collembola can have three different functions within ecosystems, as communitors, as root grazers and as fungal feeders. This primitive group of insects are probably not true root herbivores but can have a major impact on soil fungal populations including AM fungi (Gange 2000).

Further evidence comes from the effects of soil nematodes on ecosystem processes as a surrogate system (cf. Bonkowski and Scheu, Chap. 4, this Vol.) Soil nematodes can have a beneficial role in C (and other nutrients) cycling in grassland ecosystems. For example, cell content/nutrient leakage due to cell piercing by low levels of nematode root herbivory led to enhanced microbial activity, increased total microbial biomass (Denton et al. 1999), and increased the allocation of photoassimilate to roots (which may be detrimental to the individual host plant), resulting in increased root exudation and microbial activity in the rhizosphere (Bardgett et al. 1999a). If root herbivory can lead to a greater and more active soil microbial biota, then this in turn could lead to greater decomposition and mineralization resulting in increased available C and N for plant uptake (Fig. 5.2), hence decreasing the probability that the plant suffers from nutrient stress caused by root feeding, or enabling plants to resist root herbivory through buffering the effects of root consumption (Villalobos et al. 1997). However, neither Denton et al. (1999) nor Bardgett et al. (1999a) found any positive effects on plants from the increased microbial biomass caused by low levels of root herbivory. An alternative hypothesis is that a greater and more active soil biota, resulting from root herbivory, will lead to increased asymmetric competition between roots and microbes for nutrients (with microbes benefiting from the interaction). Indeed, the competitive balance between plants within a community may well be affected by root herbivory. Bardgett et al. (1999b) suggest that competing roots of plants neighbouring the host plant may gain an indirect benefit from root herbivory which may lead to an altered successional trajectory. These authors found that root herbivory on *Trifolium* by clover-specific nematodes resulted in greater uptake of nutrients (N) by co-existing grasses.

5.6 Implications of Belowground Herbivory for Multitrophic Interactions

A key aspect of ecosystem function is the role that individual species and functional groups have in driving and governing ecosystem processes. Any significant change in the balance of species, through altered performance and ultimately fitness, within an assemblage or community will lead to a shift in ecosystem function which may, or may not, be of longer-term importance (Joshi et al., Chap. 6, this Vol.; Janssen and Sabelis, Chap. 9, this Vol.; Schmitz, Chap. 14, this Vol.). Given the plethora of effects that belowground insect herbivores have on plants (e.g. growth, physiology, fecundity, architecture and diversity), there is clearly the potential for host-plant-mediated interactions with aboveground phytophages and vice versa. Such interactions can lead to changes in population dynamics and community structure, potentially leading to knock-on effects for ecosystem function.

In a controlled environment experiment, Masters and Brown (1992) demonstrated that root feeding by *Phyllopertha horticola* increased the fecundity by 20 % of the leaf-miner, *Chromatomyia syngenesiae* (Diptera: Agromyzidae). In a series of controlled environment experiments, all involving *P. horticola* at field (70 individuals m^{-2}) and infestation (210 individuals m^{-2}) densities, the performance and fecundity of the aphids *Aphis fabae* and *Myzus persicae* (both Hemiptera: Homoptera: Aphididae) and larvae of *Mamestra brassicae* (Lepidoptera: Noctuidae) were increased by 30–50 % by root feeding (Gange and Brown 1989; Masters 1992, 1995a, b; Fig. 5.3). Masters et al. (1993) suggested that belowground insect herbivory resulted in a host-plant-mediated beneficial effect on foliar-feeding insects (and an indirect negative effect of aboveground insect herbivory on root feeders through food limitation via decreased root biomass, as also shown by Crutchfield and Potter 1995). Root feeding limits the plant's ability to take up and forage for water and nutrients (Ridsdill Smith 1977), thus leading to a stress response being induced within the host plant. Such responses generally lead to the mobilization or translocation of soluble C and N to the foliage (Fig. 5.1; see Hsiao 1973; Brodbeck and Strong 1987). In all these controlled environment studies, the increase in foliar feeder performance when sharing a host plant subjected to root herbivory was associated with a physiological response consistent with the stress hypothesis (Masters et al. 1993), either a reduction in relative water content or an increase in soluble N and soluble carbohydrate (Gange and Brown 1989; Masters 1992, 1995a, b; Masters and Brown 1992). This increased food quality for herbivores leads to the

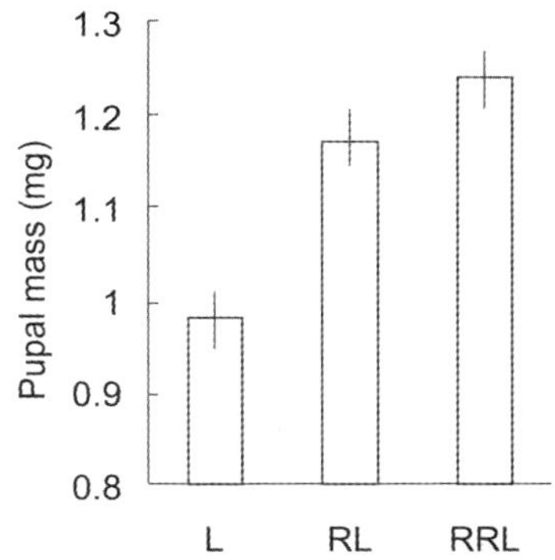

Fig. 5.3. Effects of root herbivory by the chafer *Phyllopertha horticola* (Coleoptera: Scarabaeidae) at two densities (*R* 70 individuals m^{-2} and *RR* 210 individuals m^{-2}) on fecundity (estimated by measuring pupal mass, where greater female pupal mass leads to greater fecundity; see Quiring and McNeil 1984) of the leaf-miner *Chromatomyia syngenesiae* (Diptera: Agromyzidae) feeding on the annual *Sonchus oleraceus* (Asteraceae). Root feeding leads to a significant increase in leaf-miner pupal mass (a measure of fecundity) (ANOVA $F_{3,29}=3.84$; $P<0.05$; see Masters 1995b for full details). *L* Leaf miner foliar herbivory. *Error bars* are ±1 SEM. (Modified from Masters 1995b)

recorded increased performance of foliar-feeding insects, which can result in a change in the population dynamics and community structure of the aboveground phytophages (cf. Bonkowski and Scheu, Chap. 4, this Vol.). Garden experiments, involving the reduction of insect root herbivory through the application of soil insecticide and the enhancement of root herbivory by introducing more belowground insect herbivores, and manipulative field experiments, involving the application of soil insecticide, found that root herbivory led to an increase in stem-/foliar-feeding Hemipteran populations (Moran and Whitham 1990; Masters 1992, 1995b) and changed the structure of hemipteran communities across acidic successional grasslands of different age: plots subjected to root herbivory supported more diverse and larger populations of insect herbivores even though the plant community had reduced architecture, biomass and diversity (Masters 1992; Masters et al. 1993; Masters and Brown 1997).

The evidence does suggest that root herbivory leads to an increase in the performance of foliar-feeding insects as individuals, populations and communities, mediated through physiological changes (nutrient) in the host plants, as predicted by Masters et al. (1993). Indeed, Murray and Hatch (1994) demonstrated that root herbivory by *Sitona* weevils on *Trifolium repens* led to an increase in the N levels of co-existing *Lolium perenne* (Poaceae) plants. They suggest that this was due to nodule feeding by the weevil, mediating the transfer of N between the two plants. Recently, root herbivory has been shown to lead to increased levels of seed predation by *Terellia ruficauda* (Diptera: Tephritidae) of *Cirsium palustre* (Asteraceae) (Masters et al. 2001). This suggests that it is not just foliar-/stem-feeding insects that can be affected by belowground insect herbivores. Additionally, there was a knock-on effect (see Hartley and Jones, Chap. 2, this Vol.) on the parasitoids [particularly *Torymus chloromerus* (Hymenoptera: Torymidae) and *Pteromalus elevatus* (Hymenoptera: Pteromalidae)] with a greater incidence of parasitism of this seed-pre-

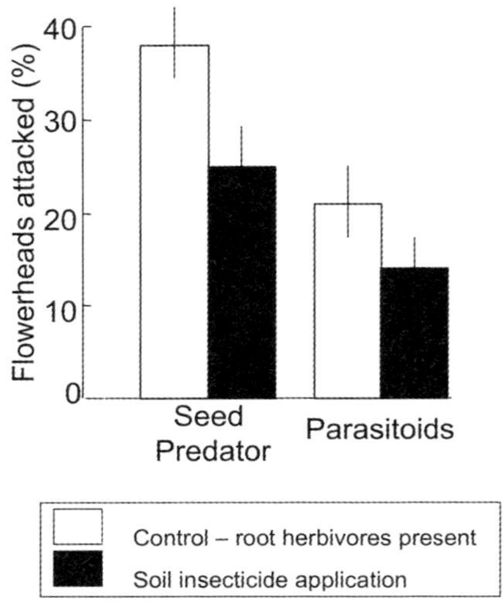

Fig. 5.4. Effect of root herbivory on mean levels of attack (±1 SEM) by the seed predator *Terellia ruficauda* (Diptera: Tephritidae) and its parasitoids on the marsh thistle *Cirsium palustre* (Asteraceae). *Open bars* represent root herbivory at field density, *solid bars* represent where soil insecticide was applied to exclude root-feeding insects. Root herbivory led to greater levels of attack of seedheads by the seed predator and on it by its parasitoids (*t*-test: $t_{predator}$=2.31, p <0.05 and *t*-test: $t_{parasitoids}$=2.08, P <0.05; see Masters et al. 2001 for full details). This increased parasitoid abundance did not, however, lead to higher levels of tephritid parasitism (parasitism of tephritids in untreated thistles = 52.7±8.1 %, treated thistles = 53.9±6.3 %; see Masters et al. 2001) (Modified from Masters et al. 2001)

dating fly where root herbivores were present (Fig. 5.4), suggesting a truly multitrophic interaction.

A counter mechanism to the nutrient-based interaction between root herbivores and their aboveground counterparts has recently been demonstrated by Wäckers and Bezemer (2003). Root herbivory was shown to induce a possible aboveground indirect defence. Cotton plants, *Gossypium herbaceum* (Malvaceae), exposed to root-feeding *Agriotes lineatus* (Coleoptera: Elateridae) had greater foliar extrafloral nectar production in comparison to undamaged plants. Extrafloral nectar production has been associated with predator attraction, thus increasing the potential protection of the plant against aboveground insect herbivores. Wäckers and Bezemer (2003) suggest that root-feeding herbivores may alter such aboveground defensive interactions. This is an area that needs further investigation, particularly a clear inclusive demonstration of root herbivory leading to an increase in defensive compounds which then, in situ, have a negative effect on aboveground insect herbivores (van Dam et al. 2003; Voelckel and Baldwin, Chap. 17, this Vol.).

Although the above studies provide insight into aboveground community structure and function, the importance of other components of the soil biota cannot be ignored when considering how the above- and belowground communities interact (e.g. Gange and Bower 1997; Bonkowski and Scheu, Chap. 4, this Vol.). A single example is that detailed by Gange (2001) where he examined the responses of root-feeding *Otiorhynchus sulcatus* (Coleoptera: Curculionidae) larvae and the leaf-feeding adults to the colonization of strawberry (*Fragaria* × *ananassa*) (Rosaceae) by one or two species of AM fungi.

Non-additive interactions were recorded where the presence of either fungal species reduced larval survival and biomass, but the effect disappeared when both fungal species were together. Mycorrhizae also mitigated the effects of larval feeding on adult weevils and root feeding only decreased plant foliar and root biomass when the fungi were absent or inoculated together. Such research illustrates the potential for AM fungi to mitigate interactions between herbivorous insects and the associated consequences for ecosystem structure and function.

There is also the reciprocal with aboveground insect herbivory affecting the performance of their root-feeding counterparts. For example, leaf galling of *Chenopodium album* (Chenopodiaceae) by *Hayhurstia atriplicis* (Hemiptera: Homoptera: Aphididae) decreased root aphid (*Pemphigus betae*) (Hemiptera: Homoptera: Pemphigidae) populations sharing the same host plant by an average of 91 % (Moran and Whitham 1990). Similarly, a 25 % reduction in root biomass of the annual *Sonchus oleraceus* by the leaf miner *Chromatomyia syngenesiae* led to a *negative* growth rate of *Phyllopertha horticola* sharing the same host plant (the chafer grub was losing weight, essentially starving) (Masters 1995b).

Root quality and quantity will be important in determining the growth and performance of any belowground insect herbivore. The quantity argument, as suggested by Masters et al. (1993), relies on the assumption that roots are of generally poor quality overall for herbivores and so it is biomass of root available for consumption that will drive the herbivore response. Foliar herbivory generally leads to a reduction in root biomass (e.g. Crawley 1983) largely due to C reallocation to sustain growth (Holland et al. 1996). However, this hypothesis suggests that plants of different life histories/growth forms could affect root herbivores in different ways. Defoliation of a perennial plant generally stimulates the restoration of the root:shoot ratio by increasing shoot growth through translocating stored assimilates from the roots, thus reducing root biomass (e.g. McNaughton 1983). This will decrease the root C:N ratio, hence increasing relative food quality for root herbivores. However, annual plants do not transport a high proportion of primary productivity to the root system for storage (Mooney 1972). When an annual is defoliated, the root:shoot ratio is likely to be restored by diverting the products of primary productivity to the shoots for foliar regrowth, thereby decreasing root biomass and quality and limiting food availability for the root-feeding insects (Masters et al. 1993).

Two general outcomes of aboveground herbivory on C flow are short-term changes in plant C allocation and root exudation or longer-term changes in root biomass and morphology (Bardgett et al. 1998). The use of C labelling techniques has provided a mechanistic understanding of how aboveground herbivory affects C allocation and root exudation. Grasshopper (*Romalea guttata*) (Orthoptera: Acrididae) herbivory led to up to 50 % of *Zea mays* (Poaceae) leaf area being removed and resulted in increased allocation of assimilate to roots and decreased allocation to shoots, as shown by $^{14}CO_2$-C pulse labelling techniques (Holland et al. 1996). Additionally, Dyer et al. (1991) used ^{11}C labelling techniques to show that feeding by *Melanoplus sanguinipes* (Orthoptera: Acrididae) on *Panicum coloratum* (a C4 grass; Poaceae) increased translocation of labile C to the roots. In both of these cases, the quality of the roots will be decreased through an increase in the C:N ratio and root herbivore performance would be predicted to decrease (Masters et al. 1993). Such changes in allocation are short term and need to be interpreted with caution when dealing with belowground insect herbivores, who generally have long life cycles (possibly because of generally poor food quality). Short-term storage of assimilates in roots following aboveground grazing enables rapid mobilization of C reserves for regrowth (Dyer et al. 1991; Holland et al. 1996).

Longer-term consequences of aboveground herbivory include reduced root biomass and altered root morphology and architecture, probably as a result of reduced allocation of C below ground (Bardgett et al. 1998). Biomass, morphology and architecture will all be important for belowground insect herbivores as resource quantity and feeding niches (distribution of the preferred fine roots). As already mentioned, changes in root biomass was the proposed mechanism for the negative impact of foliovores on root-feeding insects (Masters 1992, 1995b; Masters et al. 1993; Masters and Brown 1997). However, it should be noted that enhancement of root biomass by grazing is possible (e.g. Milchunas and Lauenroth 1993). This may lead to the opposite, where aboveground feeding can have a positive effect on root-feeding insects. For example, Ingham and Detling (1984) recorded greater populations of root-feeding nematodes (which can consume up to 25 % of annual belowground net primary productivity) in grazed areas. There could be, though, complicated feedback and source-sink relations that need further investigation. Defoliation of *Bouteloua curtipendula* (Poaceae) led to greater allocation of plant biomass to the roots, but only in plants that were supporting high densities of root-feeding nematodes (Ingham and Detling 1986). Bardgett et al. (1998) considered this to be due to a feedback, where increased root feeding in defoliated plants resulted in a feedback to the plant causing more C (biomass) to be allocated to the roots, leading to increased rates of root exudation after nematode feeding on the roots (Ingham and Detling 1986). Similar effects might be expected for plants supporting large populations of root-feeding aphids.

5.7 Conclusion

Ecosystem function can be estimated by productivity, allocation patterns, energy flow, nutrient cycling, soil processes or interactions between component species. In each case, belowground herbivory can be seen to be a strong modifying factor. These effects have been shown, and the mechanisms of interaction tested, across all scales, from the laboratory to large-scale manipulative field experiments, and the general conclusion is that root feeding by insects cannot be ignored in natural, semi-natural, managed and agricultural ecosystems as a driver or modifier of ecosystem processes. As always, more research is needed to understand fully the importance of root herbivores, but it can be suggested that they may have a keystone role in linking above- and belowground systems.

Acknowledgements. The Nuffield Foundation and Natural Environment Research Council funded elements of the work reported here. Thanks go to Bernhard Stadler and one anonymous referee for comments that improved this chapter.

References

Anonymous (1985) Dursban insecticide: technical manual. Dow Chemical, Hitchin, UK

Bardgett RD, Wardle DA, Yeates GW (1998) Linking above-ground and below-ground interactions: how plant responses to foliar herbivory influence soil organisms. Soil Biol Biochem 30:1867–1878

Bardgett RD, Cook R, Yeates GW, Denton CS (1999a) The influence of nematodes on below-ground processes in grassland ecosystems. Plant Soil 212:23–33

Bardgett RD, Denton CS, Cook R (1999b) Below-ground herbivory promotes soil nutrient transfer and root growth in grassland. Ecol Lett 2:357–360

Brodbeck B, Strong D (1987) Amino acid nutrition of herbivorous insects and stress to host plants. In: Barbosa P, Schultz JC (eds) Insect outbreaks. Academic Press, San Diego, pp 347–364

Brown MW, Schmitt JJ, Ranger S, Hogmire HW (1995) Yield reduction in apple by edaphic woolly apple aphid (Homoptera, Aphididae) populations. J Econ Entomol 88:127–133

Brown VK, Gange AC (1989a) Differential effects of above- and below-ground insect herbivory during early plant succession. Oikos 54:67–76

Brown VK, Gange AC (1989b) Root herbivory by insects depresses plant species richness. Funct Ecol 3:667–671

Brown VK, Gange AC (1990) Insect herbivory below ground. Adv Ecol Res 20:1–58

Clark SE, Van Driesche RG, Sturdevant N, Kegley S (2001) Effect of root feeding insects on spotted knapweed (*Centaurea maculosa*) stand density. Southwest Entomol 26:129–135

Clements RO (1984) Control of insect pests in grassland. Span 27:77–80

Clements RO, Bale JS (1988) The short-term effects on birds and mammals of the use of chlorpyrifos to control leatherjackets in grassland. Ann Appl Biol 112:41–47

Clements RO, Bentley BR (1983) The effect of three pesticide treatments on the establishment of white clover (*Trifolium repens*) sown with a slot-seeder. Crop Protect 2:375–378

Clements RO, Bentley BR, Jackson CA (1986) The impact of granular formulations of phorate, terbufos, carbosulphan and thiofanox on newly sown Italian ryegrass (*Lolium multiflorum*). Crop Protect 5:389–394

Coleman DC (1976) A review of root production processes and their influence on soil biota in terrestrial ecosystems. In: Anderson JM, MacFadyen A (eds) The role of terrestrial and aquatic organisms in decomposition processes. Blackwell, Oxford, pp 417–434

Crawley MJ (1983) Herbivory: the dynamics of animal–plant interactions. Blackwell, Oxford

Crutchfield BA, Potter DA (1995) Tolerance of cool-season turf grasses to feeding by Japanese-beetle and southern masked chafer (Coleoptera, Scarabaeidae) grubs. J Econ Entomol 88:1380–1387

Dawson LA, Grayston SJ, Murray PJ, Pratt SM (2002) Root feeding behaviour of *Tipula paludosa* (Meig.) (Diptera: Tipulidae) on *Lolium perenne* (L.) and *Trifolium repens* (L.). Soil Biol Biochem 34:609–615

Denton CS, Bardgett RD, Cook R, Hobbs RJ (1999) Low amounts of root herbivory positively influence the rhizosphere microbial community in a temperate grassland soil. Soil Biol Biochem 31:155–165

Dyer ML, Acra MA, Wang GM, Coleman DC, Freckman DW, McNaughton SJ, Strain BR (1991) Source-sink carbon relations in two *Panicum coloratum* ecotypes in response to herbivory. Ecology 72:1472–1483

Evans EW (1991) Experimental manipulation of herbivores in native tallgrass prairie: response of above-ground arthropods. Am Midl Nat 125:37–46

Filser J (2002) The role of Collembola in carbon and nitrogen cycling in soil. Pedobiologia 46:234–245

Fogel R (1985) Roots as primary producers in below-ground ecosystems. In: Fitter AH, Atkinson D, Reed DJ, Usher MB (eds) Ecological interactions in the soil: plants, microbes and animals. Blackwell, Oxford, pp 23–26

Gange AC (2000) Arbuscular mycorrhizal fungi, Collembola and plant growth. Trends Ecol Evol 15:369–372

Gange AC (2001) Species-specific responses of a root- and shoot-feeding insect to arbuscular mycorrhizal colonization of its host plant. New Phytol 150:611–618

Gange AC, Bower E (1997) Interactions between insects and mycorrhizal fungi. In: Gange AC, Brown VK (eds) Multitrophic interactions in terrestrial systems. Blackwell, Oxford, pp 115–132

Gange AC, Brown VK (1989) Effects of root herbivory by an insect on a foliar-feeding species, mediated through changes in the host plant. Oecologia 81:38–42

Gange AC, Brown VK (2002) Soil food web components affect plant community structure during early succession. Ecol Res 17:217–227

Gels JA, Held DW, Potter DA (2002) Hazards of insecticides to the bumble bees *Bombus impatiens* (Hymenoptera: Apidae) foraging on flowering white clover in turf. J Econ Entomol 95:722–728

Goldsmith FB, Harrison CM, Morton AJ (1986) Description and analysis of vegetation. In: Moore PD, Chapman SB (eds) Methods in plant ecology. Blackwell, Oxford, pp 437–524

Grayston SJ, Dawson LA, Treonis AM, Murray PJ, Ross J, Reid EJ, MacDougall R (2001) Impact of root herbivory by insect larvae on soil microbial communities. Eur J Soil Biol 37:277–280

Grime JP, Brown VK, Thompson K, Masters GJ, Hillier SH, Clarke IP, Askew AP, Corker D, Kielty JP (2000) The response of two contrasting limestone grasslands to simulated climate change. Science 289:762–765
Hector A (1996) Insect herbivory on herbaceous legumes. PhD Thesis, Imperial College, University of London
Holland JN, Cheng WX, Crossley DA (1996) Herbage-induced changes in plant carbon allocation: assessment of below ground carbon fluxes using carbon-14. Oecologia 107:87–94
Hsiao TC (1973) Plant responses to water stress. Annu Rev Plant Physiol 24:519–570
Ingham RE, Detling JK (1984) Plant–herbivore interactions in a North American mixed grass prairie. III. Soil nematode populations and root biomass on *Cynomys ludovicianus* colonies and adjacent uncolonized areas. Oecologia 63:307–313
Ingham RE, Detling JK (1986) Effects of defoliation and nematode consumption on growth and leaf gas exchange in *Bouteloua curtipendula*. Oikos 46:23–28
Jeffrey DW (1987) Soil–plant relationships: an ecological approach. Croom Helm, Sydney
Maron JL (1998) Insect herbivory above- and belowground: individual and joint effects on plant fitness. Ecology 79:1281–1293
Masters GJ (1992) Interactions between foliar- and root-feeding insects. PhD Thesis, Imperial College, University of London
Masters GJ (1995a) The impact of root herbivory on aphid performance – field and laboratory evidence. Acta Oecologia 16:135–142
Masters GJ (1995b) The effect of herbivore density on host-plant mediated interactions between two insects. Ecol Res 10:125–133
Masters GJ, Brown VK (1992) Plant-mediated interactions between two spatially separated insects. Funct Ecol 6:175–179
Masters GJ, Brown VK (1997) Interactions between spatially separated herbivores. In: Gange AC, Brown VK (eds) Multitrophic interactions in terrestrial ecosystems. Blackwell, Oxford, pp 217–237
Masters GJ, Brown VK, Gange AC (1993) Plant mediated interactions between aboveground and belowground insect herbivores. Oikos 66:148–151
Masters GJ, Jones TH, Rogers M (2001) Host-plant mediated effects of root herbivory on insect seed predators and their parasitoids. Oecologia 127:246–250
McNaughton SJ (1983) Compensatory plant growth as a response to herbivory. Oikos 40:329–336
Milchunas DG, Lauenroth WK (1993) Quantitative effects of grazing on vegetation and soils over a global range of environments. Ecol Monogr 63:327–366
Mooney HA (1972) The carbon balance of plants. Annu Rev Ecol Syst 3:315–346
Moran NA, Whitham TG (1990) Interspecific competition between root-feeding and leaf galling aphids mediated by host-plant resistance. Ecology 71:1050–1058
MoronRios A, Jaramillo VJ, Dirzo R (1997) Species composition of root feeding macroarthropods in a subalpine grassland associated with pine forest in Mexico. Can Entomol 129:71–80
Müller-Schärer H, Brown VK (1995) Direct and indirect effects of aboveground and belowground insect herbivory on plant density and performance of *Tripleurospermum perforatum* during early plant succession. Oikos 72:36–41
Murray PJ, Hatch DJ (1994) *Sitona* weevils (Coleoptera: Curculionidae) as agents for rapid transfer of nitrogen from white clover (*Trifolium repens* L.) to perennial ryegrass (*Lolium perenne* L.). Ann Appl Biol 125:29–33
Murray PJ, Hatch DJ, Cliquet JB (1996) Impact of insect root herbivory on the growth and nitrogen and carbon contents of white clover (*Trifolium repens*) seedlings. Can J Bot 74:1591–1595

Murray PJ, Dawson LA, Grayston SJ (2002) Influence of root herbivory on growth response and carbon assimilation by white clover plants. Appl Soil Ecol 20:97–105
Notzold R, Blossey B, Newton E (1998) The influence of below ground herbivory and plant competition on growth and biomass allocation of purple loosestrife. Oecologia 113:82–93
Nowak RS, Caldwell MM (1984) A test of compensatory photosynthesis in the field: implications for herbivory tolerance. Oecologia 61:311–318
Potter DA, Held DW (2002) Biology and management of the Japanese beetle. Annu Rev Entomol 47:175–205
Pozo C, Martineztoledo MV, Salmeron V, Rodelas B, Gonzalezlopez J (1995) Effects of chlorpyrifos on soil microbial activity. Environ Toxicol Chem 14:187–192
Preus LE, Morrow PA (1999) Direct and indirect effects of two herbivore species on resource allocation in their shared host plant: the rhizome galler *Eurosta comma*, the folivore *Trirhabda canadensis* and *Solidago missouriensis*. Oecologia 119:219–226
Quiring DT, McNeil JN (1984) Adult-larval intraspecific competition in *Agromyza frontella* (Diptera: Agromyzidae). Can Entomol 116:1385–1391
Ridsdill Smith TJ (1977) Effects of root-feeding by scarabaeid larvae on growth of perennial ryegrass plants. J Appl Ecol 14:73–80
Saner MA, Müller-Schärer H (1994) Impact of root mining by *Eteobalea* spp. on clonal growth and sexual reproduction of common toadflax, *Linaria vulgaris* Mill. Weed Res 34:199–204
Sheppard AW, Vitou J (2000) The effect of a rosette-crown fly, *Botanophila turcica*, on growth, biomass allocation and reproduction of the thistle *Carthamus lanatus*. Acta Oecol 21:337–347
Shure DJ (1971) Insecticide effects on early succession in an old-field ecosystem. Ecology 52:271–279
van Dam NM, Harvey JA, Wäckers FL, Bezemer TM, van der Putten WH, Vet LEM (2003) Interactions between aboveground and belowground induced responses against phytophages. Basic Appl Ecol 4:63–77
Villalobos FJ, Goh KM, Saville DJ, Chapman RB (1997) Interactions among soil organic matter, levels of the indigenous entomopathogenic bacterium *Serratia entomophila* in soil, amber disease and the feeding activity of the scarab larva of *Costelytra zealandica*: a microcosm approach. Appl Soil Ecol 5:231–246
Wäckers FL, Bezemer TM (2003) Root herbivory induces an above-ground indirect defence. Ecol Lett 6:9–12
Whitehouse DM (1993) Effect of root-feeding insects on establishment of selected broadleaved trees in farm woodland. PhD Thesis, Imperial College, University of London
Zhou XM, Smith DL (1996) A new technique for continuous injection into stems of field-grown corn plants. Crop Sci 36:452–456

Section III
Plant–Insect Interactions and Ecosystem Processes

6 Bottom-Up Effects and Feedbacks in Simple and Diverse Experimental Grassland Communities

J. Joshi, S.J. Otway, J. Koricheva, A.B. Pfisterer, J. Alphei, B.A. Roy, M. Scherer-Lorenzen, B. Schmid, E. Spehn and A. Hector

6.1 Summary

Finding a consistent pattern in the effects of plant diversity on higher trophic levels is a major challenge as populations at all trophic levels of an ecosystem may be regulated by a mixture of top-down and bottom-up forces. A starting point to experimentally approach the problem is to measure the effects of changing plant-species diversity on primary productivity, the basis of each food web, and to explore the potential underlying mechanisms. This was done within the European BIODEPTH project (*Biod*iversity and *E*cological *P*rocesses in *T*errestrial *H*erbaceous ecosystems). In this project, a common methodology was used at each of eight sites across Europe to experimentally assemble grassland communities of defined plant-species numbers (e.g. 1, 2, 4, 8, 32) from the local species pools. Ecosystem processes were then monitored in these different herbaceous ecosystems. Here, we report findings gathered from the sites in the UK, Switzerland, Germany and Sweden. Our data suggest that trophic levels cannot be treated as homogeneous units since the response to changes in plant diversity of individual groups within trophic levels was correlated with group-specific attributes such as host specificity, mobility or different size classes of consumers. There was also no evidence for a resource concentration effect, i.e. for a disproportionately higher specialist insect density in plots with high host density and biomass such as monocultures. Part of the diversity effects observed at higher trophic levels was indirectly driven by changes in primary productivity with changing plant diversity. However, experimental additions of a generalist insect herbivore and a plant hemiparasitic species showed that some polyphagous groups within higher trophic levels can benefit from increased diversity not only by the higher quantity, but also by the higher variety of resources.

Ecological Studies, Vol. 173
W.W. Weisser and E. Siemann (Eds.)
Insects and Ecosystem Function

6.2 Introduction

Intrigued by the large number of species in terrestrial ecosystems, ecologists have tried for a long time to predict the role of biodiversity for ecosystem functioning and for ecosystem stability by using observational studies and experiments. The first experimental plots to explore whether diverse ecosystems perform differently from simple ones were established in the 19th century (Hector and Hooper 2002). However, the question to what extent species loss might impair ecosystem functioning has only been addressed by replicated experimental work in the past decade (see, e.g., Loreau et al. 2002; Schmid et al. 2002 for reviews). Current interest was triggered by today's unprecedented speed of species-loss occurring worldwide on a local to regional and even global scale (Barbault 1995).

Effects of species diversity on ecosystem functioning can be studied both within and among the trophic levels of ecosystems. Most biodiversity–ecosystem functioning work to date, both theoretical and experimental, has concentrated on the single trophic level of producers (Loreau et al. 2001; Schmid et al. 2002; Weisser and Siemann, Chap. 1, this Vol.). Only a small amount of theory has begun to extend this work to multiple interacting trophic levels (Loreau 2001; Holt and Loreau 2002; Siemann and Weisser, Chap. 18, this Vol.), even though an interrelationship between autotrophic biomass or diversity and biomass or abundance of higher trophic levels has long been suggested by a variety of models and hypotheses (e.g. Hairston et al. 1960; May 1973; Root 1973; Oksanen et al. 1981; Abrams 1993; Schulze 1995; Carson et al., Chap. 10, this Vol.). These models differ in their assumptions and approaches to disentangling the complex and dynamic relationships between ecosystem components. Some highlight the importance of energy and material flows within ecosystems, whereas others focus on food-web processes and trophic levels or on population dynamics (Naeem and Li 1998). How changes in plant diversity and subsequent concomitant changes in ecosystem processes interact with higher trophic levels is rarely explicitly addressed in these models and experimental studies reporting effects of plant diversity on higher trophic levels are still too rare to yield an overall picture (Siemann 1998; Knops et al. 1999; Mulder et al. 1999; Koricheva et al. 2000; Symstad et al. 2000; Prieur-Richard et al. 2002; Raffaelli et al. 2002; Janssen and Sabelis, Chap. 9, this Vol.).

A reasonable starting point in unravelling the complex relationship between producer diversity and whole-ecosystem consequences is to measure the effects of changing plant-species diversity on primary productivity, the basis of each food web, and to explore the potential consequences of these effects on higher trophic levels. This general approach was chosen in several recent large-scale experiments in grassland ecosystems (e.g. Hooper and Vitousek 1997; Hector et al. 1999; Van der Putten et al. 2000; Tilman et al. 2001). In the European BIODEPTH project (Hector et al. 1999), a common methodology was used at eight experimental sites across Europe. At each site, experimental grassland communities of defined plant species numbers (e.g. 1, 2, 4, 8, 32) and of a defined number of plant functional groups but differing in plant species composition were assembled from the local species pools to monitor ecosystem processes at different plant-diversity levels and different environmental conditions at a European scale.

In this chapter, we concentrate on how the diversity of plants in experimental grassland ecosystems can affect the diversity and abundance of insect herbivores and compare the response of insect consumers with plant-diversity effects on other consumer groups such as plant pathogens, plant parasites and decomposers. We will focus on experimental results obtained at three sites within the BIODEPTH framework, the Swiss site in Lupsingen in the Jura Mountains, Silwood, one of the two UK sites, and the German site in Bayreuth (see, e.g., Diemer et al. 1997; Hector et al. 2000; Joshi et al. 2000; Spehn et al. 2000a; Scherer-Lorenzen et al. 2003), where the abundance of several groups within higher trophic levels has been monitored in response to changes in plant diversity.

6.3 Effects of Plant Diversity on Herbivorous Insects Feeding Above Ground

6.3.1 Hypotheses Predicting the Response of Herbivores to Higher Plant Diversity

Several hypotheses have been suggested to predict the response of herbivores to changes in plant-species diversity. The resource concentration or associational resistance hypothesis introduced by Tahvanainen and Root (1972) and Root (1973) predict that plant species growing in diverse communities gain protection against herbivore attack (cf. Carson et al., Chap. 10, this Vol.). Herbivores with a narrow host range are expected to find hosts more easily if they are concentrated. The host-finding process of these herbivores can be disturbed by olfactory interference of non-host plants or by visual masking (Hambäck et al. 2003). Visual masking, a general, non-species-specific mechanism (Hambäck et al. 2003), suggests that the presence of non-host plants might visually impair the host-finding behaviour of herbivores by similarities in plant stature, leaf shape and spectral quality or simply by covering the host plants (e.g. Rausher 1978, Brown and Lawton 1991, reviewed in Hambäck and Beckermann 2003). Olfactory interference, i.e. disturbed host-finding and feeding behaviour, in species-rich plant communities due to interfering chemical stimuli produced by non-host plants might not be a frequently encountered mechanism because sensory perception in insect herbivores is

Table 6.1. Hypotheses on the response of monophagous, specialist and polyphagous, generalist herbivores on increase in plant-species diversity. Bottom-up effects of herbivores on predators are not considered in these hypotheses. Polyphagous herbivores were expected to have distinct food preferences. + Increase in herbivore abundance; – decline in herbivore abundance; *ne* no effect; (–) low probability of a negative effect

	Specialists	Generalists
Resource concentration hypothesis	–	(–)
Olfactory interference	–	(–)
Visual interference	–	(–)
Enemies hypothesis	–	–
Nutrient balance and toxin dilution hypotheses	ne	+

excellent and they are not easily deceived by olfactory interference (see Hambäck and Beckermann 2003). Generalist associational refuges through physical interference, however, might cause smaller herbivore loads in species-rich plant communities (Finch and Collier 2000; Hambäck et al 2003). Smaller herbivore densities of both generalist and specialist insects in species-rich stands are also predicted by the natural enemies hypothesis (Root 1973). The mechanism behind this is an increased top-down control of herbivores in species-rich communities that provide refuges, additional resources such as pollen and nectar and offer constantly higher prey diversity, thereby supporting increased densities of predators (see Table 6.1). Thus, the enemies hypothesis is related to the trophic cascade model by Hairston et al. (1960) where predators indirectly allow plants to escape suppression by their consumers (Strong 1999).

On the other hand, higher abundance of generalist herbivores in more diverse plant communities can be expected on the basis of the nutrient balance and toxin dilution hypotheses (e.g. Pulliam 1975; Singer et al. 2002), which predict a beneficial effect of the higher diversity of secondary metabolites in diverse plant communities, making a mixed diet advantageous (Table 6.1).

Research into these hypotheses has mainly concentrated on agricultural and silvicultural crop cultivars known to suffer from attacks of specialized herbivores. These experiments usually only had two diversity-treatment levels (e.g. Risch et al. 1983; Russell 1989; Andow 1991; Altieri 1995; Wilby and Thomas 2002). There have been relatively few studies on these processes in natural or semi-natural communities, and, among those, most have concentrated on single focal plant species (e.g. Kunin 1999; Hambäck et al. 2000; Prieur-Richard et al. 2002, Hambäck et al. 2003).

6.3.2 Responses of Specialist and Generalist Herbivores in Plant Diversity Experiments

To test the resource concentration hypothesis, invertebrates were suction-sampled twice a year in May and August for 2 years at the Silwood BIODEPTH field site. We focused on abundant specialist herbivores, defined as those feeding on a single plant species and represented by at least 30 individuals. We investigated their abundance in relation to plant species diversity and host plant biomass (Otway 2000). Nine herbivore species (seven weevils, a psyllid and a leafhopper) met the criteria for inclusion in the analysis. For the reasons discussed above, the resource concentration hypothesis predicts that monophagous herbivores such as these should attain a disproportionately higher density in pure versus mixed stands (Root 1973). We would therefore expect a regression of insect density on plant biomass to have a slope greater than one.

Figure 6.1 shows these regressions for each specialist-insect/host-plant pair separately. When combined in one analysis, we found that the density of specialist insects increased with increasing host plant biomass ($F_{1,88}$=33.7594, P=9.780e^{-08}) and that differences in slope between different insect–plant pairs were not significant ($F_{8,88}$=0.5219, P=0.8371). The main impact of plant diversity on the density of insects comes through changes in the biomass of their host plants. However, counter to the resource concentration hypothesis, the overall slope (0.22342, SE: 0.03865) was significantly less than one (*t*-test=20.09, P <0.001, n=105). This 'resource dilution' effect has also been found in previous studies (reviewed in Yamamura 2002). In addition to the effect of host plant biomass there was a smaller but significant negative effect of plant species richness when host plant biomass was fitted before as a covariate (i.e. when corrected for differences in host biomass between the different plots; $F_{1,88}$=5.9934, P=0.01634; Fig. 6.1, solid vs open symbols). However, it is unclear how this secondary effect of diversity came about.

Similarly, at the Swiss BIODEPTH site, only one of the five groups of herbivores collected by sweep net, pitfalls and suction sampling – the leafhoppers (Cicadellidae) – showed a log-linear decrease in abundance as a response to an increase in plant species number, whereas the other groups (Acrididae, Aphididae, Miridae, Gastropoda) did not show the negative response predicted by the resource concentration hypothesis. Because leafhoppers are a rather specialized family, mainly feeding on grasses and reproducing sexually, it is likely that an increase in plant-species richness and hence a lower concentration of available food resources reduces the number of potential mates, resulting in a lower abundance (Koricheva et al. 2000). In contrast, wingless aphids, a group of rather sessile insect herbivores, increased in abundance with increasing number of plant functional groups (grasses, grasses + legumes or grasses + herbs, grasses + legumes + herbs) in both years sampled

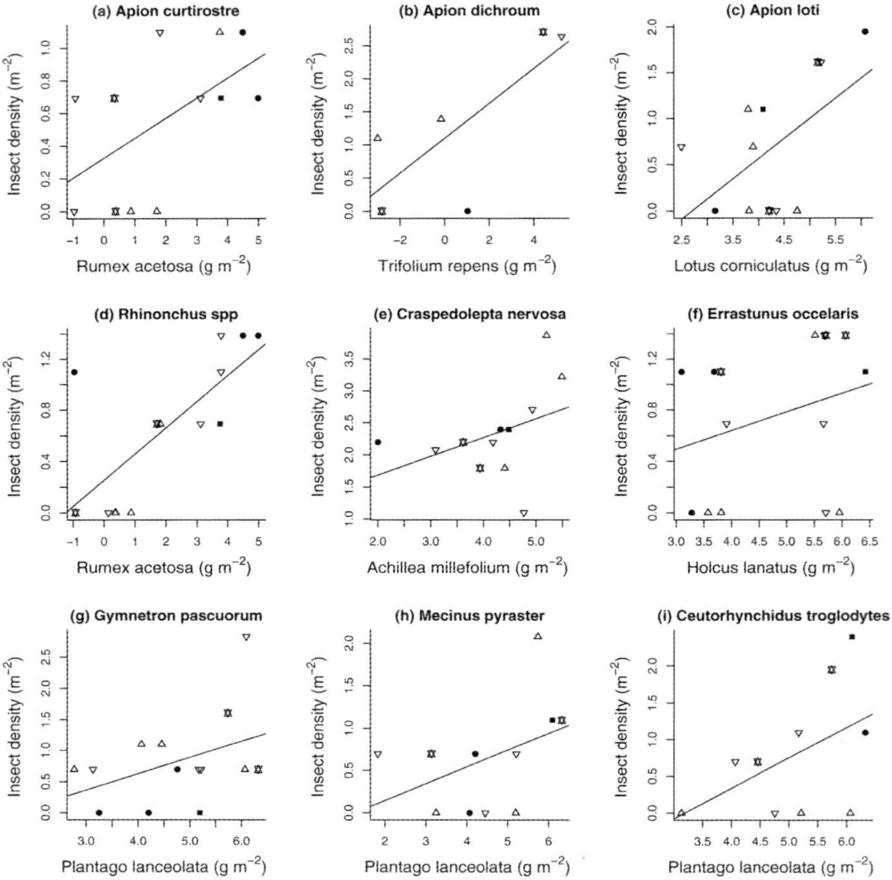

Fig. 6.1. Relationships between host-plant biomass and specialist-herbivore density in the Silwood (UK) BIODEPTH plots (both transformed with natural logs). Insect density is significantly positively related to increasing host plant biomass, but, contrary to the resource concentration hypothesis, slopes are significantly less than one rather than greater than one as predicted. Low-diversity communities are coded by *solid symbols* (*circles* one-species plots; *squares* two-species plots) and higher-diversity communities by *open symbols* (*upward-facing triangles* four-species plots; *downward-facing triangles* eight-species plots; *stars* 11-species plots)

(Koricheva et al. 2000). Aphids are a speciose group with both specialists and generalists. Hence, the higher total aphid abundance found may be due to an increased likelihood that communities with higher plant diversity sustain a higher number of aphid species and thus a higher overall aphid abundance than species-poor plots.

True generalist herbivores are expected to be more abundant in diverse plant communities because they may benefit from a mixed diet (Table 6.1), as has been found in experiments comparing diets of two to three plant species with diets consisting of just one species (Pennings et al. 1993; Bernays et al. 1994). Food mixing may allow generalists to balance intake of different nutrients (e.g. Pulliam 1975) and to avoid ingesting toxic doses of particular secondary metabolites characteristic of individual host plant species (Freeland and Saladin 1989; Singer et al. 2002). To test whether species-rich communities exert such complementary diet effects on polyphagous species, individuals of a native generalist grasshopper species (*Parapleurus alliaceus* Germar) were caged and left feeding on the Swiss BIODEPTH plots for 2 weeks (Pfisterer et al. 2003). Even though plant-biomass consumption did not significantly differ among diversity treatments and this grasshopper species is known to have a preference for grasses that occurred in a lower relative abundance in plots of higher species diversity, biomass gain of the grasshoppers increased with a greater diversity of plant species (Pfisterer et al. 2003; Fig. 6.2). A similar beneficial effect of a mixed diet was found in an experiment looking at the mutual effects of plant diversity and the performance of a hemiparasitic plant species (Joshi et al. 2000). Like generalist insect herbivores, hemiparasites can be viewed as generalist consumers of plant production (Atsatt 1977; Pennings and Callaway 2002). Although they are autotrophic with respect to carbon, they attack roots of other plants and extract water, nutrients and organic solutes from them. In the Swiss BIODEPTH plots, mature weight of sown individuals of *Rhinanthus alectorolophus* (European yellow rattle) increased linearly with the number of plant functional groups present in the host communities. Total parasite biomass per plot was almost three times as high in host communities with three functional groups as in host communities with one functional group only (Joshi et al. 2000).

All of these hypotheses that predict the impact of plant diversity on insect herbivores assume plant productivity to be fixed. However, one of the common features of plant biodiversity experiments is overyielding – plant species in mixtures are able to achieve higher biomasses than expected due to

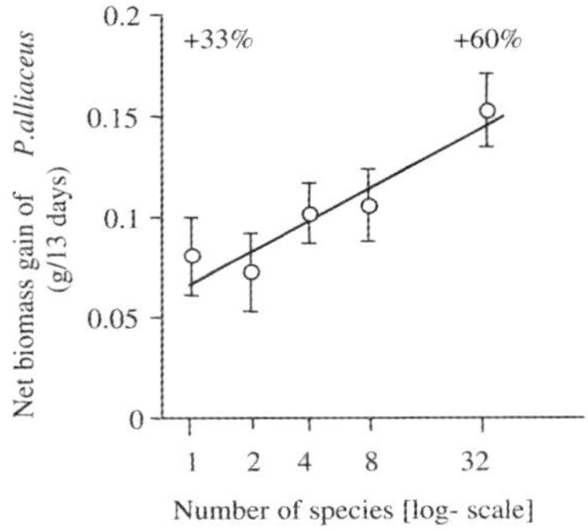

Fig. 6.2. Effect of plant species richness on biomass increases in caged female *Parapleurus alliaceus* in experimental grassland systems at the Swiss BIODEPTH site. Means of original data ±1 SE are shown with regression line, $P < 0.01$

resource partitioning and other niche differentiation effects (Hector et al. 2002). For example, in the Swiss experiment, total plant biomass log-linearly increased with increasing number of plant species or of functional groups (Fig. 6.3). The nitrogen concentration and C/N ratio of plant biomass, however, did not change with an increase in overall plant diversity. Only the presence of legumes in plant communities positively influenced N content in vegetation at the Swiss and other BIODEPTH sites (Spehn et al. 2002). Therefore, the observed effects of plant diversity on invertebrates and hemiparasites might reflect direct diversity effects or indirect diversity effects mediated by concomitant changes in plant biomass, but not by an increased N content of the vegetation with higher plant diversity. An increasing trend in herbivory and herbivore abundance with increasing net primary productivity in terrestrial systems has already been described by Cyr and Pace (1993) and by Siemann (1998). However, in the parasite experiment, the beneficial effect of plant diversity on parasite biomass still remained significant, even if host-community biomass was included as a covariate in the statistical model to correct for differences in productivity among different plant communities (Joshi et al. 2000). Hence, it can be concluded that host diversity itself, i.e. probably the complementarity of resources provided by the different host species, had a positive influence on parasite growth. Similarly, the increase in biomass gain in grasshoppers was not the result of more food consumed, but rather of more diverse food available in species-rich communities, which was indicated by the positive diversity effect remaining even after including plant biomass as a covariate (Pfisterer et al. 2003). The grasshoppers increased dietary mixing in high diversity plots by increasing the consumption of herbal plants compared to grasses, thus supporting the hypothesis that generalist insect herbivores may benefit from a higher diversity of nutrients and secondary compounds. For the majority of herbivores sampled at the Swiss site, however, path analysis (Fig. 6.4) and analysis of covariance indicated that

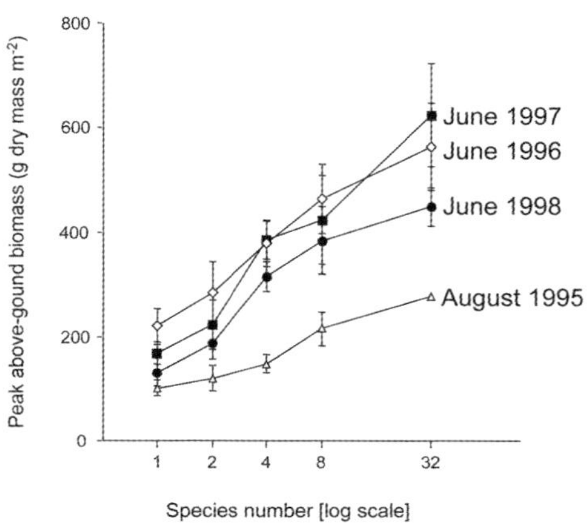

Fig. 6.3. Effect of plant species richness on aboveground biomass (g m^{-2}) at main harvest during the first 4 years of the experiment at the Swiss BIODEPTH site. Means of original data ±1 SE are shown

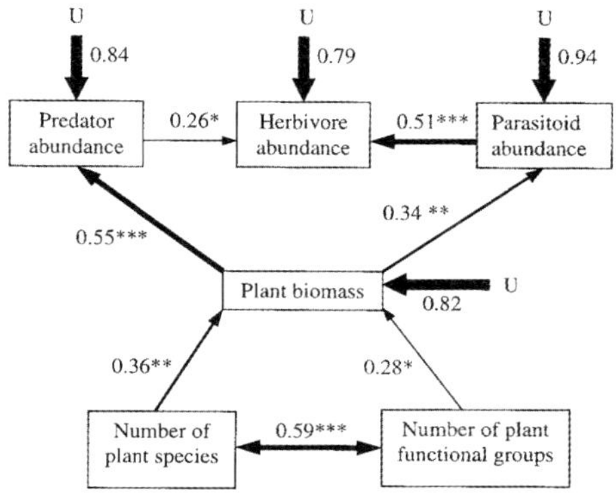

Fig. 6.4. Path diagram describing direct and indirect effects of plant diversity on abundance of different trophic groups of invertebrates based on data from the sweep net catches at the Swiss site in 1996 (n=64). *Single-headed arrows* represent causal effects of one variable on another; *double-headed arrow* represents a correlation. *U* denotes proportion of unexplained variance for each dependent variable. The *width of each line* is proportional to the strength of the relationship and *values next to arrows* indicate path-coefficient values (* $P<0.05$, ** $P<0.01$, *** $P<0.001$). For further details see Koricheva et al. (2000). (Reprinted with permission, *Oecologia*, Springer Verlag)

plant diversity effects were largely mediated by plant biomass and cover (or by some factors that may be correlated with biomass and cover, such as microclimatic conditions, or nutritive quality), or in some cases by particular plant species compositions rather than due to plant diversity per se (Koricheva et al. 2000).

Looking at aboveground herbivory from the plant side, i.e. by measuring the total herbivore damage done to nine plant species in the Swedish and Swiss BIODEPTH plots, plant biomass removed by herbivory increased with plant diversity (Mulder et al. 1999). Even though patterns of herbivore damage across species-richness levels differed between particular plant species, for all species levels, damage in monocultures was relatively low (Mulder et al. 1999).

In the BIODEPTH plots at Silwood, herbivore diversity increased log-linearly with plant diversity ($F_{32,32}$=3.96, $P<0.0001$) with on average just under five herbivore species collected in a suction sample per monoculture plot and an additional 0.91 (SE ±0.15) species per doubling of plant species richness (Otway 2000), supporting the view that more species of plants will provide more diverse resources supporting more insect herbivores, both generalists and those that specialize on particular species or plant parts.

6.3.3 Concomitant Responses of Natural Enemies of Herbivores

At the Swiss BIODEPTH site, natural enemies of herbivores were generally not affected by changes in plant species diversity (parasitoid wasps) or effects were inconsistent among years (spiders; Koricheva et al. 2000; M. Diemer, pers. comm.). Where significant effects were found, the direction of the diversity effect was negative and hence opposite to the response predicted by the enemies hypothesis (Koricheva et al. 2000). The number of generalist predators (carabid beetles and spiders) and the number of predators in pitfalls in the second and third years of the experiment decreased with plant diversity, providing little indication that herbivores were more strongly controlled by predators in more diverse plant communities. However, it is difficult to assess overall herbivore pressure, considering the many different types of insects and the sample sizes for secondary consumers, which were rather small as predicted by the various principles concerning trophic pyramids. Nevertheless, these results do not support the view that enhanced net primary productivity in species-rich plant communities translates into a clear bottom-up effect along the short linear food chain from insect herbivores to their predators.

6.3.4 Insect Herbivores as Drivers of Ecosystem Processes

Insects may be drivers of ecosystem change themselves. They may change the effect of plant species diversity on ecosystem processes by differentially altering community plant biomass at different diversity levels, by changing the relative abundance of plant species or by altering ecosystem processes directly (see Mulder et al. 1999; Weisser and Siemann, Chap. 1, this Vol.; Hartley and Jones, Chap. 2, this Vol.; Carson et al., Chap. 10, this Vol.; Schmitz, Chap. 14, this Vol.). In the Swedish BIODEPTH plots (which had a substantially lower annual biomass production than the Swiss plots; Hector et al. 1999) the presence of insects negatively affected plant evenness and the relationship between plant-species richness and biomass production, which was more strongly positive when insects were excluded (Mulder et al. 1999). In addition, a recent experimental study in old-field ecosystems revealed top-predator control of plant evenness and dominance structure (Schmitz 2003; Schmitz, Chap. 14, this Vol.). However, top-down control on aggregate ecosystem processes may depend on ecosystem productivity. Predators eat whole individuals and may thus control herbivores under some circumstances. Herbivores, however, feed on plant parts, which may regrow (at speeds depending on climatic, edaphic and biotic conditions), or select particular plant species, allowing possible compensatory growth from unattacked neighbours. As a result, we may expect aggregate ecosystem processes influenced by the producer trophic level to be relatively resilient against top-down damage imposed by herbivores.

6.4 Effects of Plant Diversity on Pathogens

The same mechanisms proposed as possible links between aboveground primary consumers and plant diversity could also be applied to plant pathogens (Kruess et al., Chap. 8, this Vol.). We expect low diversity to favour specialists when the biomass of the preferred hosts is high, whereas high diversity might favour generalists because of the variety present and the higher probability of a steady presence of host plants. In the high diversity plots, we might also expect to see more pathogen species that require more than one host for the successful conclusion of their life cycles, and there might be more pathogen species overall in the high diversity plots because of the availability of safe sites, allowing escape from their enemies. Unfortunately, most plant pathogens are specialists, and all of the 12 species we quantified were specialists, allowing no comparisons with generalists.

We were, however, able to look at the effects of plant diversity on attack rates. As with specialist herbivores, monocultures are expected to be the most susceptible to specialist pathogen attack because there a particular host species is both frequent and dense, allowing easier transmission (Zwölfer and Zimmermann, Chap. 7, this Vol.; Kruess et al., Chap. 8, this Vol.). In mixed species assemblages, not only is density of an individual species lower, but also there is interference from non-hosts; spores are trapped on species on which they cannot grow. The prediction is thus that there should be a clear relationship with diversity, with low diversity plots having more disease than high diversity plots. However, our data from the Swiss BIODEPTH plots show that plant diversity did not influence pest diversity in one predictable direction. Instead, four patterns occurred, depending on species and time of the year censused: (1) a linear decrease with increasing plant diversity, (2) a linear increase with increasing plant diversity, (3) a humped distribution in which diversity was highest at intermediate plant diversity plots, and (4) no association with diversity (Gächter 1997; Landolt 1997).

Contrary to theoretical predictions, some of the pathogens we studied were more common in the high diversity plots (e.g. *Pseudopeziza trifolii* and one of the pathogens on *Taraxacum*). Our best guess for these pathogens was that the more diverse plots had greater cover and more diverse structure and thus tended to trap moisture for longer periods. Many pathogens require free water for germination on the leaf surface (Agrios 1997; Brewer and Smith 1997), and we suspect that the higher humidity in the more diverse plots facilitated germination. Rust fungi, on the other hand, usually followed the expected pattern, with *Puccinia coronata* on *Lolium perenne* and *Puccinia* rust on *Poa pratensis* being significantly influenced by plant-species richness and number of plant functional groups causing the highest damage in low-diversity plots. These pathogens are wind-dispersed and are well known for their dependence on both host frequency and density (Roy 1993; Brunet and Mundt 2000).

6.5 Belowground Food Web

6.5.1 Plant Biomass and Microbial Response

It is likely that changes in plant diversity and concomitant changes in net primary productivity also have feedbacks to the associated soil communities. Soil heterotrophic organisms regulate soil processes such as nutrient cycling and carbon flow and thus may not only be influenced by plant diversity and the diversity and abundance of insects, but also modify the relationship between plant diversity and ecosystem processes (e.g. Bradford et al. 2002; Catovsky et al. 2002; Wardle, 2002; Wardle and Bargett, Chap. 3, this Vol.; Bonkowski and Scheu, Chap. 4, this Vol.). Subterranean communities of decomposers, detritivores and herbivores depend on plants as primary producers, i.e. on quantity (e.g. litter production, root turnover and root exudation) and quality of organic substrate (Hooper et al. 2000; Hartley and Jones, Chap. 2, this Vol.). Thus, it can be hypothesized that a decrease in plant diversity influences the biomass and also the compositional diversity of soil heterotrophic communities via reduced quantity of carbon input (Catovsky et al. 2002), via reduced biochemical diversity of root exudates or via a less constant supply of food resources, leading to a drop in decomposer diversity and possibly to a decline in decomposition (Wardle and van der Putten 2002). Altered abundances of the different constituents of the decomposer food web can be crucial for the decomposition of organic material and thus for the mineralization of nutrients (e.g. Bardgett and Shine 1999; Wardle 2002).

If a mixed diet is as beneficial for the decomposer community as it might be for generalist herbivores (Table 6.1), then an increase in producer diversity might translate into higher decomposer diversity. In turn, higher decomposer diversity may lead to enhanced ecosystem processes, i.e. decomposition and subsequent mineralization, through higher resource-use complementarity (Hector et al. 2000). On the other hand, increased net primary productivity in species-rich plant communities is also likely to influence all levels of soil organisms through increased quantities of plant litter and root exudates (Wardle and van der Putten 2002). If, however, higher net primary productivity in species-rich systems goes along with enhanced nutrient uptake by plants, then a net negative effect on microbial biomass and soil fauna might result due to increased competition for nutrients (Wardle and van der Putten 2002). To sort out the interplay between plant-species richness, ecosystem processes and the soil food web, an array of components of soil heterotrophic activity were monitored in the Swiss, German and British BIODEPTH plots (Diemer et al. 1997; Hector et al. 2000; Spehn et al. 2000b; Gastine et al. 2003). However, it was not possible to monitor changes in soil chemistry and biotic activity regularly during the whole duration of the experiment. Therefore, our results reflect only particular parameters at single points in time and might serve as an indication only of how plant-species diversity might affect the belowground black box.

In contrast to the very clear increase in aboveground biomass production with increased plant diversity at the Swiss site (Fig. 6.3), total root biomass was influenced neither by plant diversity nor by the presence or absence of legumes in the third year of the experiment (all $P > 0.3$). Fine-root biomass, however, tended to increase with species richness, in both the second and third year of the experiment ($F_{1,24}=3.19$; $P=0.087$, and $F_{1,24}=3.88$; $P=0.06$, respectively). Concomitant with the increase in aboveground biomass and fine-root biomass with plant-species richness, an increase in diversity and catabolic activity of culturable bacteria (measured by BIOLOG plates) was found in the third and fourth year of the experiment (Stephan et al. 2000). However, this increase in culturable soil bacteria was not significantly correlated with fine-root biomass (Stephan et al. 2000). Total microbial biomass also increased log-linearly with increasing plant-species diversity, with monocultures having a 15% lower biomass as compared with the most species-rich mixtures (Spehn et al. 2000b). Here again, belowground or fine-root biomass did not explain this increase if included as covariates in the analysis. In contrast, aboveground biomass explained a major part of the variation among different diversity levels. Hence, the positive effect of plant diversity on microbial biomass might have been due to enhanced input of litter supply to the soil, especially due to the increase in legume abundance and therefore higher resource quality, i.e. litter with lower C/N ratio (Spehn et al. 2000b).

Belowground processes are controlled by both biotic and abiotic factors at any given site, so we may postulate that the relationship between the diversity of primary producers and those processes varies depending on the abiotic site conditions. For example, at the German BIODEPTH site, which is characterized by low soil nitrogen availability, contrasting results to those described for the Swiss site were observed. At the German site, the presence and abundance of nitrogen-fixing legumes, which increased the nitrogen availability for the whole system (Spehn et al. 2002), had a major influence on the outcome of plant diversity–ecosystem functioning relationships. In contrast to the Swiss site, fine root biomass was not affected by plant diversity in four consecutive years despite strong diversity effects on aboveground biomass production (Gastine et al. 2003; Scherer-Lorenzen et al. 2003). Only in plots containing legumes did an increase in species number result in increased fine-root biomass and belowground processes, such as ammonification and nitrification and the decomposition of plant litter and standard test materials (Scherer-Lorenzen et al. 2003). Thus, we conclude that biodiversity effects on belowground processes are very much dependent on the nutritional status of the communities and will differ according to different plant traits present, with the ability to fix atmospheric nitrogen being one of the most important functional traits. While our sites showed an overall positive relationship between diversity and aboveground productivity (Hector et al. 1999), these differences in the details of their ecology also point out that there are major limits to our ability to generalize and predict effects of biodiversity loss on belowground processes.

6.5.2 Soil Animals That Feed on Microbes

Biogeochemical processes might also be influenced by soil animals that feed directly on microbes (Wardle and van der Putten 2002; Bonkowski and Scheu, Chap. 4, this Vol.). In the Swiss BIODEPTH plots, different functional groups of nematodes responded in various ways to enhanced plant functional diversity and the associated increase in microbial biomass. Total nematode density (individuals per gram of dry weight) did not show any significant response to changes in producer diversity, but the dominance (percentage of total nematode density) among the different nematode functional groups changed considerably. While omnivores, root-hair-feeding or predatory nematodes did not change with enhanced plant diversity, plant parasites significantly decreased with an increase in the number of plant species or functional groups, respectively (F=4.0, P=0.0311; F=6.73, P=0.0046). In contrast, bacterial and fungal feeders increased in dominance with higher plant functional diversity. The decrease of plant parasitic nematodes with increasing diversity of plant functional groups represents an especially interesting feedback that merits more experimental attention in the future.

The clearest responses to changes in producer diversity and concomitant changes in net primary productivity have been observed in feeders of fresh organic matter. Voles as well as earthworms clearly increased either their activity or in biomass, respectively, with increasing plant diversity (Spehn et al. 2000b). Voles feed on roots and rhizomes. As the total biomass of roots did not increase with plant diversity and as they showed a preference for plant communities containing legumes, they could also have been attracted by the mixed diet or a more steady food supply. The 84 % increase of earthworm biomass from monocultures to 32-plant-species communities was mainly due to a pronounced response of anecic earthworms (worms forming vertical burrows in the soil) that predominantly feed on fresh organic matter (Spehn et al. 2000b). Here again, aboveground biomass explained a considerable part of the variation in earthworm biomass – presumably due to increased litter input – whereas neither total root nor fine-root biomass showed a significant correlation with earthworm biomass (Spehn et al. 2000b).

At the German site, neither the abundance of predatory or non-predatory nematodes nor the density, biomass or species diversity of earthworms was influenced by plant diversity (Gastine et al. 2003).

6.6 Conclusions

Testing for consistent patterns in the effects of plant diversity on higher trophic levels is a major challenge because populations in all trophic levels of an ecosystem may be regulated by a mixture of top-down *and* bottom-up forces, often leading to idiosyncratic or opposing effects depending on ecosystem type, climatic and edaphic conditions, and experimental setup (e.g. Raffaelli et al. 2002; Schmid et al. 2002; Wardle 2002; Janssen and Sabelis, Chap. 9, this Vol.). In addition, the constant non-destructive monitoring of changes in species abundances and nutrient fluxes is an enormous and costly task, which is difficult to achieve in well-replicated field experiments. All our results represent snapshots as we lack continuous data on population dynamics of primary and secondary consumers in response to changes in plant-species diversity. Hence, we can draw preliminary conclusions only, based on individual observations at several points in time. Moreover, we do not consider strong feedbacks between above- and belowground components and try to find the most parsimonious explanations for the phenomena observed in our experiments (by including explanatory covariates in our statistical models or using path analysis).

Our data suggest that trophic levels cannot be treated as homogeneous units since the response to changes in plant diversity of individual groups within trophic levels was correlated with group-specific attributes such as host specificity, mobility or different size classes of consumers at our experimental scale. Among herbivores, only the abundance of the most sessile and host-specific groups (wingless aphids and leafhoppers, respectively) was significantly affected by plant diversity. We found some preliminary evidence that species-rich plots harbour a higher invertebrate diversity. There was not much support for disproportionately lower abundances of specialist herbivores in species-rich communities, as predicted by the associational resistance hypothesis. Evidence for the associational resistance hypothesis has mainly been found in agricultural, intercropping systems with a much lower range of species diversity as used in our experiments. Hence, these differences between the two kinds of studies may stem from differences in experimental design.

Part of the diversity effects observed on higher trophic levels both above ground and below ground were indirectly driven by changes in aboveground primary productivity with changing plant diversity. However, experimental additions of generalist herbivores such as grasshoppers and a plant hemiparasite showed that some polyphagous groups within higher trophic levels benefited from increased plant diversity not only by the higher quantity but also by the higher variety of food compounds.

In general, however, effects of individual plant species and individual species combinations at the Swiss, the Swedish and the British BIODEPTH sites were a more important determinant of invertebrate abundance than plant species richness per se (Mulder et al. 1999; Koricheva et al. 2000; Otway 2000). In particular, the presence of legumes in a mixture led to higher numbers of most invertebrate groups above ground. There is growing consensus that any biodiversity effect on ecosystem functioning will arise from phenotypic variation between species, i.e. from the diversity of their functional traits or attributes, and from species interactions (Lawton 2000, Díaz and Cabido 2001; Loreau et al. 2001; Hector et al. 2002). Differences in individual species effects will thus be related to the size of their functional differences (Schmid et al. 2002).

Overall, our results suggest that groups within higher trophic levels can be directly affected by plant species diversity or indirectly via enhanced net primary productivity. On the other hand, the relationship between plant diversity and ecosystem processes may be altered by a higher trophic level through direct modification of ecosystem functions like aboveground biomass production, or through altered abundances of plant species. A more exact analysis of the interaction between higher trophic levels and plant diversity will require experimental manipulations of different trophic groups – a logistically demanding job. Alternatively, we must develop and employ new methods for teasing apart the complex web of trophic interactions. Both approaches pose challenges if we want to gain a better understanding of the relationship between biodiversity and ecosystem functioning from a multitrophic perspective.

Acknowledgements. We thank John Lawton for his work on the Silwood insect sampling, Alan Stewart for help with insect identification, and all our colleagues from the BIODEPTH project for stimulating discussions and ideas. Peter Hambäck, Evan Siemann, Wolfgang Weisser and an anonymous reviewer gave constructive comments on the manuscript.

References

Abrams PA (1993) Effect of increased productivity on the abundances of trophic levels. Am Nat 141:351–371

Agrios GN (1997) Plant pathology. Academic Press, San Diego

Altieri MA (1995) Biodiversity and biocontrol: lessons from insect pest management. In: Andrews JH, Tommerup IC (eds) Advances in plant pathology. Academic Press, London, pp 191–209

Andow DA (1991) Vegetational diversity and arthropod population response. Annu Rev Entomol 36:561–586

Atsatt PR (1977) The insect herbivore as a predictive model in parasitic seed plant biology. Am Nat 111:579–586

Barbault R (1995) Generation, maintenance and loss of biodiversity. In: Heywood VH, Watson RT (eds) Global biodiversity assessment. Cambridge University Press, Cambridge, pp 197–258
Bardgett RD, Shine A (1999) Linkages between plant litter diversity, soil microbial biomass and ecosystem function in temperate grasslands. Soil Biol Biochem 31:317–321
Bernays EA, Bright KL, Gonzalez N, Angel J (1994) Dietary mixing in a generalist herbivore – tests of 2 hypotheses. Ecology 75:1997–2006
Bradford MA, Jones TH, Bardgett RD, Black HIJ, Boag B, Bonkowski M, Cook R, Eggers T, Gange AC, Grayston SJ, Kandeler E, McCaig AE, Newington JE, Prosser JI, Setala H, Staddon PL, Tordoff GM, Tscherko D, Lawton JH (2002) Impacts of soil faunal community composition on model grassland ecosystems. Science 298:615–618
Brewer CA, Smith WK (1997) Patterns of leaf surface wetness for montane and subalpine plants. Plant Cell Environ 20:1–11
Brown VK, Lawton JH (1991) Herbivory and the evolution of leaf size and shape. Phil Trans R Soc B 333:265–272
Brunet J, Mundt CC (2000) Disease, frequency-dependent selection, and genetic polymorphisms: experiments with stripe rust and wheat. Evolution 54:406–415
Catovsky S, Bradford MA, Hector A (2002) Biodiversity and ecosystem productivity: implications for carbon storage. Oikos 97:443–448
Cyr H, Pace ML (1993) Magnitude and patterns of herbivory in aquatic and terrestrial ecosystems. Nature 361:148–150
Díaz S, Cabido M (2001) Vive la difference: plant functional diversity matters to ecosystem processes. TREE 16:646–655
Diemer M, Joshi J, Körner C, Schmid B, Spehn E (1997) An experimental protocol to assess the effects of plant diversity on ecosystem functioning utilized in a European research network. Bull Geobot Inst ETH Zurich 63:95–107
Finch S, Collier RH (2000) Host-plant selection by insects – a theory based on 'appropriate/inappropriate landings' by pest insects of cruciferous plants. Entomol Exp Appl 96:91–102
Freeland WJ, Saladin LR (1989) Choice of mixed diets by herbivores – the idiosyncratic effects of plant secondary compounds. Biochem Syst Ecol 17:493–497
Gächter F (1997) The influence of plant diversity on plant diseases and herbivore damage in experimental plant communities. Diploma ETH, Zürich
Gastine A, Scherer-Lorenzen M, Leadley PW (2003) No consistent effect of plant diversity on root biomass, soil biota and soil abiotic conditions in temperate grassland communities. Appl Soil Ecol 24:101–111
Hairston NG, Smith FE, Slobodkin LB (1960) Community structure, population control, and competition. Am Nat 94:421–425
Hambäck PA, Agren J, Ericson L (2000) Associational resistance: insect damage to purple loosestrife reduced in thickets of sweet gale. Ecology 81:1784–1794
Hambäck PA, Beckerman AP (2003) Herbivory and plant resource competition: a review of two interacting interactions. Oikos 101:26–37
Hambäck PA, Pettersson J, Ericson L (2003) Are associational refuges species-specific? Funct Ecol 17:87–93
Hector A, Hooper R (2002) Ecology – Darwin and the first ecological experiment. Science 295:639–640
Hector A, Schmid B, Beierkuhnlein C, Caldeira MC, Diemer M, Dimitrakopoulos PG, Finn J, Freitas H, Giller PS, Good J, Harris R, Högberg P, Huss-Danell K, Joshi J, Jumpponen A, Körner C, Leadley PW, Loreau M, Minns A, Mulder CPH, O'Donovan G, Otway SJ, Pereira JS, Prinz A, Read DJ, Scherer-Lorenzen M, Schulze E-D, Siamantziouras A-SD, Spehn E, Terry AC, Troumbis AY, Woodward FI, Yachi S, Lawton JH (1999) Plant diversity and productivity experiments in European grasslands. Science 286:1123–1127

Hector A, Beale AJ, Minns A, Otway SJ, Lawton JH (2000) Consequences of the reduction of plant diversity for litter decomposition: effects through litter quality and microenvironment. Oikos 90:357–371
Hector A, Bazeley-White E, Loreau M, Otway S, Schmid B (2002) Overyielding in grassland communities: testing the sampling effect hypothesis with replicated biodiversity experiments. Ecol Lett 5:502–511
Holt RD, Loreau M (2002) Biodiversity and ecosystem functioning: the role of trophic interactions and the importance of system openness. In: Kinzig A, Pacala S, Tilman D (eds) Functional consequences of biodiversity: experimental progress and theoretical extensions. Princeton University Press, Princeton, pp 246–262
Hooper DU, Vitousek PM (1997) The effects of plant composition and diversity on ecosystem processes. Science 277:1302–1305
Hooper DU, Bignell DE, Brown VK, Brussaard L, Dangerfield JM, Wall DH, Wardle DA, Coleman DC, Giller KE, Lavelle P, Van der Putten WH, De Ruiter PC, Rusek J, Silver WL, Tiedje JM, Wolters V (2000) Interactions between aboveground and belowground biodiversity in terrestrial ecosystems: patterns, mechanisms, and feedbacks. Bioscience 50:1049–1061
Joshi J, Matthies D, Schmid B (2000) Root hemiparasites and plant diversity in experimental grassland communities. J Ecol 88:634–644
Knops JMH, Tilman D, Haddad NM, Naeem S, Mitchell CE, Haarstad J, Ritchie ME, Howe KM, Reich PB, Siemann E, Groth J (1999) Effects of plant species richness on invasion dynamics, disease outbreaks, insect abundances and diversity. Ecol Lett 2:286–293
Koricheva J, Mulder CPH, Schmid B, Joshi J, Huss-Danell K (2000) Numerical responses of different trophic groups of invertebrates to manipulations of plant diversity in grasslands. Oecologia 125:271–282
Kunin WE (1999) Patterns of herbivore incidence on experimental arrays and field populations of ragwort, *Senecio jacobaea*. Oikos 84:515–525
Landolt M (1997) The effect of plant species diversity on pathogen infection and herbivore damage on grassland perennials. Diploma ETH, Zürich
Lawton JH (2000) Community ecology in a changing world. In: Kinne O (ed) Excellence in ecology. Ecology Institute, Oldendorf, Germany
Loreau M (2001) Microbial diversity, producer–decomposer interactions and ecosystem processes: a theoretical model. Proc R Soc Lond Ser B-Biol Sci 268:303–309
Loreau M, Naeem S, Inchausti P, Bengtsson J, Grime JP, Hector A, Hooper DU, Huston MA, Raffaelli D, Schmid B, Tilman D, Wardle DA (2001) Biodiversity and ecosystem functioning: current knowledge and future challenges. Science 294:804–808
Loreau M, Naeem S, Inchausti P (2002) Perspectives and challenges. In: Naeem S, Loreau M, Inchausti P (eds) Biodiversity and ecosystem functioning: synthesis and perspectives. Oxford University Press, Oxford, pp 237–242
May RM (1973) Stability and complexity in model ecosystems. Princeton University Press, Princeton
Mulder CPH, Koricheva J, Huss-Danell K, Högberg P, Joshi J (1999) Insects affect relationships between plant species richness and ecosystem processes. Ecol Lett 2:237–246
Naeem S, Li SB (1998) Consumer species richness and autotrophic biomass. Ecology 79:2603–2615
Oksanen L, Fretwell SD, Arruda J, Niemelä P (1981) Exploitation ecosystems in gradients of primary productivity. Am Nat 118:240–261
Otway SJ (2000) Influences of grassland plant diversity and composition on invertebrate diversity and abundance. PhD Thesis, University of London
Pennings SC, Callaway RM (2002) Parasitic plants: parallels and contrasts with herbivores. Oecologia 131:479–489

Pennings SC, Nadeau MT, Paul VJ (1993) Selectivity and growth of the generalist herbivore *Dolabella auricularia* feeding upon complementary resources. Ecology 74:879–890

Pfisterer AB, Diemer M, Schmid B (2003) Dietary shift and lowered biomass gain of a generalist herbivore in species-poor experimental plant communities. Oecologia 135:234–241

Prieur-Richard AH, Lavorel S, Linhart YB, Dos Santos A (2002) Plant diversity, herbivory and resistance of a plant community to invasion in Mediterranean annual communities. Oecologia 130:96–104

Pulliam HR (1975) Diet optimization with nutrient constraints. Am Nat 109:765–768

Raffaelli D, van der Putten WH, Persson L, Wardle DA, Petchey OL, Koricheva J, van der Heijden MGA, Mikola J, Kennedy T (2002) Multi-trophic dynamics and ecosystem processes. In: Naeem S, Loreau M, Inchausti P (eds) Biodiversity and ecosystem functioning: synthesis and perspectives. Oxford University Press, Oxford, pp 147–154

Rausher MD (1978) Search image for leaf shape in a butterfly. Science 200:1071–1073

Risch SJ, Andow D, Altieri MA (1983) Agroecosystem diversity and pest control: data, tentative conclusions, and new research directions. Environ Entomol 12:625–629

Root RB (1973) Organization of a plant–arthropod association in simple and diverse habitats: the fauna of collards (*Brassica oleracea*). Ecol Monogr 43:95–124

Roy BA (1993) Patterns of rust infection as a function of host genetic diversity and host density in natural populations of the apomictic crucifer, *Arabis holboellii*. Evolution 47:111–124

Russell EP (1989) Enemies hypothesis: a review of the effect of vegetational diversity on predatory insects and parasitoids. Environ Entomol 18:590–599

Scherer-Lorenzen M, Palmborg C, Prinz A, Schulze E-D (2003) The role of plant diversity and composition for nitrate leaching in grasslands. Ecology 84:1539–1552

Schmid B, Joshi J, Schläpfer F (2002) Empirical evidence for biodiversity–ecosystem functioning relationships. In: Kinzig A, Pacala S, Tilman D,(eds) Functional consequences of biodiversity: experimental progress and theoretical extensions. Princeton University Press, Princeton, pp 120–150

Schmitz OJ (2003) Top predator control of plant biodiversity and productivity in an old-field ecosystem. Ecol Lett 6:156–163

Schulze ED (1995) Flux control at the ecosystem level. TREE 10:40–43

Siemann E (1998) Experimental tests of effects of plant productivity and diversity on grassland arthropod diversity. Ecology 79:2057–2070

Singer MS, Bernays EA, Carriere Y (2002) The interplay between nutrient balancing and toxin dilution in foraging by a generalist insect herbivore. Anim Behav 64:629–643

Spehn E, Joshi J, Diemer M, Schmid B, Körner C (2000a) Aboveground resource use increases with plant species-richness in experimental grassland ecosystems. Funct Ecol 14:326–337

Spehn EM, Joshi J, Schmid B, Alphei J, Körner C (2000b) Plant diversity effects on soil heterotrophic activity in experimental grassland ecosystems. Plant Soil 224:217–230

Spehn EM, Scherer-Lorenzen M, Schmid B, Hector A, Caldeira MC, Dimitrakopoulos PG, Finn JA, Jumpponen A, O'Donnovan G, Pereira JS, Schulze ED, Troumbis AY, Körner C (2002) The role of legumes as a component of biodiversity in a cross-European study of grassland biomass nitrogen. Oikos 98:205–218

Stephan A, Meyer AH, Schmid B (2000) Plant diversity affects culturable soil bacteria in experimental grassland communities. J Ecol 88:988–998

Strong DR (1999) Predator control in terrestrial ecosystems: the underground food chain of bush lupine. In: Olff H, Brown VK, Drent RH (eds) Herbivores: between plants and predators. Blackwell, Oxford, pp 577–602

Symstad AJ, Siemann E, Haarstad J (2000) An experimental test of the effect of plant functional group diversity on arthropod diversity. Oikos 89:243–253

Tahvanainen JO, Root RB (1972) Influence of vegetational diversity on the population ecology of a specialized herbivore, *Phyllotreta cruciferae* (Coleoptera: Chrysomelidae). Oecologia 10:321–346

Tilman D, Reich PB, Knops J, Wedin D, Mielke T, Lehman C (2001) Diversity and productivity in a long-term grassland experiment. Science 294:843–845

Van der Putten WH, Mortimer SR, Hedlund K, Van Dijk C, Brown VK, Leps J, Rodriguez-Barrueco C, Roy J, Len TAD, Gormsen D, Korthals GW, Lavorel S, Regina IS, Smilauer P (2000) Plant species diversity as a driver of early succession in abandoned fields: a multi-site approach. Oecologia 124:91–99

Wardle DA (2002) Communities and ecosystems; linking the aboveground and belowground components. Princeton University Press, Princeton

Wardle DA, van der Putten WH (2002) Biodiversity, ecosystem functioning and aboveground–below-ground linkages. In: Naeem S, Loreau M, Inchausti P (eds) Biodiversity and ecosystem functioning: synthesis and perspectives. Oxford University Press, Oxford, pp 155–168

Wilby A, Thomas MB (2002) Natural enemy diversity and pest control: patterns of pest emergence with agricultural intensification. Ecol Lett 5:353–360

Yamamura K (2002) Biodiversity and stability of herbivore populations: influences of the spatial sparseness of food plants. Pop Ecol 44:33–40

7 The Potential of Phytophagous Insects in Restoring Invaded Ecosystems: Examples from Biological Weed Control

H. ZWÖLFER and H. ZIMMERMANN

7.1 Summary

Biological weed control attempts to reduce population densities of noxious weeds to a level below thresholds of economic damage by introducing and establishing foreign herbivorous organisms as control agents. Successful control of target weeds using phytophagous insects has so far been achieved in more than 100 projects, but only a relatively small percentage of biologically successfully established insect species have proved capable of substantially reducing the densities of the target weed species. Comparison of the rates of biological success (establishment) with economic success (resource utilization) in biological weed control projects indicates that populations of many phytophagous insects behave not merely as a function of bottom-up and top-down processes, which could be assumed, as the introduced agents are confronted with a surplus of food resources in a relatively enemy-free space. Behavioural patterns, such as density-dependent dispersal, which prevent a full exploitation of the host resource, may explain the failure of many introduced and established phytophagous species to control weed species. We discuss weed situations and properties of phytophagous insect species which have resulted in successful biological weed control and corresponding changes of ecosystems, and list some of the prerequisites for such an insect-mediated change in vegetation cover. The biological control of *Carduus nutans* by the weevil *Rhinocyllus conicus* in North America, of *Sesbania punicea* in South Africa after the introduction of three weevil species, and of *Salvinia molesta* in Papua New Guinea are described as examples of how insect–plant interactions may strongly affect the vegetation cover and composition and ecosystem properties dependent on them. Such projects demonstrate the potential of certain phytophagous insects to restore floral diversity and food webs in ecosystems where the flow of material and energy has been blocked by pure stands of alien plants.

Ecological Studies, Vol. 173
W.W. Weisser and E. Siemann (Eds.)
Insects and Ecosystem Function

7.2 Introduction

The aim of biological control projects is to permanently reverse, usually over large areas, undesirable properties of modified invaded ecosystems by introducing and establishing foreign host-specific organisms. If properly executed and evaluated, biological control projects can therefore be considered as unique, large-scale, ecological experiments, in which certain specific, key components of an ecosystem are manipulated. This applies particularly to biological weed control (Myers 1987; Crawley 1989b; Kruess et al., Chap. 8, this Vol.) which uses key phytophagous organisms, mostly insects, to restore ecosystems modified by invading, alien plant species. As the vegetation is the most important component in structuring ecosystems, a change of vegetation cover and composition caused by alien plant invasions is also followed by drastic changes in the structure and function of the ecosystem (e.g. floral and faunal diversity, food chains and food webs, flow of energy and material, water loss through transpiration, erosion). Many of these environmental impacts worldwide are discussed by Parker et al. (1999), Van Wilgen et al. (2001) (for South Africa) and Dukes and Mooney (2004) and include excessive changes in the fire ecology and changes in soil properties.

There have also been proposals (DeLoach 1984; Pemberton 1985; Waterhouse and Norris 1987; Harris 1993) and attempts to apply biological weed control against native weeds, mainly if the latter have increased in density as a consequence of overgrazing or other human influences, but these have seldom been carried out and so far have not been successful. In this chapter, we present data from three projects directed against introduced weeds which have succeeded in restoring the invaded ecosystems.

We start with the available data on the success rates of projects and agents, and discuss those weed properties and qualities of agents that have led to the control of dense weed populations. As an example of a successful project, we describe the biological control of *Carduus nutans* in North America by the introduced seed weevil, *Rhinocyllus conicus*. We then compare the North American situation to that in the country of origin. Another case study deals with the biological control of a small tree, *Sesbania punicea*, in South Africa. This was successfully controlled by the interaction of three introduced weevils of South American origin. The third example is the successful control of the waterweed *Salvinia molesta* in Papua New Guinea by the weevil *Cyrtobagous salviniae*. The chapter ends with a general discussion of the potential role of phytophagous insects as a means to restore invaded ecosystems and ecosystem functioning.

7.3 Success Rates and Successes in Biological Weed Control

Worldwide, there is a steady increase in the number of biological control projects against weeds. Crawley (1989a) lists a total of 627 projects carried out before 1981 and 130 in the following decade. The success rate of these projects allows an estimation of the proportion of specialized phytophagous insects having the potential to restore habitats invaded by an alien weed species. According to a global analysis by Julien (1989), 24 % of the phytophagous species released are effective biocontrol agents. Laing and Hamai (1976) evaluated 57 at least partially successful control projects and reported that 13 (22.8 %) were completely successful. Harris (1984a) estimates that one third of the 30 biocontrol agents introduced into Canada inflicted major damage to their target weed species. Crawley (1989a) examined 572 biological weed control projects and found that only 17 (~3 %) were completely successful while another 106 (20 %) showed that the introduced agents had a marked impact. More than 85 species of biocontrol agents have been released against 47 weed species in South Africa and at least 22 species (47 %) are considered to be under complete or substantial control (Olckers and Hill 1999).

Figures on the success rates of biological weed control projects vary greatly and are often based on subjective estimates. Thus, the mass of information accumulated in *Biological Control of Weeds, A World Catalogue of Agents and Their Target Weeds* by Julien and Griffiths (1998) shows that success usually varies greatly in different regions, and sometimes even in different years in the same region. They also refer to case studies in which an initial success was transitory, particularly when the biocontrol agents acquired native enemies (e.g. *Procecidochares utilis* on *Ageratina adenophora* in Australia; Wilson 1969). Another difficulty is the variation in the amount of time required by single control agents before they adapt to the local conditions in the country of introduction and become effective, and the general long-term nature of many projects. Harris (1997) demonstrated this delay effect for three introduced phytophagous species (*Chrysolina quadrigemina* released in British Columbia against *Hypericum perforatum*; *Tyria jacobaeae* released in eastern Canada against *Senecio jacobaeae*; and *Aphthona cyparissiae* released in eastern Canada against *Euphorbia cyparissias*). It can take many years before seed-destroying biological control agents show an effect on weed populations, especially with perennial weeds which have a long-lived seed bank (Neser and Kluge 1986). Additionally, in many control projects, not a single agent but a complex of phytophagous species has to become established before an effect on the target weed population can be achieved (Harris 1991).

However, in spite of these unavoidable inadequacies, the history of biological weed control shows that even single phytophagous insect species may have the potential to eliminate dense covers of single alien species and to restore the original situation. Some striking examples are the control of

Opuntia stricta by *Cactoblastis cactorum* in Australia (Dodd 1940), of *Hypericum perforatum* by *Chrysolina* spp. (Briese 1997), of *Carduus nutans* by *Rhinocyllus conicus* in North America (Andres et al. 1976) and of the shrub *Clidemia hirta* by *Liothrips urichi* in Fiji (Simmonds 1933). In these examples the success of the biocontrol agent led to the re-establishment of the original pasture ecosystems. The control of *Salvinia molesta* by *Cyrtobagous salviniae* in Australia and Papua New Guinea (Thomas and Room 1986), and most recently the dramatic control of *Azolla filiculoides* by the weevil *Stenopelmus rufinatus* in South Africa (Hill 1999), cleared water surfaces formerly completely covered by alien aquatic weeds and restored the former conditions of aquatic ecosystems.

About one third of the introduced agents [in Canada 34% (Harris 1997), worldwide 34% (Julien 1989) or 37% (Crawley 1989a)] fail to become established. By far not all of the two thirds of successfully established insect species in biocontrol projects are effective control agents, and many have only a limited impact at the ecosystem level. The percentage of effective control agents varies according to the different authors and their interpretation of success, and lies between 3 and 30%. If all phytophagous insects would follow the bottom-up vs top-down control model, the rate of economic success in biological weed control should be higher since the agents are carefully selected and introduced without their specific enemies into a new ecosystem where food abounds and specialized enemies (particularly parasitoids) are absent or rare. Based on an analysis of Canadian weed projects and supported by their own investigations, Zwölfer and Völkl (1997) proposed, for phytophagous insects, the replacement of the dichotomy of a bottom-up and top-down process by a triangle scheme of population control. In addition to the two components resource availability and mortality factors (parasitoids, predators, pathogens), specific stabilizing behavioural patterns, such as density-dependent dispersal, determine the population densities of phytophagous insects as the third component. Zwölfer and Völkl emphasized that the underutilization of local resources can be adaptive in the sense of 'spreading the risk', i.e. of escaping enemies operating in a density-dependent manner or as a trade-off between host quantity and host quality.

7.4 Weed Characteristics and Positive Traits of Insects in Biological Control

The following two sections discuss plant types likely to be affected by insects with respect to ecosystem functioning and life history features of phytophagous insects, which are likely to have the greatest influence on plant populations.

7.4.1 Weed Species

There is a broad consensus among biological control practitioners that the chances of success are greatest against weed species that are confronted with many natural enemies in their country of origin, and which have escaped these in their country of introduction, resulting in serious invasions (Huffaker 1964; Harris 1973b; Huffaker and Messenger 1976; Waterhouse and Norris 1987). In such cases, biological control attempts to substantially reduce the density of the alien invader and to restore the competitive balance with the native vegetation (e.g. range grasses), resulting in a shift back to the original status of the ecosystem. Some new and spectacular successes against aquatic weeds (Thomas 1985; Julien and Griffiths 1998) show, however, that under certain circumstances an invading weed can even be controlled by a phytophagous insect in situations in which it is not subject to interspecific competition.

Crawley (1989a) analysed successes and failures in biological weed control and lists among weeds successfully controlled by insect agents host species such as *Opuntia* (cacti), a monocarpic thistle (*Carduus nutans*), a rhizomatous, polycarpic herb (*Hypericum perforatum*) and a sterile polyploid floating fern (*Salvinia molesta*). This shows that different types of plants can be controlled by insects. An exception are annual weeds of arable land where the short life span of the plants in combination with crop rotation and cultivation measures reduces the impact of phytophagous insects. Many failures have come from perennials with substantial powers of regrowth (e.g. *Cirsium arvense*) or perennials with relatively low leaf-nitrogen content (e.g. members of the Poaceae or Cyperaceae). Myers (1987) emphasized the role of the nutritional status of the host plant and suggested that fertilization of weeds may facilitate the initial establishment of control agents.

7.4.2 Insect Species

A general review of the influence of phytophagous insects on plant population dynamics is given by Crawley (1989b). With regard to biological weed control, for which only insect species with a narrow and predictable host range can be used, the impact of monophagous or oligophagous specialists is of interest. Several authors (e.g. Huffaker 1964; Harris 1973a; Goeden 1983; Myers 1987; Crawley 1983, 1989a) have dealt with the problem of why among phytophagous specialists certain species are successful weed control agents, but others failed. They have proposed that mainly phytophagous insects with the following attributes may have a substantial impact on target hosts:

- Species with a high intrinsic rate of increase (Harris 1973a; Myers 1987; Crawley 1987, 1989a), or high voltinism (Harris 1973a).

- Species that by their feeding activity cause substantial damage per individual (e.g. destruction of the vascular or mechanical support system; Harris 1973a).
- Gregarious feeders (Harris 1973a).
- Species with a prolonged attack or which attack the host plant in a particularly vulnerable phase (Harris 1973a).
- Species that damage their host by feeding and thus open the way for subsequent infection by fungi and bacteria (Harris 1973a; Moran and Zimmermann 1984; Crawley 1989a).
- Internal feeders that may suffer relatively high rates of attack from parasitoids in their native environments but which are relatively safe from generalist, external natural enemies (Crawley 1989a).
- Species that are subject to extensive mortality from specialized enemies, including diseases, and are relatively immune to non-specific enemies (Harris 1973a).
- Species among the native guild of insects attacking a particular plant species that have evolved under conditions of high mortality from interspecific competition and/or specialized natural enemies, a handicap that makes them good invaders (Zwölfer 1973; Myers 1987).

With regard to host specificity, Harris (1973a) mentioned that oligophagous species (associated with a number of closely related host species or host genera) have a greater potential than monophagous species. This is particularly relevant when the target weed is genetically variable and consists of a polyploid complex of varieties or morphotypes. In addition, species, mainly leaf feeders, that induce acquired immunity in their perennial hosts are usually less successful biocontrol agents (Mattson et al. 1988).

7.5 Three Examples of Successful Weed Control

7.5.1 *Rhinocyllus conicus* on *Carduus nutans*

Carduus nutans (nodding thistle, musk thistle) and its subspecies are biennial thistles of European origin which form dense stands on dry, uncultivated grasslands in many part of North America. On pastureland they can reduce the grazing area drastically. Stands tend to be self-perpetuating as the death of the flowering stems in August creates a seedbed largely devoid of competing vegetation (Harris 1984b). The weevil *Rh. conicus*, an agent that Crawley (1989a) gives rank 4 in his list of the 11 most successful individual cases of weed biocontrol, was introduced from Europe to control *C. nutans* in Canada and the USA (Julien and Griffiths 1998). It is a suitable species to illustrate the latent potential of certain phytophagous insects to control populations of their host plant. Zwölfer and Harris (1984) give details on this species that belongs to a monophyletic group of weevil genera which co-evolved with the flower heads of 'thistles' (i.e. members of the Asteraceae tribe Cynareae). The species is a cluster of biotypes associated with hosts belonging to the genera *Carduus*, *Cirsium*, *Silybum* and *Onopordum* (Zwölfer and Preiss 1983). Screening tests demonstrated that adult feeding responses were restricted to members of the Cardueae subtribes Carduinae and Centaureinae, and that oviposition and full larval development of the biotype from *C. nutans* was only possible on some *Carduus* and *Cirsium* spp. Additional tests in Canada confirmed these results and showed that the *C. nutans* strain may occasionally oviposit on native North American *Cirsium* spp. In such cases, however, larval mortality was high. Thus, Zwölfer and Harris (1984) concluded that it was unlikely that the normally scattered native thistles would be adopted as prime hosts by the *C. nutans* strain of *Rh. conicus*. Between 1992 and 1996, *Rh. conicus* has increasingly attacked native thistles in the USA (Louda et al. 1997). As more than one ecotype of *Rh. conicus* was introduced (Goeden et al. 1984), it is so far not clear whether the attack on non-target thistles occurred as a consequence of host-range extension or of the introduction of another *Rh. conicus* biotype. The larvae of the weevil develop gregariously within the immature flower heads, where they feed on receptacle and callus tissue and destroy the ovarioles. In our European study areas (Austria, Germany, France), the impact of *Rh. conicus* on *C. nutans* was significantly weakened by the highly specialized egg parasitoid *Pterandrophysalis levantina*, three larval and pupal parasitoid species, intra-guild predation by larvae of pyralids (*Homoeosoma* spp.) and by interspecific competition with larvae of the weevil *Larinus sturnus* and the gall-forming tephritid *Urophora solstitialis*. Nevertheless, *Rh. conicus* does survive within this complex food web. Adults are relatively long-lived, have a high egg potential and are able to find scattered thistles (Harris 1986). Another advantage of *Rh. conicus* is its relatively high capacity for dispersal (Kok and Surles 1975). Rees (1977) reports flight distances between 7 and 19 km/year. An analysis of the oviposition pattern of *Rh. conicus* showed that this species laid its eggs on a much broader proportion of available flower heads than the other guild members and was therefore able to find competitor-free resources (Zwölfer 1979). As long as its population densities are not too high, *Rh. conicus* can adapt its egg load to the varying sizes of the flower heads chosen for oviposition (Zwölfer 1985). We therefore assume that the high pressure of mortality factors and its competitive inferiority in its guild of flower-bud feeders is a selective force which has promoted the efficiency of host utilization in *Rh. conicus*.

In Canada and the USA, the introduced *Rh. conicus* is not exposed to intra-guild predation, interspecific competition or attack by specialized parasitoids. It could therefore make full use of its potential as a herbivore and utilize a much higher proportion of the tissues in flower heads of *C. nutans*. As an example, we compare in Fig. 7.1 the food web studied in 1971 at Munchhouse, a locality near Mulhouse, Haut Rhin, France (one of the main collection areas for the export of *Rh. conicus* to North America) and the situation in the Galatin valley (Montana, USA) where *Rh. conicus* was introduced in 1969 and was annually monitored by Rees (1977). For our comparison, we use his data from 1977 (Rees 1978), describing the impact of the *Rh. conicus* 7 years after its introduction on stands of *C. nutans*. As a measure of the efficiency of *Rh. conicus*, we use the relation of its larval biomass (in milligrams) per dry weight (grams) of *C. nutans* flower heads. The autochthonous *Rh. conicus* population is integrated into a complex food web, whereas in the allochthonous population in the Galatin valley *Rh. conicus* is the only consumer in an extremely simplified food chain. If losses due to parasitoids are corrected (hatched areas in Fig. 7.1), the Mulhouse food web produced an average of 17.5 mg biomass of surviving third-instar larvae of *Rh. conicus* per gram of head dry weight. For corresponding values in the Galatin valley, the data of

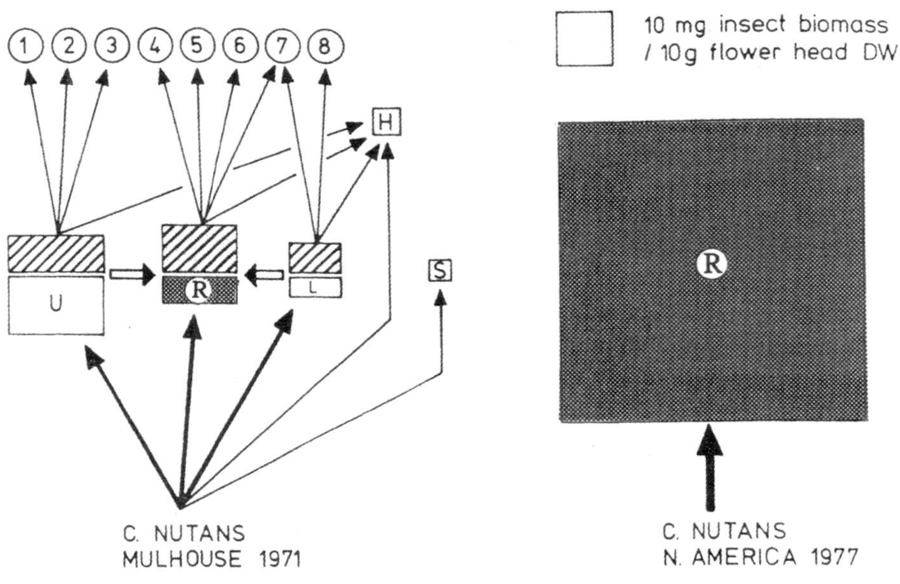

Fig. 7.1. The food web in heads of *Carduus nutans*, in 1971, near Mulhouse, Haut Rhin, France (*left*) and in 1977, the Galatin valley, Montana, USA (*right*). Size of *squares* represents the biomass of living mature larvae of primary consumers (in mg) per flower head dry wt. (*DW*) (in g). *Hatched areas* show proportion of consumer biomass lost to parasitoids and a facultative predator (*H, Homoeosoma* spp.). *U, Urophora solstitialis*; *L, Larinus sturnus*; *R, Rhinocyllus conicus*; *S*, additional phytophagous insects of minor importance. *Numbers* represent the parasitoids: *1, Eurytoma tibialis*; *2, E. robusta*; *3, Torymus* sp.; *4, Pterandrophysalis levantina*; *5, Pteromalus* sp.; *6, Bracon minuator*; *7, Bracon urinator*; *8, Tetrastichus crassicornis*. *Arrows* show the flow of energy and nutrients and indicate competitive interactions between the consumers. (Modified from Zwölfer 1980)

Rees (1978) allow an estimated 306 mg biomass for 1977. In the Mulhouse area, the average number of mature larvae/head was 1.6, in the Galatin valley this increased steadily (1974: 15.6; 1975: 17.8; 1976: 21.3 (Rees 1977) and reached 28 in 1977 (Rees 1978).

Rh. conicus is attacked by a number of native parasitoids in North America (Dowd and Kok 1982), but their impact is so far insignificant. The substantial reduction of seed production in the attacked heads, and of flower production later in the season, caused by *Rh. conicus* provided 'good to excellent control in many areas, resulting in 80–99 % reduction in the weed' (Julien and Griffiths 1998). Many authors (e.g. Kok and Surles 1975; Rees, 1977, 1978; Harris 1984a; Zwölfer and Harris 1984) showed that *Rh. conicus* is capable of restoring the original vegetation in rangeland completely occupied by *C. nutans*. On recently disturbed sites denuded of vegetation, *C. nutans* populations have survived the attack by *Rh. conicus* because of a drastic reduction in the competition with other plant species. This shows that the outcome of biological control can be severely influenced by other ecological processes.

7.5.2 Interactions Between Three Weevil Species in the Biocontrol of the Invader *Sesbania punicea* in South Africa

Monitoring the impact of introduced phytophagous insect species on the dynamics of alien invading trees or shrubs has not been attempted often because of the long-term nature of these studies. The case of the successful biocontrol of the leguminous tree *Sesbania punicea* revealed some important aspects that have consequences for other similar projects involving trees. The most recent review of this project is by Hoffmann and Moran (1991b). *S. punicea* has invaded many parts of South Africa, in particular rivers and watercourses where it increases erosion and, through excessive transpiration, depletes valuable water resources (Hoffmann and Moran 1999). *S. punicea* is native to the Parana Delta of South America. It is a small deciduous tree that lives for about 10 years, and which produces copious numbers of winged pods containing about ten seeds per pod. Seeds disperse mainly by water. The seed bank is relatively short-lived, up to about 4 years (Graaf and Van Staden 1984), and maximum germination from seeds occurs after 2 years. In South America it is utilized by at least 18 insect herbivore species (Erb 1980). The plant has become scarce in its native range mainly because of extensive development in that region. In South Africa, *S. punicea* suffered negligible damage from local generalist insect feeders. Four species from its native range were selected for further studies in South Africa with the intention of controlling large infestations of the invader, mainly along rivers and wetlands. One species was rejected, but the remaining three species were eventually released and included: a bud-destroying and leaf-feeding apionid, *Trichapion lativentre*, a seed-feeding curculionid, *Rhyssomatus marginatis*, and a stem-boring weevil,

Neodiplogrammus quadrivittatus (Hoffmann and Moran 1999). A post-release monitoring programme spanning more that 10 years was implemented to evaluate the impact of each biocontrol agent, alone and in combination with the others, at different time intervals (Hoffmann and Moran 1998). Values for plant and seed survival, seed-set and germination rates for the various scenarios were used to describe the population dynamics of the weed and its three natural enemies (Hoffmann 1988; Hoffmann and Moran 1989, 1992a, 1992b). These data were eventually used in simulation models to predict the outcome of the project and to help in planning a long-term biological and integrated control programme (for a detailed account of the model see Hoffmann 1990).

7.5.2.1 The Seed-Destroying Agents: *Trichapion lativentre* and *Rhyssomatus marginatus*

The bud-feeding weevil *T. lativentre* consistently destroys 98 % or more of the annual pod production when plants are growing in dense aggregations. A smaller proportion of pods are destroyed in areas where the plants grow isolated (Hoffmann and Moran 1992a). In spite of this significant impact on seed production by *T. lativentre*, *S. punicea* still produces some seed pods that replenished the seed bank and this ensures that the population persists. The seed-weevil *R. marginatus* was then released to reduce seed production even further. The weevil is able to locate and destroy between 49 and 86 % of the isolated seed pods that have escaped *T. lativentre*. Its unique oviposition behaviour ensures that eggs are well dispersed over the available food resource, which minimizes intraspecific encounters (Hoffmann and Moran 1992b). It is equally as effective on isolated as on aggregated trees. In field observations, the bud-feeding weevil *T. lativentre* alone had no significant effect on the density of mature *S. punicea* plants, although the density of seedlings declined significantly with the number of years that *T. lativentre* was present (Hoffmann and Moran 1999). However, in combination with *R. marginatus*, the seed set was reduced by up to 99.7 % (Hoffmann and Moran 1992b), with one site showing a highly significant decline of mature plants. When applying these data to the simulation model (Hoffmann 1990) with and without these two seed-destroying agents, the population dynamics showed that at year 25 the weed population equilibrates at a mean density of about 25 plants/10 m^2 and that the weevils have no impact on plant density for the first 3 years after release. However, after 3 years, when the seed bank is reduced because of reduced germination in combination with a reduction of the seed-rain and natural mortality, the recruitment of new plants into the population drops and the population declines as the older trees die. The model shows that a reduction of less than 95 % does not adequately control the weed within 25 years after release of the weevils and that satisfactory control is only achieved within 10 years when seed-set is reduced by 99% (Hoffmann 1990). The model reflects long-term field observations on *S. punicea*. Similar conclusions were also made in areas where two seed-reducing agents were used in the biological control of another perennial invading tree in South Africa, *Acacia longifolia*. In this case, a consistent reduction of seed production of between 85 and 100% since the late 1980s has been achieved (Dennill and Donnelly 1991), and, in addition, the slow decline of populations has recently been noted. The slow rate of decrease of weed populations with such high levels of seed reduction was disconcerting at times, especially with plants with long-lived seed banks. However, in combination with the mechanical clearing of mature plants and/or in combination with fire in the case of *A. longifolia*, the process of population decline was considerably accelerated.

7.5.2.2 The Stem-Borer: *Neodiplogrammus quadrivittatus*

Initially, this univoltine weevil did not seem to be a promising biocontrol agent because of its slow rate of increase and its poor dispersal abilities. Females oviposit a single egg in a young branch or stem. The larvae feed underneath the bark, destroying the vascular tissues as they move downward, often reaching the root crown. Large *S. punicea* plants are often killed by a small number of larvae. Hoffmann and Moran (1991a) have described the impressive control of *S. punicea* caused by this weevil throughout the country, particularly with large plants in dense infestations. The introduction of *N. quadrivittatus* had a drastic effect on the population dynamics of the weed. In long-term field observations (Hoffmann and Moran 1992b) the combination of *T. lativentre* and *N. quadrivittatus* induced a significant reduction in the density of mature plants. In the presence of all three weevils, there was a highly significant decline in the overall density of plants. Although it was not possible to measure the effect of either *R. marginatus* or *N. quadrivittatus* alone on the population dynamics of the weed, evidence from the simulation model (Hoffmann 1990) and from manual clearing operations showed that neither agent on its own would have provided satisfactory control of the invader. In combination, however, *T. lativentre* and *N. quadrivittatus* resulted in a significant reduction of mature plants as was confirmed in 10-year-old population studies (Hoffmann and Moran 1999). The killing of the older plants by *N. quadrivittatus* in the simulation model enhanced the survival of the younger plants, which more than compensated for the loss of the older plants. Plant population levels consequently increased but only to be reduced again by subsequent attacks of the stem-boring weevil. This accounts for the initial large fluctuations in the model, which did not persist, as the population of the weed eventually declined because of the destruction caused by the weevil (Hoffmann 1990).

When all three natural enemies were present in a weed population, there was little seedling recruitment following the destruction of the larger plants by the stem-borer and control was achieved much earlier and was more successful. In practice, the combined effect of these biocontrol agents has reduced populations of the weed to levels below economic thresholds. Even where the seed-set is reduced by only 75 % by the two seed-destroying weevils, control would still be achieved as indicated by the model. Hoffmann and Moran (1991b) showed that predictions of the model were reflected in the results of two detailed, long-term studies on *S. punicea*. The model as described by Hoffmann (1990) allows prediction of the degree of control that can be expected from a single census of the age structure of the weed population approximately 4–5 years after the release of the weevils. Early detection of effectiveness of biological control will be crucial to design overall management operations and to decide whether other control methods, beside biological control, will be needed.

Considerably more effort has been spent on post-release evaluations of this biological control programme against *S. punicea* in South Africa than is normally the case, and it can now be stated that *S. punicea* is under complete control. Large scattered infestations have now changed to isolated plants and occasional small clumps, which do not pose a threat to ecosystem functioning or to agricultural resources. Unfortunately, the study did not include the successive processes that lead to ecosystem restoration after biological control of the invasive species.

Compared to other control methods, biological control has the advantage that the course of control is gradual, allowing natural processes to drive rehabilitation without further intervention. The remarkable recovery of the Fynbos vegetation in the western Cape after the successful control of *Acacia saligna* using the rust fungus, *Uromycladium tepperianum*, (Morris 1999) after a period of 10 years was significant. This study also supports the contention that "agents that reduce seed set or destroy seeds of their host plants should be used routinely as a first line of attack in the biological control of weeds" (Hoffmann and Moran 1998). South African biocontrol programmes rely heavily on agents that prevent seeding. It was more by good fortune rather than by deliberate design that this particular sequence of releases of biological control agents happened to be the ideal sequence. Ideally all new biocontrol projects should be carefully designed around the biology of the weed.

7.5.3 Aquatic Weeds

Over the last 20 years, some remarkable successes in biocontrol have been achieved against aquatic weeds that form floating mats. An example is *Salvinia molesta*, a water fern of South American origin, which is a serious weed outside its native range. Here, floating mats can reduce or prevent the use of water for transport, fishing and watering of stock. They may cause flooding where *Salvinia* reduces the flow of water after heavy rain. Mats of *Salvinia* inhibit photosynthesis and oxygenation of submersed plants, they taint water for human consumption and are a risk for the survival of native plant and animal species (Harley and Mitchell 1981). During several surveys in South America, where *Salvinia* is native, a total of 12 insect species were found to feed on it. The release of the weevil *Cyrtobagous salviniae* in Australia and Papua New Guinea resulted in the spectacular control of *Salvinia molesta*. Between 1972 and 1980, continuous mats of *Salvinia molesta* occupied about 250 km^2 of the River Sepik area in Papua New Guinea, which led to the abandonment of many villages. Releases of *C. salviniae* started in 1982. By August 1985, the total area covered by the weed in the Sepik region had dropped from about 250 to 2 km^2 and village life had returned to normal (Waterhouse and Norris 1987).

Whereas in Australia *C. salviniae* multiplied immediately after release, in Papua New Guinea the nitrogen level in the *Salvinia* tissues was too low for the nutritional requirements of the weevil. To start a population outbreak, an initial and locally limited fertilization of *Salvinia* with urea was necessary. The activities of *Cyrtobagous* then caused an abrupt rise in the nitrogen level of any undamaged *Salvinia* buds and this enriched food enabled more weevils to thrive (Waterhouse and Norris 1987). It is also remarkable that a closely related sibling species, *Cyrtobagous singularis*, which in its native area is associated with *Salvinia auricularia*, failed to build up populations on *Salvinia molesta*, in spite of a very similar biology (Waterhouse and Norris 1987). This shows the importance of subtle biological differences between agents which can determine the success or failure of a biocontrol project.

7.6 Discussion and Conclusions

The influence of phytophagous insects on the structure and function of ecosystems differs greatly from the clear impact of large herbivorous mammals such as many ungulates. Thus, there is little doubt that the coevolution of megaherbivores with the vegetation (McNaughton 1991) was one of the determinants in the evolution of grassland ecosystems. In phytophagous insects, the interactions with the plant community of terrestrial ecosystems are much more subtle and are integrated into a much more complex system of 'checks and balances'. With regard to undisturbed ecosystems, only a few cases are known in which native phytophagous insects have an obvious and profound impact on ecosystem structure and function. An example is the North American spruce budworm, *Choristoneura fumiferana*, whose spectacular outbreaks are a driving force in the dynamics of *Abies balsamea* forest ecosystems (Clark and Holling 1979). As projects of biological weed control operate in ecological systems disturbed by man, they do not allow conclusions concerning to what extent phytophagous insects are able to influence the ecology and evolution of undisturbed ecosystems. However, the successes in the biological control of weeds demonstrate that certain allochthonous insect species have the potential to clear sites from an invading allochthonous weed species or to reduce the density of a weed to such an extent that the former vegetation (and hence the former community structure) or a cleared water surface is reestablished.

Invasive alien weed species usually possess an increased reproductive and competitive potential, as they have no specialized antagonists in their new environment (Andres et al. 1976). Biological control projects use a similar principle for the suppression of such species: the foreign agent is introduced without specialized enemies and/or pathogens, which theoretically allows an unrestricted utilization of its resource. This measure, however, is often not sufficient, as incomplete adaptation to the abiotic and biotic conditions of the new ecosystem may prevent the establishment of the agent. Furthermore, even if an agent is successfully established, behavioural stabilizing mechanisms (such as density-dependent dispersal) may reduce its impact on its host plant to such an extent that it does not become a control factor. The example of *Rh. conicus* on *C. nutans* shows that phytophagous species that have evolved under extreme conditions of competition and enemy pressure may be particularly successful in a new environment.

There is an important difference in the insect–plant interactions in the biological control of weeds in terrestrial ecosystems and of floating aquatic weeds. The usual goal in biological control projects against terrestrial weeds is to substantially reduce the density of the invader by weakening its reproductive potential and its competitive stress on the vegetation of the ecosystem. This can restore and stabilize the former community together with residual individuals of the invading weed species. In this way, the energy flow of the ecosystem, which was formerly blocked by the invading population, is restored and a multitude of food webs can form anew. Former pasture grounds became again accessible and could be used again by grazing animals (Holloway 1964; Zimmermann et al. 1986). Injury to livestock by the poisonous alien weed *Hypericum perforatum* ceased, as the weed was reduced to 1 % of its former density by *Chrysolina* spp. In California, such biological control was successful on over 2 million acres of range land (Holloway 1964). If the targets of a successful biological weed control were invasive shrubs or trees, valuable water resources were regained (Hoffmann and Moran 1999), soil properties restored and species diversity increased. Several studies in South Africa have demonstrated these impacts and their effect on ecosystem functioning caused by woody alien plant invaders (Musil and Midgley 1990; Witkowski 1991; Le Maitre et al. 1996; Holmes and Cowling 1997). In contrast, the biological control of floating aquatic weeds attempts simply to clear the water surface from an invader. In spite of initial skepticism, the recent successes in the biological control of aquatic weeds (Waterhouse and Norris 1987; Julien and Griffiths 1998) show that certain phytophagous insect species can eliminate dense mats of an invading floating weed species even if there is no competing vegetation. This means that biocontrol agents exist which are powerful enough to control a weed permanently by direct effects and not indirectly by reducing its competitive capacity. The effect of initial fertilization with urea on the control of *Salvinia molesta* corresponds with Australian experiences during the control of prickly pear with *Cactoblastis cactorum*. There, in an area with poor soil, successful control was achieved only after nitrogen fertilizer improved the quality of the weed (Myers 1987). The results of these fertilizer experiments are interesting examples of the ecological 'paradox of enrichment' (Rosenzweig 1971).

One of the major drawbacks in biocontrol of weed projects is the large number of failures and the unpredictability of project outcomes. Some guidelines that could increase the chances of finding 'winners' were listed above, but there are many exceptions that question the validity of each generalization. The high cost of developing a biocontrol agent to the point of release is a compelling reason why more research is warranted into the methodologies and selection processes of new biocontrol agents. The new tools offered by DNA analyses that allow us to do exact matching of host and predator will certainly increase the success rate, as is the ability to select climatically adapted biotypes of a species.

Reuniting neophytes with their key associated phytophages provides unique and one-off opportunities to study plant-insect interactions, even more so if the phytophages are released in sequence and in various combinations, and with or without predation at higher tropic levels. The value of these studies increases further when they can be supported with detailed studies of the plant dynamics as affected by the complete set of associated phytophages in the invader's country of origin. The long-term project on the biological control of thistles is one example (Harris 1984a; Kruess et al., Chap. 8, this Vol.).

Well-monitored biological control projects of weeds have certainly contributed more than is generally realized to our knowledge on the impact of insect phytophages on plant populations. Unfortunately, many countries, particularly in the developing world, do not have the capacity and means to capture all the processes in play after an alien plant has been reunited with one of its key natural enemies. Most of the funding bodies for biocontrol projects are also reluctant to finance long-term monitoring effects and usually insist on quick results from 'winners'. An increased emphasis on follow-up studies of successful projects of biological weed control and systematically conducted analyses of failures would certainly not only facilitate decision-making processes for practitioners of biological weed control but also extend our understanding of the impact of phytophagous insects on ecosystem functioning.

Acknowledgements. We thank two anonymous reviewers for reading an earlier version of the manuscript and making extensive comments, which helped to improve the chapter.

References

Andres LA, Davis CJ, Harris P, Wapshere AJ (1976) Biological control of weeds. In: Huffaker CB, Messenger PS (eds) Theory and practice of biological control. Academic Press, New York, pp 481–499

Briese DT (1997) Biological control of St John's wort: past, present and future. Plant Protect Q 12:73–80

Clark WC, Holling CS (1979) Process models, equilibrium structures, and population dynamics: on the formulation and testing of realistic theory in ecology. Fortschr Zool 25:29–52

Crawley MJ (1983) Herbivory. The dynamics of animal–plant interactions. Blackwell, Oxford

Crawley MJ (1987) What makes a community invasible? In: Gray AJ, Crawley MJ, Edwards PJ (eds) Colonization, succession and stability. Blackwell, Oxford, pp 429–453

Crawley MJ (1989a) The successes and failures of weed biocontrol using insects. Biocontrol News Info 10:213–223

Crawley MJ (1989b) Insect herbivores and plant population dynamics. Annu Rev Entomol 34:531–564

DeLoach CJ (1984) Conflicts of interest over beneficial and undesirable aspects of mesquite (*Prosopis* spp.) in the United States as related to biological control. In: Delfosse ES (ed) Proc 6th Int Symp on Biological Control of Weeds, 19–25 Aug, Vancouver, Canada, Agriculture Canada, pp 301–340

Dennill GB, Donnelly D (1991) Biological control of *Acacia longifolia* and related weed species (Fabaceae). Entomophaga 36:259–306

Dodd AP (1940) The biological campaign against prickly pear. Commonwealth Prickly Pear Board Bulletin, Brisbane, pp 1–177

Dowd PF, Kok LT (1982) Parasitism of *Rhinocyllus conicus* in Virginia. Environ Entomol 11:71–77

Dukes JS, Mooney HA (2004) Biological invaders disrupt ecosystem processes in western North America. In: Bradshaw GA, Mooney HA, Alaback P (eds) Disruptions and variability in ecosystem processes and patterns. Springer, Berlin Heidelberg New York (in press)

Erb HE (1980) The natural enemies and distribution of *Sesbania punicea* (Cav.) Benth. in Argentina. In: Neser S, Cairns ALP (eds) Proc 3rd National Weeds Conf of South Africa, AA Balkema, Cape Town, pp 205–210

Goeden RD (1983) Critique and revision of Harris' scoring system for selection of insect agents in biological control of weeds. Protect Ecol 5:287–301

Goeden RD, Ricker DW, Hawkins BA (1984) Ethological and genetic differences among three biotypes of *Rhinocyllus conicus* (Coleoptera: Curculionidae) introduced into North America for the biological control of asteraceous thistles. In: Delfosse ES (ed) Proc 6th Int Symp Biological Control of Weeds, 19–25 Aug, Vancouver, Canada, Agriculture Canada, pp 181–189

Graaf JL, Van Staden J (1984) The germination characteristics of two *Sesbania* species. S Afr J Bot 3:59–62

Harley KLS, Mitchell DS (1981) The biology of Australian weeds. 6. *Salvinia molesta*. J Aust Inst Agric Sci 47:67–76
Harris P (1973a) The selection of effective agents for the biological control of weeds. Can Entomol 105:1495–1503
Harris P (1973b) Insects in population dynamics of plants. In: van Emden HF (ed) Proc Symp Insect/Plant Relationships, Royal Entomological Society of London. Blackwell, Oxford, pp 201–209
Harris P (1984a) Current approaches to biological control of weeds. In: Kelleher JS, Hulme MA (eds) Biological control programmes against insects and weeds in Canada 1969–1980. Commonwealth Agricultural Bureaux, Farnham Royal, UK, pp 95–103
Harris P (1984b) *Carduus nutans* L., nodding thistle, *C. acanthoides*, plumeless thistle (Compositae). In: Kelleher JS, Hulme MA (eds) Biological control programmes against insects and weeds in Canada 1969–1980. Commonwealth Agricultural Bureaux, Farnham Royal, UK, pp 115–126
Harris P (1986) Biological control of nodding thistle and bull thistle with *Rhinocyllus conicus* (Froel.). Canadex, Agriculture Canada, Communication branch, Ottawa, pp 1–4
Harris P (1991) Classical biocontrol of weeds: its definition, selection of effective agents, and administrative-political problems. Can Entomol 123:827–849
Harris P (1993) Effects, constraints and the future of weed biocontrol. Agric Ecosyst Environ 46:289–303
Harris P (1997) Monitoring and impact of weed biological control agents. In: Andow DA, Ragsdale DW, Nyvall RE (eds) Ecological interactions and biological control. Westview Press, Boulder, pp 215–223
Hill MP (1999) Biological control of red water fern, *Azolla filiculoides* Lamarck (Pteridophyta: Azollaceae), in South Africa. Afr Entomol Mem 1:119–124
Hoffmann JH (1988) An early assessment of *Trichapion lativentre* (Coleoptera: Apionidae) for biological control of the weed *Sesbania punicea* (Fabaceae) in South Africa. J Entomol Soc S Afr 51:265–273
Hoffmann JH (1990) Interactions between three weevil species in the biocontrol of *Sesbania punicea* (Fabaceae): the role of simulation models in evaluation. Agric Ecosyst Environ 32:77–87
Hoffmann JH, Moran VC (1989) Novel graphs for depicting herbivore damage on plants: the biological control of *Sesbania punicea* (Fabaceae) by an introduced weevil. J Appl Ecol 26:353–360
Hoffmann JH, Moran VC (1991a) Biocontrol of a perennial legume, *Sesbania punicea*, using a florivorous weevil, *Trichapion lativentre*: weed population dynamics with a scarcity of seeds. Oecologia 88:574–576
Hoffmann JH, Moran VC (1991b) The biological control of *Sesbania punicea* (Fabaceae) in South Africa. Agric Ecosyst Environ 37:157–174
Hoffmann JH, Moran VC (1992a) The influence of host-plant location in the incidence and impact of *Trichapion lativentre* (Coleoptera: Apionidae), a biocontrol agent of *Sesbania punicea* (Fabaceae). J Entomol Soc S Afr 55:258–287
Hoffmann JH, Moran VC (1992b) Oviposition patterns and the supplementary role of a seed-feeding weevil, *Rhyssomatus marginatus* (Coleoptera: Curculionidae), in the biological control of a perennial leguminous weed, *Sesbania punicea*. Bull Entomol Res 82:343–347
Hoffmann JH, Moran VC (1998) The population dynamics of an introduced tree, *Sesbania punicea*, in South Africa, in response to long-term damage caused by different combinations of three species of biological control agents. Oecologia 114:343–348
Hoffmann JH, Moran VC (1999) A review of the agents and factors that have contributed to the successful biological control of *Sesbania punicea* (Cav.) Benth. (Papilionaceae) in South Africa. Afr Entomol Mem 1:75–79

Holloway JK (1964) Projects in biological control of weeds. In: DeBach P (ed) Biological control of insect pests and weeds. Chapman and Hall, London, pp 650–670

Holmes PM, Cowling RM (1997) The effects of invasion by *Acacia saligna* on the guild structure and regeneration capabilities of South African fynbos shrublands. J Appl Ecol 34:317–332

Huffaker CB (1964) Fundamentals of biological weed control. In: DeBach P (ed) Biological control of insect pests and weeds. Chapman and Hall, London, pp 631–647

Huffaker CB, Messenger PS (1976) (ed) Theory and practice of biological control. Academic Press, New York

Julien MH (1989) Biological control of weeds worldwide: Trends, rates of successes and the future. Biocontrol News Info 10:299–306

Julien MH, Griffiths MW (eds) (1998) Biological control of weeds, a world catalogue of agents and their target weeds, 4th edn. CAB International, Wallingford, UK

Kok LT, Surles WW (1975) Successful biocontrol of musk thistle by an introduced weevil, *Rhinocyllus conicus*. Environ Entomol 4:1025–1027

Laing JE, Hamai J (1976) Biological control of insect pests and weeds by imported parasites, predators, and pathogens. In: Huffaker CB, Messenger PS (eds) Theory and practice of biological control. Academic Press, New York, pp 685–743

Le Maitre DC, van Wilgen BW, Chapman RA, McKelly DH (1996) Invasive plants and water resources in the western Cape Province, South Africa: modelling the consequences of a lack of management. J Appl Ecol 33:161–172

Louda SM, Kendall D, Connor J, Simberloff D (1997) Ecological effects of an insect introduced for the biological control of weeds. Science 277:1088–1090

Mattson WJ, Levieux J, Bernard-Dagan C (eds) (1988) Mechanisms of woody plant defenses against insects: search for pattern. Biosis Book no 37842. Springer, Berlin Heidelberg New York

McNaughton SJ (1991) Evolutionary ecology of large tropical herbivores. In: Price PW, Lewinsohn TM, Wilson Fernandes G, Benson WW (eds). Plant–animal interactions: evolutionary ecology in tropical and temperate regions. Wiley, New York, pp 509–522

Moran VC, Zimmermann HG (1984) The biological control of cactus weeds: achievements and prospects . Biocontrol News Info 5:297–320

Morris MM (1999) The contribution of the gall-forming rust fungus *Uromycladium tepperianum* (Sacc.) McAlp. to the biological control of *Acacia saligna* (Labill.) Wendl. (Fabaceae) in South Africa. Afr Entomol Mem.1:125–128

Musil CF, Midgley GF (1990) The relative impact of invasive Australian acacias, fire and season on the soil chemical status of a sand plain lowland fynbos community. S Afr J Bot 56:419–427

Myers JH (1987) Population outbreaks of introduced insects: lessons from the biological control of weeds. In: Barbosa P, Schultz JC (eds) Insect outbreaks. Academic Press, San Diego, pp 173–193

Neser S, Kluge RL (1986) The importance of seed-attacking agents in the biological control of invasive alien plants. In: Macdonald IAW, Kruger FJ, Ferrar AA (eds) The ecology and management of biological invasions in southern Africa. Oxford University Press, Cape Town, pp 285–294

Olckers T, Hill MP (eds) (1999) Biological control of weeds in South Africa (1990–1998). African Entomology Memoir 1, Entomological Society of Southern Africa, Pretoria

Parker IM, Simberloff D, Lonsdale WM, Goodell K, Wonham M, Kareiva PM, Williamson MH, Von Holle B, Moyle PB, Byers JE, Goldwasser L (1999) Impact: towards a framework for understanding the ecological effect of invaders. Biol Invasions 1:3–19

Pemberton RW (1985) Native weeds as candidates for biological control research. In: Delfosse ES (ed) Proc 6th Int Symp Biological Control of Weeds, 19–25 Aug, Vancouver, Canada, Agriculture Canada, Ottawa, Ontario, pp 869–877

Rees NE (1977) Impact of *Rhinocyllus conicus* on thistles in southwestern Montana. Environ Entomol 6:839–842

Rees NE (1978) Interaction of *Rhinocyllus conicus* and thistles in the Galatin valley. In: Frick KE (ed) Biological control of thistles in the genus *Carduus* in the United States. Science and Education Administration, US Department of Agriculture, Washington, DC, pp 31–38

Rosenzweig ML (1971) Paradox of enrichment: destabilization of exploitation ecosystems in ecological time. Science 171:385–387

Simmonds FJ (1933) The biological control of the weed *Clidemia hirta* D. Don, in Fiji. Bull Entomol Res 24:345–348

Thomas PA (1985) Management of *Salvinia molesta* in Papua New Guinea. FAO Plant Protect Bull 33:50–56

Thomas PA, Room PM (1986) Towards biological control of *Salvinia* in Papua New Guinea. In: Delfosse ES (ed) Proc 6th Int Symp Biological Control of Weeds, 19–25 Aug, Vancouver, Canada, pp 567–574

Van Wilgen BW, Richardson DM, Le Maitre DC, Marais C, Magadlela D (2001) The economic consequences of alien plant invasions: examples of impacts and approaches to sustainable management in South Africa. Environ Dev Sustainability 3:145–168

Waterhouse DF, Norris KR (1987) Biological control, Pacific prospects. Inkata Press, Melbourne

Wilson CL (1969) Use of plant pathogens in weed control. Annu Rev Phytopathol 7:411–434

Witkowski RTF (1991) Effects of invasive alien acacias on nutrient cycling in the coastal lowlands of the Cape Fynbos. J Appl Ecol 28:1–15

Zimmermann HG, Moran VC, Hoffmann HJ (1986) Insect herbivores as determinants of the present distribution and abundance of invasive cacti in South Africa. In: MacDonald IAW, Kruger FJ, Ferrar AA (eds) The ecology and management of biological invasions in southern Africa. Oxford University Press, Cape Town, pp 269–274

Zwölfer H (1973) Competition and coexistence in phytophagous insects attacking the heads of *Carduus nutans*. In: Dunn PH (ed) Proc 2nd Int Symp Biological Control of Weeds, 1971, Rome, Misc Publ 6. Commonwealth Institute of Biological Control, Farnham Royal, UK, pp 74–81

Zwölfer H (1979) Strategies and counterstrategies in insect population systems competing for space and food in flower heads and plant galls. Fortschr Zool 25:331–353

Zwölfer H (1980) Distelblütenköpfe als ökologische Kleinsysteme: Konkurrenz und Koexistenz in Phytophagenkomplexen. Mitt Dtsch Ges Allg Angew Entomol 2:21–37

Zwölfer H (1985) Energieflußsteuerung durch informationelle Prozesse – ein vernachlässigtes Gebiet der Ökosystemforschung. Verh Ges Ökol 13:285–294

Zwölfer H, Harris P (1984) Biology and host specificity of *Rhinocyllus conicus* (Froel.) (Col. Curculionidae), a successful agent for biocontrol of the thistle, *Carduus nutans* L. Z Angew Entomol 97:36–62

Zwölfer H, Preiss M (1983) Host selection and oviposition behaviour in West-European ecotypes of *Rhinocyllus conicus* Froel. (Col. Curculionidae). Z Angew Entomol 95:113–122

Zwölfer H, Völkl W (1997) Einfluß des Verhaltens adulter Insekten auf Ressourcen-Nutzung und Populationsdynamik: Ein Drei-Komponenten-Modell der Populationsdichte-Steuerung. Entomol Gen 21:129–144

8 Plant–Insect–Pathogen Interactions on Local and Regional Scales

A. Kruess, S. Eber, S. Kluth and T. Tscharntke

8.1 Summary

In many plant communities, weeds are important drivers of ecosystem processes, but natural enemies may control their growth and population dynamics. These enemies may directly reduce plant growth by grazing or infection, or indirectly enhance plant susceptibility to other enemy species. The role of insects as vectors of plant pathogens has thereby often been underestimated but appears to be widespread and important. Crucial for the understanding of such plant–enemy interactions is the consideration of processes on local and regional scales. We analysed the effects of herbivore and pathogen attack and their interactions using the weed creeping thistle, *Cirsium arvense*, as a model system. Local thistle patches were limited by the thistles' low competitive ability in secondary succession, by pathogens and insects and by the disturbance or management of habitats. The regional dynamics of *C. arvense* were mainly driven by human activities and showed a mosaic of increasing and decreasing local patches with low persistence. High patch turnover rates and subsequent small average patch sizes limited the success of herbivore populations.

Furthermore, we analysed thistle–herbivore–pathogen–parasitoid interactions on different spatial scales in a landscape context. Interactions between *C. arvense* and either insect herbivores or rust pathogens were related to landscape context at large spatial scales (3,000 m radius of landscape sector), whereas herbivore–parasitoid interactions were influenced at smaller scales (750 m). Hence, species at higher trophic levels appeared to have a smaller range of dispersal than those at lower levels. These marked spatial differences emphasize the need to consider both local and landscape management in biological control. In general, such an approach may help to explain the dynamics of plant populations in dependence on possible control by insects and pathogens which can affect the plants' influence on ecosystem processes. However, the potential of facilitation among the plants' antagonists via plant-mediated indirect interactions or with insects as vectors of pathogen spores

Ecological Studies, Vol. 173
W.W. Weisser and E. Siemann (Eds.)
Insects and Ecosystem Function

and their dependence on species-specific spatial scales need much more experimental evidence.

8.2 Introduction

No organism lives isolated, and the interactions between organisms and their biotic and abiotic environment are the ecological functions determining patterns and processes in ecosystems. The numerous complex, direct and indirect, interactions often connect taxonomically remote organisms, which is well known from a natural-history perspective. This real-world complexity, however, has only recently been acknowledged by ecologists (Janssen and Sabelis, Chap. 9, this Vol.). Thus the focus in ecology is switching from the traditional study of simple communities and interactions to approaches considering the effects of multiple trophic levels for the distribution and abundance of populations (Gange and Brown 1997; Schmitz et al. 2000; Tscharntke and Hawkins 2002; Schmitz, Chap. 14, this Vol.). Along with the incorporation of the complexity found in nature goes the realization that local interactions and community structure are directly affected by landscape structure, and that populations therefore need to be studied at different spatial scales (Kareiva and Wennergren 1995; Wiegand et al. 1999).

In this chapter, we analyse plant populations and their complex interactions with natural enemies. In plant communities, weeds are well known to greatly affect ecosystem processes. In Germany, more than half of the sprayed pesticides are herbicides, indicating how fundamentally weeds influence processes in man-made ecosystems. In the USA, invasive species have been implicated for 49 % of the cases of endangered species and cause high costs of about US$ 30 billion per year. Invasive plants contribute substantially to these figures with 57 % of imperilled plants threatened by invasive species (Wilcove et al. 1998; Zwölfer and Zimmermann, Chap. 7, this Vol.).

Plant competition is usually thought to be the dominant driving force of plant community dynamics, but top-down effects by herbivory have also been shown to alter plant community structure and ecosystem processes (Hulme 1996; Mulder et al. 1999; Wilby and Brown 2001). Although weeds are important drivers of ecosystem processes, little is known about how their natural antagonists feeding on vegetative or reproductive parts of the plant affect plant growth and its overall population dynamics. The influence of these antagonists, such as insects and pathogens, on weed performance will be an important topic of this chapter. Starting with a more general overview on biological weed control and ecosystem processes we will give examples from classical biological weed control and then shift to plant–pathogen–herbivore interactions. The latter will lead to our own case studies on weed–herbivore–pathogen interactions, which focus on one of the most important weeds worldwide, creeping thistle, *Cirsium arvense* (Asteraceae). We will deal with the influence of weed population dynamics on local processes and with dispersal patterns and population dynamics of weed antagonists. Since local processes depend not only on habitat factors, our focus will be on both local and landscape scales. Data of a long-term study on the temporal dynamics of a weed and weed antagonists are presented. This will also illustrate the importance of long-term studies – still in shortage in ecological research – for analysing regional patterns and processes.

8.3 Biological Weed Control, Interactions and Ecosystem Processes

8.3.1 Classical Biological Control

Examples from classical biological weed control show that insects and pathogens may have an important impact on individual plant species (Zwölfer and Zimmermann, Chap. 7, this Vol.). The suppression of weed populations by such antagonists can restore formerly weed-degraded ecosystems, as was shown by the local eradication of *Opuntia* cactus in Australia through feeding damage of the introduced moth *Cactoblastis cactorum* (Pyralidae) and subsequent infection with bacteria and pathogens (Dodd 1940).

Given that insects are ubiquitous and numerous in plant communities, a negative influence of insects on primary production is to be expected. However, their effects on net primary production (NPP) are controversially debated (Crawley 1989a; Weisser and Siemann, Chap. 1, this Vol.). One argument for a limited effect of herbivores on NPP was the presumed population limitation of herbivores by natural enemies (Hairston et al. 1960). In this context, it has been suggested that the outcome of herbivory and herbivore-control may depend on productivity (Oksanen et al. 1981; Coley et al. 1985). More recently, this debate has shifted from top-down vs. bottom-up effects towards hypotheses on how these different forces may work together in structuring communities, altering complex food webs and maintaining biodiversity (Leibold 1996; Persson 1999; Dunne et al. 2002; Dyer and Letourneau 2003).

Concerning weeds and invasive plants, there is clear evidence that both can have negative effects on biodiversity and ecosystem processes (Gordon 1998; Wardle 2002). For instance, the salt cedar *Tamarix ramosissima*, a Mediterranean invader into the USA, can lower the water table due to its deep roots and hence drain desert oases (Vitousek 1986). The nitrogen-fixing shrub *Myrica faya* was introduced from the Azores to Hawaii, where it colonizes lava flows. By enriching the soil with nitrogen, this species affects soil processes, productivity and the structure of the subsequent early-successional plant communities (Vitousek and Walker 1989).

On the other hand, weeds and invasive plants can function as beneficials in agro-ecosystems by providing nectar for insect pollinators, by attracting natural enemies (Root 1973), by reducing erosion, by conserving soil moisture and enhancing nitrogen content in the soil (Gliessman et al. 1981; Marshall et al. 2003). High quality biomass and litter produced by invasive plants can improve decomposition or alter decomposer community structure, thereby influencing the availability of nutrients. The European grass *Agrostis stolonifera*, introduced to the subantarctic Marion Islands, affected several components in the decomposer community. Thereby some species were suppressed, others enhanced (Gremmen et al. 1998).

In most cases, herbivores and pathogens affect plant growth, reproduction or other fitness parameters in a negative way. This is the case for most of the economically important insect and pathogen pests attacking crops. Weeds also suffer from their insect and pathogen load, but these effects are less recognized. Frequently the impact of natural enemies on weeds only becomes apparent when the weed is released from enemy attack. The fact that many alien plant species became important weeds after introduction was attributed to this release from enemy pressure (Blossey and Nötzold 1995; Klironomos 2002) and was stated as the 'enemy release hypothesis' (Keane and Crawley 2002). The great success of some classical approaches of biological weed control by re-establishing weed–enemy interactions supports this idea. The control of klamath weed (*Hypericum perforatum*) in California after the introduction of two leaf beetles, *Chrysolina hyperici* and *Chrysolina quadrigemina*, is one of the most impressive examples in the recent past, where insect herbivores were able to suppress a weed (DeBach 1964). Another example is the successful control of skeleton weed (*Chondrilla juncea*) by the rust *Puccinia chondrillae* in Australia (Burdon and Marshall 1981), which was the first intentional use of a plant pathogen for biological weed control.

Hence, depending on various factors, such as the spatio-temporal dynamics of weeds and their antagonists, and on the outcome of species interactions, biological control of weeds and invasive plants by their antagonists will result in either positive or negative effects on biodiversity and ecosystem processes.

8.3.2 Plant–Pathogen–Herbivore Interactions

Low-level herbivory by herbivores or pathogens can induce compensatory plant growth and thereby enhance both plant productivity and reproduction (McNaughton 1983; Trumble et al. 1993). Although compensatory responses tend to counteract the harmful effects of herbivory, there is no clear experimental evidence for perfect compensation or even overcompensation (as has been, for example, discussed with respect to grass–grazer mutualism: Owen and Wiegert 1976; Detling 1988). Symptom-less endophytic fungi are known to attack grasses without reducing host vitality, and may even improve host-plant tolerance to abiotic and biotic stress (Siegel 1993). There is evidence, however, that acquired resistance against herbivory by endophytes is costly to the plant, and that the production of alkaloids may be nutrient-limited (Cheplick et al. 1989; Faeth and Bultman 2002).

When herbivores and pathogens share the same host, the overall negative effect on plant fitness will depend on the partial impact of each antagonist and also on the outcome of interactions between them. These herbivore–pathogen interactions can be classified as direct and indirect effects, whereby indirect interactions are mediated by the shared host plant (Hatcher 1995). For instance, plant growth changes in response to the attack of one natural enemy may affect the performance of the other antagonists. A positive indirect interaction occurs when a pathogen infection leads to an increase in the nutritional quality of the host, for example when aphids on infected thistles profit from increased levels of free amino-acids in the phloem (Kluth et al. 2002). Hatcher and Paul (2001) found a negative indirect interaction between herbivory and pathogen infection: feeding of the leaf beetle *Gastrophysa viridula* on *Rumex obtusifolius* induced a systemic resistance that reduced the subsequent infection by the rust *Uromyces rumicis*. Stem-boring insects such as flea beetles of the genus *Phyllotreta* can advance infection with *Phoma lingam* (blackleg) because they create entry holes for the pathogen which is hardly able to penetrate undamaged live tissue (Hoffmann et al. 1994).

Examples of direct interactions are the grazing of insects on fungal tissue or spores which reduces plant disease caused by the pathogen. High densities of larvae of the beetle *Phalacropis dispar* can completely consume the aecia of the western pine stem rust fungus (Steiner 1984).

The combined attack of weed antagonists usually results in negative fitness consequences. Hatcher (1995) classified four categories of possible interaction effects: synergistic, additive, equivalent and inhibitory. Hence, the combination of weed antagonists in biological control does not generally lead to better results than the use of a single agent, depending on the species-specific interaction types. As pointed out by Hatcher and Paul (2001), the understanding of these interactions is still limited and thus the success of biological weed control is hardly predictable (Crawley 1989b).

In addition to this lack of local studies on plant–enemy interactions, valuable information gets even scarcer where larger spatial scales are concerned. Little is known about landscape effects on biotic interactions (but see Didham et al. 1996; With et al. 1997; Thies and Tscharntke 1999; Östman et al. 2001; Kruess 2003a, b; Thies et al. 2003), except that populations of higher trophic levels appear to be more susceptible to landscape changes (Kruess and Tscharntke 1994; Roland and Taylor 1997; Holt et al. 1999; Steffan-Dewenter et al. 2002). Success in the management of weeds requires more detailed analyses of the mechanisms driving plant population dynamics (Müller-Schärer et al. 2000; Hatcher and Paul 2001) and needs to include studies on much larger spatial and temporal scales. In the following chapter we present recent case studies on the weed *Cirsium arvense* and its antagonists. We will illustrate the complexity of this plant–herbivore–pathogen system and its spatio-temporal dynamics with respect to local and regional habitat management in agricultural landscapes.

8.4 Creeping Thistle, Insects, Pathogens and Processes

8.4.1 The Creeping Thistle (*Cirsium arvense*)

The notorious creeping thistle [*Cirsium arvense* (L.) Scop. (Asteraceae)] is one of the most persistent, competitive and widespread weeds in Europe (Schroeder et al. 1993). It is a perennial, dioecious, clonal plant typical of early successional, ruderal communities (Hegi 1987). Its extensive seed bank (Thompson et al. 1997), huge seed production and ability for clonal growth promote a wide distribution in agricultural landscapes. Vegetative reproduction can arise from adventitious root buds or from old stem bases (Donald 1994), and may extensively expand horizontally. *C. arvense* can produce fertile seeds in high abundance, but the importance of sexual reproduction and seed dispersal has been discussed controversially (Heimann and Cussans 1996; McClay et al. 2001). Population dynamics of *C. arvense* strongly depend on management and disturbance frequency. In the first 3 years after successful colonization, net primary production of *C. arvense* is very high so that the thistle can build up large and dense populations within a few years. This is followed by a gradual decline of the populations during the following years (Denys and Tscharntke 2002; Kruess et al., unpubl. data) due to a shift from sexual to asexual reproduction and a low competitive ability of the thistle (Donald 1994). Thereby, grazing or mowing will promote the decline of *C. arvense* in grasslands, along with increasing competition by perennial grasses. In arable fields, the abundance of the thistle is influenced by crop rotation and tends to be stabilized by continuing disturbances (Donald 1994).

The creeping thistle has been accidentally introduced into many countries (Moore 1975), causing dramatic weed problems, particularly on arable land in the temperate regions (Holm et al. 1977). In North America it is listed as a noxious weed in 35 US states (Skinner et al. 2000). However, in a long history of North-American trials of classical biological control introducing insects and pathogens, little has been achieved (e.g. Peschken 1981, 1984; McClay et al. 2001). In Europe, problems with *C. arvense* are on the rise because changes in land management and restrictions in chemical weed control promote the spread of the thistle in ecological compensation areas and on organic farms.

With little European tradition in the use of natural enemies for weed control (Müller-Schärer and Scheepens 1997), the scientific interest in the biological control of *C. arvense* has been limited (Schroeder 1980), but is now increasing. Thereby, the main focus is on the use of indigenous insects and pathogens (Frantzen 1994a, b; Frantzen and Van der Zweerde 1994; Völker and Boyle 1994; Guske et al. 1996; Friedli and Bacher 2001a, b; Kluth et al. 2001, 2002, 2003; Kruess 2002). One of the main questions is now how the biological control of *C. arvense* is influenced by spatio-temporal dynamics of the weed and its antagonists on different spatial scales.

8.4.2 Interactions Between Pathogens and Insect Vectors on a Local Scale

Weed antagonists may show inhibitory or facilitating interactions mediated by species-specific plant responses. Kruess et al. (unpubl. data) studied the effects of the two pathogens *Puccinia punctiformis* and *Phoma destructiva* and the leaf-feeding beetle *Cassida rubiginosa*. They found that leaf feeding and subsequent infection with *P. punctiformis* or *P. destructiva* decreased the weight of thistle flower heads, whereas in the non-infected, beetle-only controls their weight even increased. This indicated a negative synergistic effect of these antagonists on the development of reproductive thistle structures. A systemic infection with thistle rust can also cause a reduction in the number of flower buds of more than 90 % (Watson and Keogh 1980). Thereby, seed-set is dramatically reduced. In addition to these negative effects on reproductive structures, systemic rust infections alter the growth of thistle shoots: infected shoots were up to 50 % thinner and smaller (Kluth et al. 2001), so that aboveground biomass was significantly reduced. A systemic infection with the pathogen *P. destructiva* also decreased thistle growth and biomass. Infected shoots were significantly thinner and the number of leaves per shoot was more than 30 % reduced (Kruess 2002).

Enhancement of pathogen infection by insect attack is a promising idea for biological weed control. In a common-garden experiment, simulated herbivory (wounding of leaves) increased infection of the rust *P. punctiformis*, evaluated by the number of uredosori on the leaves (Kluth et al. 2001). As wounding of plants was carried out immediately before spore inoculation, the impact of possible plant defence mechanisms might have been reduced.

These results show that herbivore attack may not only stress a plant, causing growth reductions, but also facilitate pathogen infection. In addition to this mechanism of pathogen facilitation, insects may be effective vectors of rust spores. The aphids *Uroleucon cirsii* L. and *Aphis fabae* ssp. *cirsiiacanthoidis* Scop. and the leaf beetle *C. rubiginosa* caused infection on formerly healthy thistles when transferred from infected leaf tissue to healthy plants (Fig. 8.1; Kluth et al. 2002). The aphids profited by an infection of their host plant and formed larger colonies than on control plants, presumably due to an

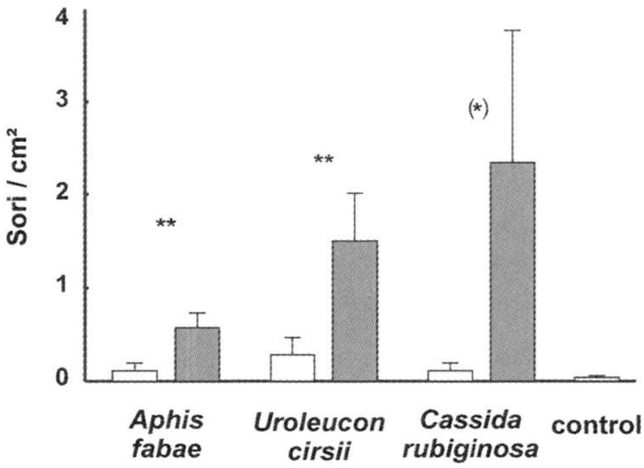

Fig. 8.1. Insects as vectors of rust spores. Sorus density on thistle leaves with herbivores originating from healthy plants (*white bars*) or thistle leaves infected with uredin-iospores of *Puccinia punctiformis* (*shaded bars*); Mann-Whitney *U*-test: *Aphis fabae*: U=52, n=32, P=0.003; *Uroleucon cirsii*: U=42, n=30, P=0.003; *Cassida rubiginosa*: U=68, n=30, P=0.067. Comparison of control with treatments with *A. fabae* ssp. *cirsiiacanthoidis*, *U. cirsii* and *C. rubiginosa* taken from healthy leaves; Kruskal-Wallis-ANOVA, not significant. (After Kluth et al. 2002)

increase in free amino-acid concentrations in the phloem (Kluth et al. 2002). However, insect facilitation of pathogens can also result in an antagonistic interaction. The leaf beetle *C. rubiginosa* served as a good spore vector, but was negatively affected by pathogen infections causing enhanced mortality and development time (Kluth et al. 2002; Kruess 2002). Similarly, Friedli and Bacher (2001a) observed insects as vectors of thistle rust. When females of the weevil *Apion onopordi* were placed on potted experimental thistles, the percentage of systemic rust infection increased. Weevils profited from infecting thistles, reaching higher weights when they hatched from the roots of infected plants.

Hence, phytophagous insects may contribute to biological control in different ways. First, they often reduce plant growth by grazing. Second, they may be important seed predators, also shown for thistles (Kruess et al., unpubl. data). Third, insects may weaken plant defences or enhance food quality of plants, thereby facilitating further antagonists such as pathogens (or vice versa, pathogens may facilitate insect populations). Fourth, they are often vectors of plant pathogens, thereby enhancing infection rates, as shown for systemic thistle rust which greatly reduces thistle fitness. However, depending on the species involved, the plant-mediated interactions between herbivores and pathogens may range from mutualistic to antagonistic, causing higher as well as lower damage of creeping thistle compared to the attack of a single agent (see Sect. 8.3.2).

8.4.3 Regional Dynamics of *Cirsium arvense* and an Associated Herbivore

The spatial and temporal dynamics of *C. arvense* in an agricultural landscape typical of central Europe have been described in Eber and Brandl (2003). Patches of *C. arvense* were found to be dynamic in space and time, whereby both the population and spatial dynamics were related to patch size. There was a characteristic patch size distribution, estimated as the number of plants per patch, with a marked skew towards many more small than large patches. Plant density within patches was fairly constant. The rate of increase of patches decreased with patch size, whereby large patches tended to have rates below one, and small ones rates above one with a net increase in their plant number. There was considerable turnover of thistle patches with a mean annual extinction rate of 28 %, which was negatively correlated with patch size (Fig. 8.2) and mainly caused by mowing and other human influences, and by apparently random events. Despite high extinction rates *C. arvense* patches were common in the study area, suggesting a successful recruitment strategy which provided about 36 % of the total number of patches found in any one year. This ability to persist regionally is due to a high seed production, and to a large and persistent seed bank (Thompson et al. 1997). Frequent disturbances, mostly caused by human activities, provide sites for recruitment from seeds and establishment of new patches. Established *C. arvense* clones can spread up to 12 m per year and 'forage for good conditions' (Chancellor 1970). Hence the distributional pattern of *C. arvense* in the study area was very dynamic, whereby large patches represented more persistent elements in the landscape, whilst small patches appeared and disappeared with high turnover rates.

These temporal and spatial dynamics of *C. arvense* typical for semi-natural areas in central Europe are crucial for associated endo- and ectophagous herbivores. Following the spatial distribution of its host plant, the specialized

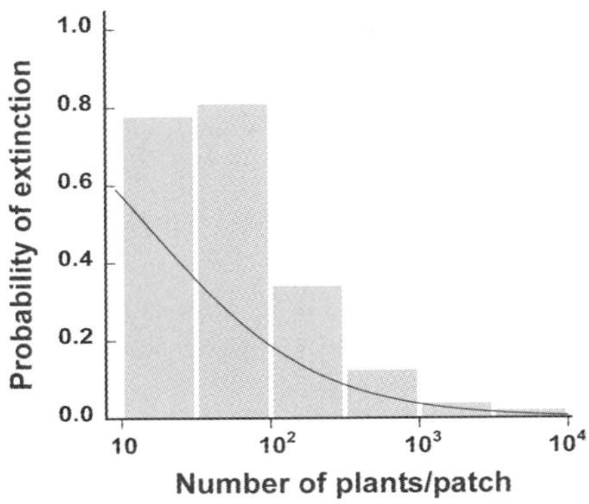

Fig. 8.2. Annual extinction probability of *C. arvense* patches with increasing patch size estimated from a logistic regression ($P<0.001$, n=1,264 in 5 years). *Histogram* gives the distribution of patch size for all patches recorded during the 5-year surveys. (After Eber and Brandl 2003)

tephritid stem gall *U. cardui* forms a metapopulation consisting of local subpopulations on spatially distinct thistle patches (Seitz and Komma 1984; Jeltsch et al. 1992; Eber and Brandl 1994, 1996). Due to high rates of parasitism and to stochastic events, subpopulations go extinct at an annual rate of around 30 %. The risk of extinction showed a negative relationship with local population size, as well as with patch size and neighbourhood density. Extinctions are balanced by similar rates of colonization of unoccupied thistle patches. The colonization rate was highest in years with low rates of parasitism, and was positively correlated with patch size and neighbourhood density.

The influence of the dynamics of *C. arvense* on the density and occurrence of its herbivore was most obvious during a period of patch fragmentation (Eber and Brandl 2003). As the number of *C. arvense* patches in the study area almost doubled with a three-fold decrease in the mean patch size (Fig. 8.3A,

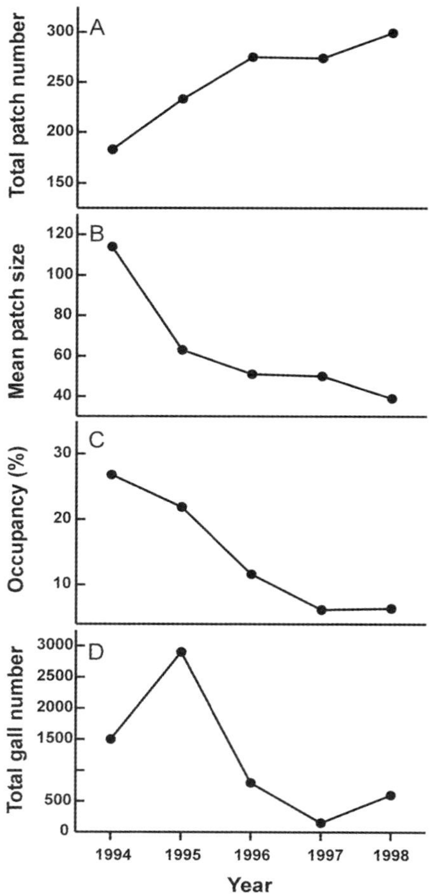

Fig. 8.3. Number of *Cirsium arvense* patches (**A**), mean patch size (**B**), as well as occupancy (**C**) and total gall number (**D**) of its specialist herbivore *Urophora cardui* in the study area during the 5-year survey period. (After Eber and Brandl 2003)

B), there was a simultaneous decline in the occurrence and density of the herbivore. Occupancy of the tephritid fly *U. cardui* dropped from 26.8% to around 6% (Fig. 8.3C) and the total number of galls decreased from 2,900 galls to a minimum of 150 galls (Fig. 8.3D). Patch size was the key factor in this decline of the herbivore, reducing the colonization probability and increasing the extinction rates of *U. cardui* populations. The dynamics of the host plant exacerbated this effect in that small and newly founded patches had a higher extinction probability and therefore increased turnover rates and extinction of local thistle populations.

Consideration of such spatio-temporal dynamics and cycling processes is important for the understanding of ecosystem dynamics and processes. Local and regional patch dynamics have been already described in the classic papers by Watt (1947) with respect to the so-called *hummock-and-hollow-cycle*, and by Whittaker (1953) in his *climax-pattern-hypotheses.*

8.4.4 The Influence of Landscape Context at Different Spatial Scales

Landscape context at different spatial scales may play an important role in the trophic interactions between weeds and their antagonists. In many agricultural areas, the main thistle habitats are crop fields, field margin strips and large perennial habitats such as grasslands and fallows (Kruess 2003a). Crop fields are nutrient-rich but ephemeral habitats with a high disturbance frequency, whereas perennial fallows are characterized by poorer soils but fewer disturbances. Field margins are intermediate but differ from the other habitats by a more linear shape and a smaller area. *C. arvense* density, diversity of stem-boring thistle insects and their parasitoids and plant–insect interactions were studied in these three habitat types and related to landscape context, which was estimated as the percentage of non-crop area within 15 differently structured landscape sectors. Patch size of *C. arvense* was largest in large perennial habitats (34,000±15,300 shoots, mean±SE, n=91) and smallest in crop fields (1,080±300 shoots, n=68) and field margins (390±140 shoots, n=186). Species richness and abundance of stem-boring herbivores were positively correlated with thistle patch size, and total herbivore abundance was also related to landscape context (Kruess 2003a). The correlation coefficients of the relationship between percent non-crop area and thistle attack were calculated for eight spatial scales (radius of the landscape sector: 250, 500, 750, 1,000, 1,500, 2,000, 2,500 and 3,000 m). The result showed the strongest influence of landscape context on herbivore abundance at the largest scale (3,000 m radius, Fig. 8.4A) with an increase of the strength of this correlation with increasing spatial scales (Fig. 8.4D).

Habitat type and landscape context also had a significant influence on the diversity of parasitoids associated with the insect herbivores. Parasitism of the agromyzid fly *Melanagromyza aeneoventris* was affected by landscape

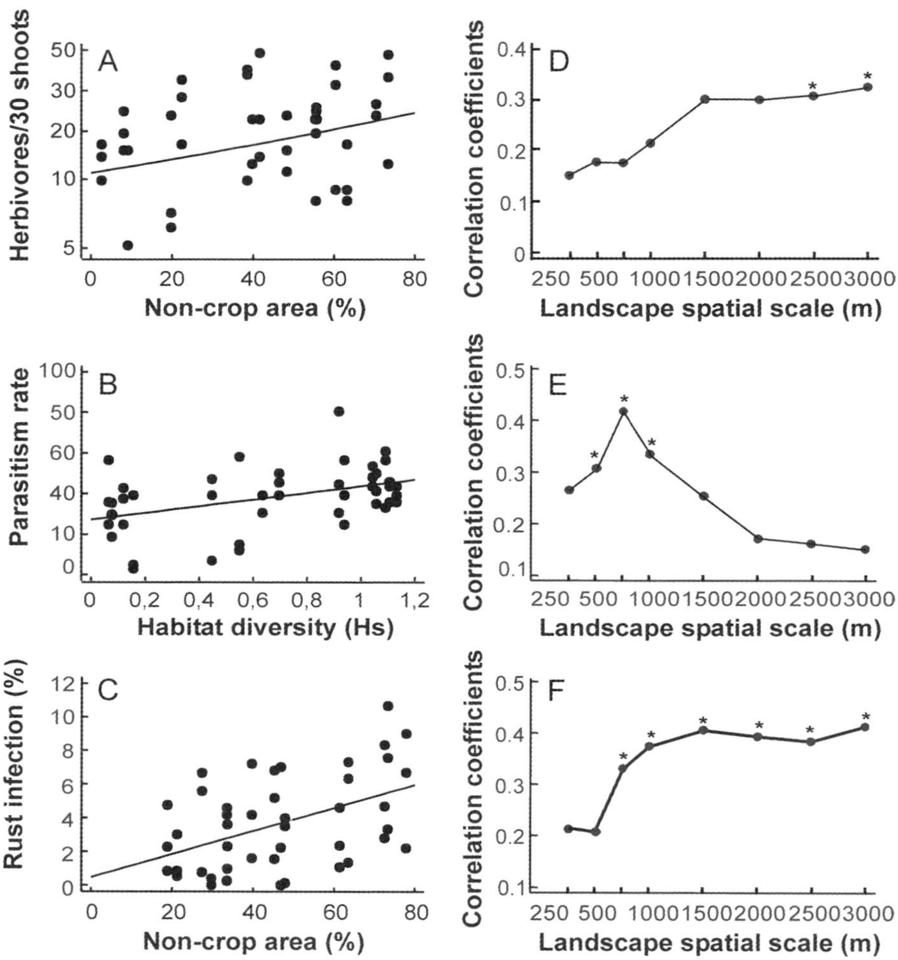

Fig. 8.4. Landscape effects on stem-boring insects on *C. arvense*: **A** correlation between total herbivore abundance and percentage of non-crop area in a landscape area with 3,000 m radius ($F_{1,43}$=5.2, r=0.32, P=0.03, n=45); **B** correlation between percent parasitism of the agromyzid fly *M. aeneoventris* and diversity of habitats in a landscape area with 750 m radius ($F_{1,43}$=9.5, r=0.42, P=0.004, n=45); **C** correlation coefficients of the regression between percentage of thistle shoots systemically infected with *Puccinia punctiformis* and percentage of non-crop area in a landscape area with 3,000 m radius ($F_{1,43}$=10.7, r=0.43, P=0.002, n=45); **D–F** correlation coefficients from the regressions shown in **A–C** respectively, calculated separately for eight landscape scales (250–3,000 m radius). *Asterisks* indicate significant correlations. (**D** after Kruess 2003b; **E** after Kruess 2003a)

structure on a much smaller spatial scale (Kruess 2003a). Non-crop area and percent parasitism were best correlated at a landscape scale with 750 m radius (Fig. 8.4B) with a strong decline in correlation strength for larger spatial scales (Fig. 8.4E). Parasitoids as an important group of the third trophic level have smaller populations and tend to be more restricted in dispersal ability than those of their hosts. Local extinction and recolonization processes appeared to be driven by the landscape structure within a much smaller radius (500–750 m) than their hosts. These results indicate that species from different trophic levels may respond to landscape factors at different spatial scales (Thies et al. 2003). Many theoretical and empirical studies have found that phytophagous insects are better dispersers than their specific parasitoids (Kruess and Tscharntke 1994, 2000; Holt et al. 1999; Tscharntke and Kruess 1999; Hassell 2000). However, mobile parasitoids may also block the spatial spread of hosts (Roland and Taylor 1995; Brodmann et al. 1997; Maron and Harrison 1997).

Fungal spores dispersed by wind are expected to spread further than larger insects, but there is a general lack of information (Jules et al. 2002). As insects are known to be important vectors of plant pathogens (Hoffmann et al. 1994; Kluth et al. 2002, see above), both the local insect-mediated dispersal and the regional dispersal as aerial plankton appear to be important.

In a recent study (Kruess, unpubl. data) the percentage of systemically rust-infected shoots was <1 % in crop fields, 4 % in field margins and 6 % in old perennial habitats. This indicated an influence of habitat type on the local incidence of thistle rust, with an increase in rust infections from frequently disturbed ephemeral habitats to near-natural and less disturbed perennial habitats. Moreover, the percentage of systemically infected shoots per thistle patch was positively correlated with patch size. Insect density on thistles, which may be an indicator of local pathogen dispersal, also increased from crop fields and field margins to fallows, and from small to large thistle patches (Kruess 2003a). The analysis on a landscape scale showed that percent non-crop area was positively related to infection rate (Fig. 8.4C). The correlation was best at large spatial scales with radii between 1,500 and 3,000 m, obviously due to the long-distance dispersal of this species (Fig. 8.4F). In both cases landscape effects on thistle rust and thistle herbivores concerned large spatial scales. It remains unclear whether this is due to similar dispersal abilities or to mutualistic pest–insect interactions. Overall the results showed that the studied thistle antagonists were affected not only by host plant traits and habitat characteristics but also by landscape context. Thereby, mainly large spatial scales seemed to be crucial for herbivores and pathogens, whereas insect parasitoids, which may limit the biocontrol success of thistle herbivores, tended to respond to landscape context at smaller scales.

8.5 Conclusions and Future Outlook

In many plant communities, weeds are important drivers of ecosystem processes, but their growth and population dynamics may be controlled by natural enemies. These enemies may directly reduce plant growth by grazing or infection, and thereby change the plant's competitive ability (Hulme 1996; Mulder et al. 1999; Wilby and Brown 2001; Zwölfer and Zimmermann, Chap. 7, this Vol.). However, they may also indirectly enhance plant susceptibility to further enemy species. The role of insects as vectors of plant pathogens has often been underestimated but appears to be widespread. Further, understanding of plant–enemy interactions needs the consideration of processes on local and regional scales.

In the creeping thistle *Cirsium arvense*, plant–herbivore, plant–pathogen and herbivore–parasitoid interactions were affected by landscape context at different spatial scales. Such differences in how species experience the surrounding landscape may be one reason for the low predictability of the effectiveness of biological weed control (Hatcher and Paul 2001). There is limited evidence that combinations of herbivore and pathogen species may yield better control of weed populations and better reduce their impact on ecosystems than single-agent approaches (Hatcher and Paul 2001). The results in this chapter provide evidence for such complementary and mutualistic effects in that insects enhance pathogen infection by (1) the wounding of leaves as an indirect, plant-mediated effect and (2) the direct interplant dispersal of spores. Wilson's (1969) demand for more research on the combined effects of insects and diseases has only recently been taken up by several authors, so the role of insect–pathogen interactions in weed control is still little known (Hatcher and Paul 2001).

In addition, interactions among belowground and aboveground processes are almost a 'black box'. Belowground interactions in detritus-based food webs are relatively well studied (Brown and Gange 1990; DeRuiter et al. 1995; Mortimer et al. 1999), but few data exist for soil pathogen interactions with plants and their effects on aboveground food webs (Van der Putten 1999; Van der Putten et al. 2001). Only few recent studies show the importance of belowground processes on plant diversity, invasion success of weeds and aboveground plant–insect interactions (Masters and Brown 1997; Van der Putten et al. 2001; De Deyn et al. 2003; Wardle and Bardgett, Chap. 3, this Vol.; Bonkowski and Scheu, Chap. 4, this Vol.; Masters, Chap. 5, this Vol.). Roots of invasive plants (such as *C. arvense* in North America) were found to accumulate more symbionts than pathogens in non-native habitats, resulting in better plant performance. In contrast, rare native plants suffered from higher pathogen accumulation (Klironomos 2002). It is likely that interactions between aboveground and belowground processes are a key to many still unexplained or contradictory results.

In conclusion, we still lack a good understanding of the many possible factors driving plant population dynamics. Both herbivores and pathogens may exert effective control, but little is known of their manifold interactions (including the role of insects in pathogen infections). Further, the great differences in the spatial scale experienced by each species (with respect to both below- and aboveground processes; Ettema and Wardle 2002) emphasize the need to take local as well as landscape management into account. Such an approach may help to better explain the complexity of plant population dynamics in terms of a possible control by insects and pathogens which can affect the plants' influence on ecosystem processes. Possible facilitation among the plants' antagonists and their dependence on species-specific spatial scales need much more experimental evidence from field studies.

References

Blossey B, Nötzold R (1995) Evolution of increased competitive ability in invasive non-indigenous plants: a hypothesis. J Ecol 83:887–889

Brodmann PA, Wilcox CV, Harrison S (1997) Mobile parasitoid may restrict the spatial spread of an insect outbreak. J Anim Ecol 66:65–72

Brown VK, Gange AC (1990) Insect-herbivory below-ground. Adv Ecol Res 20:1–58

Burdon JJ, Marshall DR (1981) Biological control and the reproductive mode of weeds. J Appl Ecol 18:649–658

Chancellor RJ (1970) Biological background to the control of three perennial broad-leaved weeds. In: Proc 10th British Weed Control Conf, British Weed Control Council, London, pp 1114–1120

Cheplick GP, Clay K, Marks S (1989) Interactions between infection by endophytic fungi and nutrient limitation in the grasses *Lolium perenne* and *Festuca arundinacea*. New Phytol 111:89–97

Coley PD, Bryant JP, Chapin FS (1985) Resource availability and plant antiherbivore defense. Science 230:895–899

Crawley MJ (1989a) The relative importance of vertebrate and invertebrate herbivores in plant population dynamics. In: Bernays E (ed) Insect–plant interactions. CRC Press, Boca Raton, pp 45–71

Crawley MJ (1989b) The success and failures of weed biocontrol using insects. Biocontrol News Inf 10:213–223

DeBach P (1964) Biological control of insect pests and weeds. Reinhold, New York

De Deyn GB, Raaijmakers CE, Zoomer HR, Berg MP, de Ruiter PC, Verhoef HA, Bezemer TM, van der Putten WH (2003) Soil invertebrate fauna enhances grassland succession and diversity. Nature 422:711–713

Denys C, Tscharntke T (2002) Plant–insect communities and predator–prey ratios in field margin strips, adjacent crop fields, and fallows. Oecologia 130:315–324

DeRuiter PC, Neutel AM, Moore JC (1995) Energetics, patterns of interaction strength, and stability in real ecosystems. Science 269:1257–1260

Detling JK (1988) Grasslands and savannas: regulation of energy flow and nutrient cycling by herbivores. In: Pomeroy LR, Alberts JJ (eds) Concepts of entomology. Springer, Berlin Heidelberg New York, pp 131–148

Didham RK, Ghazoul J, Stork NE, Davis AJ (1996) Insects in fragmented forests: a functional approach. Trends Ecol Evol 11:255–260

Dodd AP (1940) The biological control of prickly-pear in Australia. In: Whyte RO (ed) The control of weeds. Herbage Publ Ser Bull 27, Imperial Bureau of Pastures and Forage Crops, Aberystwyth, pp 131–143

Donald WW (1994) The biology of Canada thistle (*Cirsium arvense*). Rev Weed Sci 6:77–101

Dunne JA, Williams RJ, Martinez ND (2002) Network structure and biodiversity loss in food webs: robustness increases with connectance. Ecol Lett 5:558–567

Dyer LA, Letourneau D (2003) Top-down and bottom-up diversity cascades in detrital vs. living food webs. Ecol Lett 6:60–68

Eber S, Brandl R (1994) Ecological and genetic spatial patterns of *Urophora cardui* (Diptera: Tephritidae) as evidence for population structure and biogeographical processes. J Anim Ecol 63:187–199

Eber S, Brandl R (1996) Metapopulation dynamics of the tephritid fly *Urophora cardui*: an evaluation of incidence-function model assumptions with field data. J Anim Ecol 65:621–630

Eber S, Brandl R (2003) Regional patch dynamics of the weed *Cirsium arvense*, and possible implications for plant animal interactions. J Veg Sci 14:259–266

Ettema CH, Wardle DA (2002) Spatial soil ecology. Trends Ecol Evol 17:177–183

Faeth SH, Bultman T (2002) Endophytic fungi and interactions among host plants, herbivores, and natural enemies. In: Tscharntke T, Hawkins BA (eds) Multitrophic level interactions. Cambridge University Press, Cambridge, pp 89–123

Frantzen J (1994a) An epidemiological study of *Puccinia punctiformis* (Str.) Röhl as a stepping stone to the biological control of *Cirsium arvense* (L.) Scop. New Phytol 127:147–154

Frantzen J (1994b) The role of clonal growth in the pathosystem *Cirsium arvense–Puccinia punctiformis*. Can J Bot 72:832–836

Frantzen J, Van der Zweerde W (1994) Quantitative resistance of *Cirsium arvense* to root bud infection by *Puccinia punctiformis*. Biocontrol Sci Tech 4:223–228

Friedli J, Bacher S (2001a) Mutualistic interaction between a weevil and a rust fungus, two parasites of the weed *Cirsium arvense*. Oecologia 129:571–576

Friedli J, Bacher S (2001b) Direct and indirect effects of a shoot-base boring weevil and plant competition on the performance of creeping thistle, *Cirsium arvense*. Biol Control 22:219–226

Gange AC, Brown VK (eds) (1997) Multitrophic level interactions in terrestrial systems. In: Proc 36th Symp of the British Ecological Society, Blackwell Science, Oxford

Gliessman SR, Garcia ER, Amador AM (1981) The ecological basis for the application of traditional agricultural technology in the management of tropical agro-ecosystems. Agro-ecosystems 7:173–185

Gordon DR (1998) Effects of invasive, non-indigenous plant-species on ecosystem processes: lessons from Florida. Ecol Appl 8:975–989

Gremmen NJM, Chown SL, Marshall DJ (1998) Impact of the introduced grass *Agrostis stolonifera* on vegetation and soil fauna communities at Marion Island, sub-Antarctic. Biol Cons 85:223–231

Guske S, Boyle C, Schulz B (1996) New aspects concerning biological control of *Cirsium arvense*. IOBC WPRS Bull 19:281–290

Hairston NG, Smith FE, Slobodkin LB (1960) Community structure, population control, and competition. Am Nat 94:421–425

Hassell MP (2000) The spatial and temporal dynamics of host–parasitoid interactions. Oxford University Press, London

Hatcher PE (1995) Three-way interactions between plant pathogenic fungi, herbivorous insects and their host plants. Biol Rev 70:639–694

Hatcher PE, Paul ND (2001) Plant pathogen–herbivore interactions and their effects on weeds. In: Jeger MJ, Spence NJ (eds) Biotic interactions in plant–pathogen associations. CAB International, Wallingford, pp 193–226
Hegi G (1987) Illustrierte Flora von Mitteleuropa. Band VI, Teil 4. Paul Parey, Berlin
Heimann B, Cussans GW (1996) The importance of seeds and sexual reproduction in the population biology of *Cirsium arvense* – a literature review. Weed Res 36:493–503
Hoffmann GM, Nienhaus F, Poehling H-M, Schönbeck F, Weltzien HC, Wilbert H (1994) Lehrbuch der Phytomedizin. 3. Auflage. Blackwell, Berlin
Holm LG, Plucknett DL, Pancho JV, Herberger JP (1977) The world's worst weeds. Distribution and biology. University Press of Hawaii, Honolulu
Holt RD, Lawton JH, Polis GA, Martinez ND (1999) Trophic rank and the species-area relationship. Ecology 80:1495–1504
Hulme PE (1996) Herbivores and the performance of grassland plants: a comparison of arthropod, mollusc and rodent herbivory. J Ecol 84:43–51
Jeltsch F, Wissel C, Eber S, Brandl R (1992) Oscillating dispersal patterns of tephritid fly populations. Ecol Mod 69:53–75
Jules ES, Kauffman MK, Ritts WD, Carroll AL (2002) Spread of an invasive pathogen over a variable landscape: a nonnative root rot on Port Orford cedar. Ecology 83:3167–3181
Kareiva P, Wennergren U (1995) Connecting landscape patterns to ecosystem and population processes. Nature 373:299–302
Keane RM, Crawley (2002) Exotic plant invasions and the enemy release hypothesis. Trends Ecol Evol 17:164–170
Klironomos JN (2002) Feedback with soil biota contributes to plant rarity and invasiveness in communities. Nature 417:67–70
Kluth S, Kruess A, Tscharntke T (2001) Interactions between the rust fungus *Puccinia punctiformis* and ectophagous and endophagous insects on creeping thistle. J Appl Ecol 38:548–556
Kluth S, Kruess A, Tscharntke T (2002) Insects as vectors of plant pathogens: mutualistic and antagonistic interactions. Oecologia 133:193–199
Kluth S, Kruess A, Tscharntke T (2003) Influence of pathogen application and cutting on the performance and nutrient storage of *Cirsium arvense*. J Appl Ecol 40:334–343
Kruess A (2002) Indirect interaction between a fungal plant pathogen and a herbivorous beetle of the weed *Cirsium arvense*. Oecologia 130:563–569
Kruess A (2003a) Effects of landscape structure and habitat type on a plant–herbivore–parasitoid community. Ecography 26:283–290
Kruess A (2003b) Effects of landscape and habitat type on potential biological control agents against creeping thistle and their natural enemies. IOBC WPRS Bull 26(4):83–88
Kruess A, Tscharntke T (1994) Habitat fragmentation, species loss, and biological control. Science 264:1581–1584
Kruess A, Tscharntke T (2000) Effects of habitat fragmentation on plant–insect communities. In: Ekbom B, Irvin ME, Robert Y (eds) Interchanges of insects between agricultural and surrounding landscapes. Kluwer, Dordrecht, pp 53–70
Leibold MA (1996) a graphical model of keystone predators in food webs: trophic regulation of abundance, incidence, and diversity patterns in communities. Am Nat 147:784–812
Maron JL, Harrison S (1997) Spatial pattern formation in an insect host–parasitoid system. Science 278:1619–1621
Marshall EJP, Brown VK, Boatman ND, Lutman PJW, Squire GR, Ward LK (2003) The role of weeds in supporting biological diversity within crop fields. Weed Res 43:77–89
Masters GJ, Brown VK (1997) Host-plant mediated interactions between spatially separated herbivores: effects on community structure. In: Gange AC, Brown VK (eds) Multitrophic interactions in terrestrial systems. Blackwell, Oxford, pp 217–237

McClay AS, Bourchier RS, Butts RA, Peschken DP (2001) *Cirsium arvense*. In: Mason PG, Huber JT (eds) Biological control programmes in Canada 1981–2000. CAB International, Wallingford, pp 318–330
McNaughton SJ (1983) Compensatory plant growth as a response to herbivory. Oikos 40:329–336
Moore RJ (1975) The biology of Canadian weeds. 13. *Cirsium arvense* (L.) Scop. Can J Plant Sci 55:1033–1048
Mortimer SR, Van der Putten WH, Brown VK (1999) Insect and nematode herbivory below-ground: interactions and role in vegetation development. In: Olff H, Brown VK, Drent RH (eds) Herbivores: between plants and predators. Blackwell, Oxford, pp 205–238
Mulder CPH, Koricheva J, Huss-Danell K, Hogberg P, Joshi J (1999) Insects affect relationships between plant species richness and ecosystem processes. Ecol Lett 2:237–246
Müller-Schärer H, Scheepens PC (1997) Biological control of weed in crops: a coordinated European research programme (COST 816). Integrated Pest Manage Rev 2:1–6
Müller-Schärer H, Scheepens PC, Greaves MP (2000) Biological control of weeds in European crops: recent achievements and future work. Weed Res 1:83–98
Oksanen L, Fretwell SD, Arruda J, Niemela P (1981) Exploitation ecosystems in gradients of primary productivity. Am Nat 118:240–261
Östman Ö, Ekbom B, Bengtsson J (2001) Landscape complexity and farming practice influence the condition of polyphagous carabid beetles. Ecol Appl 11:480–488
Owen DF, Wiegert RD (1976) Do consumers maximize plant fitness? Oikos 27:488–492
Persson L (1999) Trophic cascades: abiding heterogeneity and the trophic level concept at the end of the road. Oikos 85:385–397
Peschken DP (1981) Biological control of Canada thistle. Proc N Cent Weed Conf 36:169–173
Peschken DP (1984) *Cirsium arvense* (L.) Scop., Canada thistle (Compositae). In: Kelleher JS, Hulme MA (eds) Biological control programmes against insects and weeds in Canada 1969–1980. Commonwealth Agricultural Bureau, Slough, pp 139–146
Roland J, Taylor PD (1995) Herbivore–natural enemy interactions in fragmented and continuous forests. In: Cappucino N, Price PW (eds) Population dynamics: new approaches and synthesis. Academic Press, San Diego, pp 195–208
Roland J, Taylor PD (1997) Insect parasitoid species respond to forest structure at different spatial scales. Nature 386:710–713
Root RB (1973) Organization of a plant–arthropod association in simple and diverse habitats: the fauna of collards (*Brassica oleracea*). Ecol Monogr 43:95–125
Schmitz OJ, Hambäck PA, Beckerman AP (2000) Trophic cascades in terrestrial ecosystems: a review of the effects of carnivore removal on plants. Am Nat 155:141–153
Schroeder D (1980) The biological control of thistles. Biocontrol News Inf 1:9–26
Schroeder D, Müller-Schärer H, Stinson CSA (1993) A European weed survey in 10 major crop systems to identify targets for biological control. Weed Res 33:449–458
Seitz A, Komma M (1984) Genetic polymorphism and its ecological background in tephritid populations (Diptera: Tephritidae). In: Wöhrmann K, Löschke V (eds) Population biology and evolution. Springer, Berlin Heidelberg New York, pp 143–158
Siegel MR (1993) *Acremonium* endophytes: our current state of knowledge and future directions for research. Agric Ecosyst Environ 44:301–321
Skinner K, Smith L, Rice P (2000) Using noxious weed lists to prioritize targets for developing weed management strategies. Weed Sci 48:640–644
Steffan-Dewenter I, Münzenberg U, Bürger C, Thies C, Tscharntke T (2002) Scale-dependent effects of landscape context on three pollinator guilds. Ecology 83:1421–1432

Steiner WE (1984) A review of the biology of the phalcrid beetles. In: Wheeler Q, Blackwell M (eds) Fungus–insect relationships, perspectives in ecology and evolution. Columbia University Press, New York, pp 424–445

Thies C, Tscharntke T (1999) Landscape structure and biological control in agroecosystems. Science 285:893–895

Thies C, Steffan-Dewenter I, Tscharntke T (2003) Effects of landscape context on herbivory and parasitism at different spatial scales. Oikos 101:18–25

Thompson K, Bakker JP, Bekker RM (1997) The soil seed banks of north west Europe: methodology, density and longevity. Cambridge University Press, Cambridge

Trumble JT, Kolodny-Hirsch DM, Ting IP (1993) Plant compensation for arthropod herbivory. Annu Rev Entomol 38:93–119

Tscharntke T, Hawkins BA (2002) Multitrophic level interactions. Cambridge University Press, Cambridge

Tscharntke T, Kruess A (1999) Habitat fragmentation and biological control. In: Hawkins BA, Cornell HV (eds) Theoretical approaches to biological control. Cambridge University Press, Cambridge, pp 190–205

Van der Putten WH (1999) Interactions of plants, soil pathogens and their antagonists in natural ecosystems. In: Jeger MJ, Spence NJ (eds) Biotic interactions in plant pathogen associations. CAB International, Wallingford, pp 285–305

Van der Putten WH, Vet LEM, Harvey JA, Wäckers FL (2001) Linking above- and belowground multitrophic interactions of plants, herbivores, pathogens, and their antagonists. Trends Ecol Evol 16:547–554

Vitousek PM (1986) Biological invasions and ecosystem properties: can species make a difference? In: Mooney HA, Drake JA (eds) Ecology of biological invasions of North America and Hawaii. Ecological Studies 58. Springer, Berlin Heidelberg New York, pp 163–176

Vitousek PM, Walker LR (1989) Biological invasions by *Myrica faya* in Hawaii: plant demography, nitrogen fixation, ecosystem effects. Ecol Monogr 59:247–265

Völker K, Boyle C (1994) Bean rust as a model system to evaluate efficiency of teliospore induction, especially in the potential mycoherbicide *Puccinia punctiformis*. Weed Res 34:275–281

Wardle DA (ed) (2002) Communities and ecosystems: linking the aboveground and belowground components. Princeton University Press, Princeton, NJ

Watson AK, Keogh WJ (1980) Mortality of Canada thistle due to *Puccinia punctiformis*. In: Delfosse ES (ed) Proc 5th Int Symp on Biological Control of Weeds, Brisbane, CSIRO, Melbourne, pp 325–332

Watt AS (1947) Pattern and process in the plant community. J Ecol 35:1–22

Whittaker RH (1953) A consideration of climax theory: the climax as a population and pattern. Ecol Monogr 23:41–78

Wiegand T, Molony KA, Naves J, Knauer F (1999) Finding the missing link between landscape structure and population dynamics: a spatially explicit perspective. Am Nat 154:605–627

Wilby A, Brown VK (2001) Herbivory, litter and soil disturbance as determinants of vegetation dynamics during early old-field succession under set-aside. Oecologia 127:259–265

Wilcove DS, Rothstein D, Dubow J, Phillips A, Losos E (1998) Quantifying threats to imperiled species in the United States. Bioscience 48:607–615

Wilson CL (1969) Use of plant pathogens in weed control. Annu Rev Phytopath 7:411–434

With KA, Gardner RH, Turner MG (1997) Landscape connectivity and population distributions in heterogeneous environments. Oikos 78:151–169

9 Food Web Interactions and Ecosystem Processes

A. Janssen and M.W. Sabelis

9.1 Summary

The effects of higher trophic levels such as herbivores and predators are usually not considered in studies of biodiversity and ecosystem processes. However, plants and organisms of higher trophic levels interact in many different ways, resulting in effects of plants on biodiversity of higher trophic levels as well as effects of higher trophic levels on plant biomass and diversity. We review the effects of higher trophic levels on plant biomass, plant diversity and ecosystem processes. Food web interactions such as apparent competition, omnivory, intraguild predation, interactions among plants, indirect plant defences and behavioural effects are important for ecosystem processes. Given the large variety of food web structures, generalizing rules relating food web interactions to ecosystem processes probably do not exist. Moreover, such predictions will also be impeded by the occurrence of multiple steady states in ecological systems.

9.2 Introduction

Most work on biodiversity and ecosystem functioning focuses on the correlations between plant biodiversity and plant biomass and nutrient cycling (Tilman et al. 1996; Hooper and Vitousek 1997; Hector et al. 1999; Hector et al. 2002; Tilman et al. 2002; Weisser and Siemann, Chap. 1, this Vol.). The effects of higher trophic levels such as herbivores and predators are usually not considered. The underlying assumption is thus that herbivore and predator diversity are the result and not the cause of plant diversity (Siemann et al. 1998; Haddad et al. 2001). However, this assumption is probably false; plants and organisms of higher trophic levels interact in many different ways, resulting in effects of plants on biodiversity of higher trophic levels as well as effects of higher trophic levels on plant biomass and diversity (Carson and Root

Ecological Studies, Vol. 173
W.W. Weisser and E. Siemann (Eds.)
Insects and Ecosystem Function

1999; Schmitz et al. 2000; Bradford et al. 2002; Duffy 2002; Joshi et al., Chap. 6, this Vol.; Schmitz, Chap. 14, this Vol.). It is therefore time to specifically study the effects of trophic levels higher than plants on ecosystem processes such as biomass production and nutrient cycling (Loreau et al. 2001; Raffaelli et al. 2002; Naeem and Wright 2003). By now, there are several examples of such studies and we give a non-exhaustive review. We will first address the effects of the selective removal of entire trophic levels and then discuss the various indirect interactions that occur within trophic levels as well as their possible consequences for ecosystem processes.

What kind of effects on ecosystem processes can we expect from the addition or removal of (species from) higher trophic levels? This question is difficult to answer. The many studies to date show that even the effects of *plant* diversity on ecosystem processes are still not well understood, let alone the effect of higher trophic levels. It is clear that biodiversity matters, but it is not obvious how it matters (Naeem and Wright 2003). Many hypotheses and mechanisms through which plant diversity affects ecosystem processes have been proposed, but almost no manipulative experiments have been done to test them (Schmid et al. 2002 but see Ruesink and Srivastava 2001). Ruesink and Srivastava (2001) propose alternative scenarios for the effects of species loss on ecosystem processes, based on the idea that remaining species may or may not compensate for the effects of the lost species on ecosystem processes (Naeem and Li 1997). If the species that is lost is responsible for the processing of some resource, two things might happen. First, the remaining species cannot process the resource that was utilized by the lost species. Hence, the loss of a species leads to an irreversible change in ecosystem processes. Second, the remaining species may be capable of compensating for the loss of resource processing by the excluded species, and thus the ecosystem process may be restored to some extent (Ruesink and Srivastava 2001). Ruesink and Srivastava distinguish a short-term per capita compensation, caused by the remaining species increasing their individual rate of resource processing, and a longer-term numerical compensation, where the numbers of remaining species increase, thus increasing the community rate of resource processing after the decline caused by the loss of a species (McGrady-Steed and Morin 2000). However, the response of communities to species loss may well be a combination of both short-term per capita compensation and longer-term numerical compensation. Nevertheless, Ruesink and Srivastava (2001) offer a hypothesis that can explain the variation in results among experimental studies. Hence, it now becomes important to assess the existence and the strength of compensation within communities.

So when can we expect compensation of ecosystem processes after species loss by remaining species? As long as there are other members of the functional group of the lost species, there will be some level of compensation within communities. However, Ruesink and Srivastava (2001) showed that compensation does not always occur, not even within the same functional group. Compensation will be strongest when one or a few of the remaining species affect ecosystem processes in a similar manner as the lost species did. However, in general, species that are more similar are less likely to coexist within communities. Local communities consist of coexisting species that are drawn from a regional species pool; coexistence is the result of migration, extinction and local interspecific interactions (Holt and Loreau 2001; Bond and Chase 2002). Differences in characteristics among species may facilitate coexistence of species because of reduced interspecific competition, but the level to which species in a local community differ in characteristics will depend on the type of population dynamics of the species in the local community (Armstrong and McGehee 1980; Huisman and Weissing 1999) and/or on the relative importance of local interspecific competition versus immigration from the regional species pool (Loreau 2000; Holt and Loreau 2001; Bond and Chase 2002). Nevertheless, we expect that species in a local community will often possess different characteristics because exclusion of species with similar characteristics by interspecific competition leads to communities of species that differ at least somewhat. The idea behind this is that each species is adapted in its own unique niche and there is a trade-off between being adapted to different niches. Hence, we expect that compensation will not be perfect in naturally formed local communities because species removal leads to vacant niches that are initially occupied by species that are adapted to different niches. Only subsequent adaptation to the newly occupied niche could restore ecosystem processes in the long run. This does not mean that the loss of any single species will always lead to a significant decrease in ecosystem processes, but we expect that, in general, ecosystem processes do initially decrease and not increase with species loss (Loreau 2000).

The above argument does not apply to artificial communities; any pattern may occur within communities that are randomly assembled from species that do not coexist in nature. Because selection leading to niche segregation did not occur within artificial communities, it is important to study the effects of species diversity on ecosystem processes in natural communities in their natural environment.

In conclusion, because all species in natural communities cannot completely compensate for the change in ecosystem processes resulting from species loss, the general expectation is that species loss will lead to a decrease in ecosystem functioning. This does not mean that communities with higher diversity are more productive. Indeed, many ecosystems with high productivity have low plant diversity (Huston and McBride 2002; Schmid et al. 2002). The maximum number of species that can co-occur locally is set by the number of species in the regional species pool (Huston 1997; Loreau et al. 2001). Hence, within each community in its natural environment, we would expect a decrease in ecosystem processes with the loss of species (although the loss of a single species may not have significant effects under all circumstances).

Having assessed that production of plant biomass, nutrient cycling and other ecosystem processes in co-evolved local communities will decrease with the loss of plant species, the question now becomes how species of higher trophic levels will affect such ecosystem processes. Undoubtedly, higher trophic levels will affect plant biomass, nutrient cycling and other processes directly. For example, herbivores reduce the aboveground plant biomass by grazing and affect nutrient cycling through the production of faeces. However, higher trophic levels can also affect ecosystem processes *indirectly* whenever they affect the local plant community (Tilman et al. 1996, 2002; Hooper and Vitousek 1997; Hector et al. 1999, 2002). Therefore, we will focus on effects of higher trophic levels on ecosystem processes such as plant biomass production, but also on effects on plant species composition.

9.3 Interactions Among Entire Trophic Levels

Including higher trophic levels in studies of ecosystem processes can in theory be done in a variety of ways, depending on the sort of question to be addressed. The most obvious way would be to assess the effects of entire trophic levels, such as herbivores, omnivores, predators (including parasitoids) and hyperpredators (including hyperparasitoids) on ecosystem processes. Hence, a first set of experiments would consist of a manipulation of entire trophic levels, starting with the full natural diversity of a co-evolved community of plants and all other trophic levels, and subsequently removing the higher trophic levels, starting with hyperpredators and hyperparasitoids and ending with herbivores. This will result in assessment of the effect of entire trophic levels on plant species diversity and ecosystem processes. Such manipulations could lead to trophic cascades, with the effect on plant biomass, plant species composition and ecosystem processes depending on the number of trophic levels (Oksanen et al. 1981; Hairston and Hairston 1997; Schmitz, Chap. 14, this Vol.).

Although to date the full set of manipulations of trophic levels has not been included in studies of ecosystem processes, there are experimental data on the joint effects of several trophic levels on ecosystem processes. In aquatic microcosms, the algal and bacterial biomass decreased when more trophic levels were added (Naeem and Li 1997). In terrestrial communities, exclusion of insect herbivores and higher trophic levels reduced plant species richness (Brown and Gange 1992; Carson and Root 1999; Mulder et al. 1999), while it increased plant biomass (Carson and Root 1999; Mulder et al. 1999). The few experiments in which predators and higher trophic levels were excluded while herbivores were included mostly showed a decrease in plant biomass upon predator removal, whereas some showed positive or negative effects on plant diversity (see Schmitz et al. 2000 for a review; Schmitz, Chap. 14, this Vol.). One study showed that a group of predator species suppressed a herbivore population and increased plant biomass, while each predator separately did not have a significant effect (Cardinale et al. 2003).

9.4 Effects of Diversity Within Trophic Levels

Having assessed the effect of entire trophic levels, the next step is to assess how diversity within different trophic levels affects the diversity of other trophic levels and ecosystem processes. Several studies have shown that plant diversity is positively correlated to herbivore and predator abundance as well as herbivore and predator species richness (Siemann et al. 1998; Haddad et al. 2001), although the effect per herbivore and predator species was highly variable (Haddad et al. 2001). Herbivore diversity was also positively correlated with diversity of predators and parasitoids (Siemann et al. 1998).

No experiments, to our knowledge, have addressed the effect of species richness and species composition of herbivores, predators and hyperpredators, either separately or together, on plant diversity and ecosystem processes. So what effects can we expect? The loss of some species of herbivores or predators may cascade directly down to the plant level (Duffy 2002), thus affecting plant biomass production, one of the ecosystem processes. However, due to indirect interactions among species, more complex, reticulate food webs show essentially different dynamics from linear food chains (Hulot et al. 2000; Mylius et al. 2001; Kuijper et al. 2003) and this may at times prevent the occurrence of cascades. Even if removal of species of herbivores and predators or of higher trophic levels did not affect total plant biomass, the distribution of biomass over the different plant species could change, resulting in changes in ecosystem processes.

Because interactions within trophic levels become important at this level of detail, it will matter much for the outcome of experiments whether generalist or specialist herbivores and predators are included or excluded. Hence, all trophic levels need to be subdivided into functional groups, with the danger that defining such groups will be rather arbitrary (Naeem and Wright 2003). For example, depending on the quality of plant tissue, omnivores mainly feed on prey or plant tissue, and should thus be treated as either predators or herbivores (Coll and Ridgway 1995; Coll and Izraylevich 1997; Agrawal et al. 1999; Agrawal and Klein 2000; Eubanks and Denno 2000). Another problem of varying the diversity within trophic levels is that the effects of indirect interactions within trophic levels become more important. The effects of indirect interactions on species coexistence are often difficult to predict and depend on much biological detail. This probably precludes the formulation of general rules relating trophic interactions to ecosystem processes (Holt and Loreau 2001). Yet, it is at this level of detail where the loss or addition of species is studied, and advancement of our knowledge of the relationships between diversity and ecosystem processes is impossible without considering such indirect interactions. We will therefore briefly review the various types of indirect interactions that occur within trophic levels and discuss their possible consequences for food web dynamics and ecosystem processes. The emphasis will be on aboveground interactions on terrestrial plants (see de Ruiter et al. 2002; Wardle and van der Putten 2002 for recent reviews of belowground interactions and van der Putten et al. 2001 for a discussion of the links between aboveground and belowground interactions).

9.4.1 Apparent Competition

Apparent competition is the indirect interaction between different prey species via a shared predator (Holt 1977, 1984). The mechanism behind this interaction is that the density of one species of prey affects the density of a predator, which subsequently affects the density of a second prey species. The effects are comparable to, and under field situations even difficult to distinguish from, those of resource competition between two species; some of the examples of competitive interactions among plants are likely to be attributable to apparent competition (Gurevitch et al. 2000). In theory, apparent competition may lead to exclusion of prey species (Holt 1977, 1984; Holt and Lawton 1994, van Rijn et al. 2002), and can therefore affect diversity. Apparent competition can occur at all levels: between plant species that are attacked by the same herbivore species, but also between herbivores that are attacked by the same predator and predators that are attacked by the same hyperpredator. This shows how species within the same trophic level may differentially affect interactions between species: in the presence of specialist herbivores, apparent competition among plants will be unimportant, while the presence of generalist herbivores may result in strong apparent competition (Sessions and Kelly 2002), with the possible outcome of plant species exclusion (Grover and Holt 1998) and changes in ecosystem processes. On the other hand, when apparent competition between herbivores leads to exclusion of a herbivore species, this will have a positive effect on the host plants of the excluded species. Hence, the effects of apparent competition can cascade down to the plant level where it affects species composition and consequently ecosystem processes.

9.4.2 Omnivory

Omnivory is a widespread phenomenon in terrestrial communities of arthropods (Polis et al. 1989; Rosenheim et al. 1995; Polis and Winemiller 1996; Coll and Guershon 2002). It blurs the distinction among trophic levels and may prevent trophic cascades (Morin and Lawler 1995; Polis and Strong 1996). As remarked above, depending on diet choice, omnivores can be damaging to a plant or act as a predator of herbivores. Their diet choice depends on food quality (Coll and Ridgway 1995; Coll and Izraylevich 1997; Agrawal et al. 1999; Agrawal and Klein 2000; Eubanks and Denno 2000; Janssen et al. 2003). When plant quality is low, omnivores will feed more on non-plant food. It is therefore expected that the effect of omnivores will depend on the plant species; when a plant is of poor quality to the omnivores, these may feed on herbivores (Agrawal et al. 1999) or even on predators (Sabelis et al. 1999b; Janssen et al. 2003), thus functioning as predators or even hyperpredators. On the other hand, when plant quality is high, omnivores may function as herbivores. Hence, omnivores will differentially affect biomass of plant species that differ in quality and this could affect competitive interactions among plant species, leading to changes in plant species composition.

9.4.3 Intraguild Predation

Intraguild predation is defined as the killing and eating of species that otherwise use similar resources, and are thus potential competitors (Polis et al. 1989). Intraguild predation is widespread, also among arthropod food webs (Polis et al. 1989). Addition of an intraguild predator to a food web may result in exclusion of other predators, the intraguild prey. In theory, the occurrence of intraguild predation may lead to alternative steady states, to complex dynamics or to exclusion of the intraguild prey or predators, and it thus affects community structure (Holt 1977). Intraguild predation can also affect ecosystem processes; if the intraguild prey is a herbivore, it is predicted to be excluded at high ranges of productivity, while there are two alternative steady states at intermediate levels of productivity, one with and one without herbivores (Mylius et al. 2001; Kuijper et al. 2003). Exclusion of the herbivore leads to increased plant biomass (Mylius et al. 2001; Kuijper et al. 2003) and may also affect plant species composition and thereby ecosystem processes.

9.4.4 Plant-Mediated Indirect Interactions Between Herbivores

Plants defend themselves directly against herbivores by producing toxins, spines, glandular hairs and digestion inhibitors. Some of these defences are inducible, causing the first-arriving herbivore to experience gradually increasing levels of plant defence, whereas herbivores that arrive later will find an alerted and well-defended host plant (Voelckel and Baldwin, Chap. 17, this Vol.). Attack by one herbivore species induces plant resistance, which decreases the population growth rate not only of that herbivore, but also of herbivores that attack the plant at a later date (English-Loeb et al. 1993; Hougen-Eitzman and Karban 1995; Karban et al. 1997). Thus, plant defences lead to indirect interactions between herbivore species and these interactions are different from resource competition and apparent competition. Moreover, these defences can lead to indirect interactions among plants; plants can become unpalatable as a result of attacks by herbivores and these may move to less-defended neighbouring plants (van Dam et al. 2000). Hence, direct plant defences can induce changes in diet choice of herbivores, resulting in shifts in the distribution of biomass or changes in plant species composition, which in turn lead to changes in ecosystem processes.

Because induced resistance decreases plant quality for omnivores, it can also change the diet and functional role of omnivores in food web interactions (Agrawal et al. 1999; Agrawal and Klein 2000). For example, the omnivorous western flower thrips feeds more on eggs of herbivores on plants that are induced by previous herbivory (Agrawal and Klein 2000). Hence, the effects of this omnivore on plant biomass strongly depend on plant quality, which in turn depends on the feeding activity of omnivores. Such feedback loops preclude general predictions of effects of omnivores on plant biomass and plant species composition.

9.4.5 Indirect Plant Defences

Plants can also defend themselves by facilitating the third trophic level (Price et al. 1980; Voelckel and Baldwin, Chap. 17, this Vol.). For example, many species of plants are known to provide shelter such as domatia that are inhabited by predators (ants and mites) and fungivores (Beattie 1985; Turner and Pemberton 1989; Walter and O'Dowd 1992; Walter 1996; O'Dowd and Willson 1997; Norton et al. 2000). Plants provide food, such as pollen and nectar, thus increasing the survival and/or reproduction of predators (Bentley 1977; Pemberton and Lee 1996; van Rijn et al. 2002). Plants also convey information on the presence of prey to predators by producing herbivore-induced volatiles (Dicke and Sabelis 1988; Dicke et al. 1990; Turlings et al. 1990; Tumlinson et al. 1993; Dicke 1994; Sabelis et al. 1999a). This facilitates the searching of predators for prey and results in higher numbers of predators and lower numbers of herbivores on plants (Drukker et al. 1995; Shimoda et al. 1997; De Moraes et al. 1998; Janssen 1999; Kessler and Baldwin 2001).

Although it is clear that plants interact with the third trophic level, the overall effects of these interactions on plant fitness are by no means clear (Sabelis et al. 1999a; van der Meijden and Klinkhamer 2000; Janssen et al. 2002). This is partly because, once produced, the volatile signals, the food, and the protective structures are not under the control of the plant and can be used by other species, such as herbivores, hyperpredators and parasites. Nevertheless, there are many examples of such interactions among plants and the third trophic level and evidence is accumulating that the third trophic level can have a positive effect on plant fitness (Thaler 1999; Turlings and Fritzsche

1999; van Loon et al. 2000; Kessler and Baldwin 2001). Since plant fitness is often linked to biomass of the plant and ultimately to the number of offspring, indirect plant defences will affect the species composition and biomass of the plant community.

The existence of interactions among plants and predators also suggests that plants may compete for natural enemies by luring natural enemies from neighbours, thus increasing their indirect defence while decreasing that of plants nearby. On the other hand, if the foraging range of predators spans several plants, neighbours may also profit from the attraction and/or arrestment of predators (Sabelis et al. 1999a). In conclusion, plants affect the abundance and possibly even the occurrence of herbivore species as well as predator species through their indirect defences. This will result in changes in plant biomass, and affects other plants in the neighbourhood as well.

9.4.6 Interactions Among Plants

While most of the theory on diversity and ecosystem processes has focused on resource competition among plants, it is clear that plants can affect other plants through shared herbivores (apparent competition), that induced resistance of a plant can cause herbivores to feed on undefended neighbours and that plants may compete for predators as their bodyguards. These indirect effects of plants on neighbours will probably vary with spatial scale and with the herbivores and predators involved. Because of the higher mobility of herbivores and predators relative to plants, the spatial scale of these indirect interactions among plants is likely to be larger than the scale over which plants typically compete for water, nutrients and light. Such openness of local plant communities to invasion of herbivores and higher trophic levels will obscure the rules of community structure derived from local exclusion, because immigration allows for persistence of species inside the community even when they would go extinct locally (Holt and Loreau 2001).

Another way in which plants can interact is plant-to-plant communication (Bruin et al. 1992; Arimura et al. 2000, 2001; Dolch and Tscharntke 2000; Bruin and Dicke 2001; Bruin and Sabelis 2001; Dicke and Bruin 2001a, b; Karban 2001). This phenomenon has not been taken seriously for a long time, but has gained a steady foothold recently (Dicke and Bruin 2001a, b). In theory, plants could benefit from receiving signals from neighbours that are attacked by herbivores by switching on their own defence mechanisms, or, alternatively, they could rely entirely on the defence mechanisms of their neighbours (Augner 1994; Augner et al. 1991; Tuomi et al. 1994; Sabelis et al. 2002). What is the best option probably depends on the relative mobility of herbivores and higher trophic levels. How communication among plants affects species diversity and biomass at various trophic levels is still an open question, let alone how this affects nutrient cycling.

9.4.7 Behavioural Effects

Species can also affect each other through the induction of changes in each other's behaviour. Although this is not an indirect interaction in a strict sense (there is no third, intermediate species), it can also have strong effects on food web dynamics. A good example is the predator-induced behaviour of prey that results in lower predation risk (Sih 1980; Lima and Dill 1990; Lima 1998). Such antipredator behaviour usually comes at a cost because time and energy is spent in avoiding predation at the expense of other vital tasks (Lima and Dill 1990; Lima 1998; Dugatkin and Godin 2002). Although such interactions seem short-lasting and therefore possibly unimportant for species diversity and ecosystem processes, an elegant study by Schmitz et al. (1997) shows the contrary: a change in behaviour of grasshoppers that is induced by the presence of spider predators led to changes in consumption of the herbivores, resulting in a shift in biomass distribution over grasses and herbs!

Arthropods avoid not only predation, but also competition (Janssen et al. 1995, 1997; Pallini et al. 1997). In particular, the presence of competing herbivores on a plant can deter herbivores (Pallini et al. 1997) that can subsequently concentrate their attack on other plants of the same or other species.

In conclusion, the response of animals to the presence of their predators and competitors on certain plants will result in the attack of other plants, leading to shifts in herbivory and this can affect plant species abundances and plant biomass. Such shifts are expected to change ecosystem processes because different plant species cannot completely compensate for changes in ecosystem processes that result from the decrease or loss of other plant species.

9.5 Conclusions and Perspectives

This short overview shows that a suite of interactions occurs in terrestrial food webs of plant-inhabiting arthropods and these interactions may well feed back to the plant level. Although there are few studies concerning the effect of these interactions on plant species composition, plant biomass, nutrient cycling and other ecosystem processes, it would be surprising if these interactions would not affect ecosystem processes (Holt and Loreau 2001; Duffy 2002). The lesson to be learned from this overview is that it is impossible to predict effects of species loss on ecosystem processes without much more insight into the strength and magnitude of these interactions (Hulot et al. 2000). Given the large variety of food web structures, generalizing rules relating food web interactions to ecosystem processes probably do not exist (Holt and Loreau 2001). Hence, predictions of how changes in biodiversity at any trophic level affect ecosystem processes are bound to be specific for the system under study. Moreover, such predictions will also be impeded by the occurrence of alternative steady states in ecological systems (Scheffer et al. 2001). Effects of such hysteresis are two-fold: first, a decline in numbers of a species may initially not result in large shifts in biomass production, nutrient cycling or other ecosystem processes, but a small further decline may result in a sudden catastrophic shift in the state of the system (Scheffer et al. 2001, de Roos and Persson 2002). Second, such shifts are often very difficult to reverse; reintroduction of a species after its loss is often only possible after large changes in conditions such as nutrient concentration in the environment (Scheffer et al. 2001). As outlined above, several of the food web interactions, such as intraguild predation and apparent competition, may lead to alternative steady states. Hence, a proper prediction of effects of removal of species from a community as well as of the success of reintroducing species in a community is probably impossible without detailed knowledge of the interspecific interactions in the community. Such knowledge can only be obtained by a combination of experimental and theoretical work.

Whereas it is difficult to predict the effects of species loss or species addition, prediction of the effects of entire trophic levels on ecosystem processes is more straightforward and can be assessed relatively easily both theoretically and experimentally. The theoretical analysis of Holt and Loreau (2001) is a good example. They studied the effect of herbivory on ecosystem processes, while varying the migration of species from the regional species pool into the local community and concluded that the relationship between biodiversity and ecosystem processes is determined by trophic interactions, biodiversity within the trophic levels and on the migration of species into the system. Extension of this type of model with higher trophic levels may prove complex but is the way forward. Another promising way is the combination of practically oriented studies on the effect of biodiversity on biological control of pests (Thies and Tscharntke 1999; Cardinale et al. 2003) with studies on biodiversity and ecosystem processes (Cardinale et al. 2003). The obvious advantage of studying agroforestry food webs is their reduced complexity compared to natural communities; the disadvantage is that agroecological communities are often not coevolved and the effects of food web interactions in artificial communities may differ from those in natural communities.

Acknowledgements. We thank Martijn Egas for stimulating discussions and Wolfgang Weisser for the invitation to participate in the workshop 'Insects and Ecosystem Function' and to write this chapter. We benefited from remarks of the editors and two anonymous reviewers. A.J. was employed by the University of Amsterdam, within the framework of a PIONIER grant (no. 030-84-469) from the Netherlands Organization for Scientific Research (NWO) awarded to A.M. de Roos.

References

Agrawal AA, Klein CN (2000) What omnivores eat: direct effects of induced plant resistance on herbivores and indirect consequences for diet selection by omnivores. J Anim Ecol 69:525–535
Agrawal AA, Kobayashi C, Thaler JS (1999) Influence of prey availability and induced host-plant resistance on omnivory by western flower thrips. Ecology 80:518–523
Arimura G, Ozawa R, Shimoda T, Nishioka T, Boland W, Takabayashi J (2000) Herbivory-induced volatiles elicit defense genes in lima bean leaves. Nature 406:512–515
Arimura G, Ozawa R, Horiuchi J, Nishioka T, Takabayashi J (2001) Plant–plant interactions mediated by volatiles emitted from plants infested by spider mites. Biochem Syst Ecol 29:1049–1061
Armstrong RA, McGehee R (1980) Competitive exclusion. Am Nat 115:151–170
Augner M (1994) Should a plant always signal its defence against herbivores? Oikos 70:322–332
Augner M, Fagerström T, Tuomi J (1991) Competition, defence and games between plants. Behav Ecol Sociobiol 29:231–234
Beattie AJ (1985) The evolutionary ecology of ant–plant mutualisms. Cambridge University Press, Cambridge
Bentley BL (1977) Extra-floral nectaries and protection by pugnacious bodyguards. Annu Rev Ecol Syst 8:407–427
Bond EM, Chase JM (2002) Biodiversity and ecosystem functioning at local and regional spatial scales. Ecol Lett 5:467–470
Bradford MA, Jones TH, Bardgett RD, Black HIJ, Boag B, Bonkowski M, Cook R, Eggers T, Gange AC, Grayston SJ, Kandeler E, McCaig AE, Newington JE, Prosser JI, Setala H, Staddon PL, Tordoff GM, Tscherko D, Lawton JH (2002) Impacts of soil faunal community composition on model grassland ecosystems. Science 298:615–618
Brown VK, Gange AC (1992) Secondary plant succession – how is it modified by insect herbivory? Vegetatio 101:3–13
Bruin J, Dicke M (2001) Chemical information transfer between wounded and unwounded plants: backing up the future. Biochem Syst Ecol 29:1103–1113
Bruin J, Sabelis MW (2001) Meta-analysis of laboratory experiments on plant–plant information transfer. Biochem Syst Ecol 29:1089–1102
Bruin J, Dicke M, Sabelis MW (1992) Plants are better protected against spider-mites after exposure to volatiles from infested conspecifics. Experientia 48:525–529
Cardinale BJ, Harvey CT, Gross K, Ives AR (2003) Biodiversity and biocontrol: emergent impacts of a multi-enemy assemblage on pest suppression and crop yield in an agroecosystem. Ecol Lett 6:857–865
Carson WP, Root RB (1999) Top-down effects of insect herbivores during early succession: influence on biomass and plant dominance. Oecologia 121:260–272
Coll M, Guershon M (2002) Omnivory in terrestrial arthropods: mixing plant and prey diets. Annu Rev Entomol 47:267–297
Coll M, Izraylevich S (1997) When predators also feed on plants: effects of competition and plant quality on omnivore–prey population dynamics. Annu Entomol Soc Am 90:155–161
Coll M, Ridgway RL (1995) Functional and numerical responses of *Orius insidiosus* (Heteroptera, Anthocoridae) to its prey in different vegetable crops. Annu Entomol Soc Am 88:732–738
De Moraes CM, Lewis WJ, Pare PW, Alborn HT, Tumlinson JH (1998) Herbivore-infested plants selectively attract parasitoids. Nature 393:570–573

de Roos AM, Persson L (2002) Size-dependent life-history traits promote catastrophic collapses of top predators. Proc Natl Acad Sci Am 99:12907–12912
de Ruiter P, Griffiths B, Moore JC (2002) Biodiversity and stability in soil ecosystems: patterns, processes and the effects of disturbance. In: Loreau M, Naeem S, Inchausti P (eds) Biodiversity and ecosystem functioning: synthesis and perspectives. Oxford University Press, Oxford, pp 102–113
Dicke M (1994) Local and systemic production of volatile herbivore-induced terpenoids – their role in plant–carnivore mutualism. J Plant Physiol 143:465–472
Dicke M, Bruin J (2001a) Chemical information transfer between damaged and undamaged plants – preface. Biochem Syst Ecol 29:979–980
Dicke M, Bruin J (2001b) Chemical information transfer between plants: back to the future. Biochem Syst Ecol 29:981–994
Dicke M, Sabelis MW (1988) How plants obtain predatory mites as bodyguards. Neth J Zool 38:148–165
Dicke M, van Beek TA, Posthumus MA, Ben Dom N, van Bokhoven H, de Groot Æ (1990) Isolation and identification of volatile kairomone that affects acarine predator–prey interactions – involvement of host plant in its production. J Chem Ecol 16:381–396
Dolch R, Tscharntke T (2000) Defoliation of alders (*Alnus glutinosa*) affects herbivory by leaf beetles on undamaged neighbours. Oecologia 125:504–511
Drukker B, Scutareanu P, Sabelis MW (1995) Do anthocorid predators respond to synomones from *Psylla*-infested pear trees under field conditions? Entomol Exp Appl 77:193–203
Duffy JE (2002) Biodiversity and ecosystem function: the consumer connection. Oikos 99:201–219
Dugatkin LA, Godin JGJ (2002) Prey approaching predators – a cost-benefit perspective. Ann Zool Fenn 29:233–252
English-Loeb GM, Karban R, Hougen-Eitzman D (1993) Direct and indirect competition between spider mites feeding on grapes. Ecol App 3:699–707
Eubanks MD, Denno RF (2000) Host plants mediate omnivore–herbivore interactions and influence prey suppression. Ecology 81:936–947
Grover JP, Holt RD (1998) Disentangling resource and apparent competition: realistic models for plant–herbivore communities. J Theor Biol 191:353–376
Gurevitch J, Morrison JA, Hedges LV (2000) The interaction between competition and predation: a meta-analysis of field experiments. Am Nat 155:435–453
Haddad NM, Tilman D, Haarstad J, Ritchie M, Knops JMH (2001) Contrasting effects of plant richness and composition on insect communities: a field experiment. Am Nat 158:17–35
Hairston NG, Hairston NG (1997) Does food web complexity eliminate trophic-level dynamics? Am Nat 149:1001–1007
Hector A, Schmid B, Beierkuhnlein C, Caldeira MC, Diemer M, Dimitrakopoulos PG, Finn JA, Freitas H, Giller PS, Good J, Harris R, Hogberg P, Huss-Danell K, Joshi J, Jumpponen A, Korner C, Leadley PW, Loreau M, Minns A, Mulder CPH, O'Donovan G, Otway SJ, Pereira JS, Prinz A, Read DJ, Scherer-Lorenzen M, Schulze ED, Siamantziouras ASD, Spehn EM, Terry AC, Troumbis AY, Woodward FI, Yachi S, Lawton JH (1999) Plant diversity and productivity experiments in European grasslands. Science 286:1123–1127
Hector A, Loreau M, Schmid B, Biodepth Project (2002) Biodiversity manipulation experiments: studies replicated at multiple sites. In: Loreau M, Naeem S, Inchausti P (eds) Biodiversity and ecosystem functioning: synthesis and perspectives. Oxford University Press, Oxford, pp 36–46
Holt RD (1977) Predation, apparent competition, and structure of prey communities. Theor Popul Biol 12:197–229

Holt RD (1984) Spatial heterogeneity, indirect interactions, and the coexistence of prey species. Am Nat 124:377–406

Holt RD, Lawton JH (1994) The ecological consequences of shared natural enemies. Annu Rev Ecol Syst 25:495–520

Holt RD, Loreau M (2001) Biodiversity and ecosystem functioning: the role of trophic interactions and the importance of system openness. In: Kinzig A, Tilman D, Pacala SW (eds) Functional consequences of biodiversity: experimental progress and theoretical extensions. Princeton University Press, Princeton, pp 246–262

Hooper DU, Vitousek PM (1997) The effects of plant composition and diversity on ecosystem processes. Science 277:1302–1305

Hougen-Eitzman D, Karban R (1995) Mechanisms of interspecific competition that result in successful control of Pacific mites following inoculations of Willamette mites on grapevines. Oecologia 103:157–161

Huisman J, Weissing FJ (1999) Biodiversity of plankton by species oscillations and chaos. Nature 402:407–410

Hulot FD, Lacroix G, Lescher-Moutoue FO, Loreau M (2000) Functional diversity governs ecosystem response to nutrient enrichment. Nature 405:340–344

Huston MA (1997) Hidden treatments in ecological experiments: re-evaluating the ecosystem function of biodiversity. Oecologia 110:449–460

Huston MA, McBride AC (2002) Evaluating the relative strengths of biotic versus abiotic controls on ecosystem processes. In: Loreau M, Naeem S, Inchausti P (eds) Biodiversity and ecosystem functioning: synthesis and perspectives. Oxford University Press, Oxford, pp 47–60

Janssen A (1999) Plants with spider-mite prey attract more predatory mites than clean plants under greenhouse conditions. Entomol Exp Appl 90:191–198

Janssen A, van Alphen JJM, Sabelis MW, Bakker K (1995) Odour-mediated avoidance of competition in *Drosophila* parasitoids – the ghost of competition. Oikos 73:356–366

Janssen A, Bruin J, Jacobs G, Schraag R, Sabelis MW (1997) Predators use volatiles to avoid prey patches with conspecifics. J Anim Ecol 66:223–232

Janssen A, Sabelis MW, Bruin J (2002) Evolution of herbivore-induced plant volatiles. Oikos 97:134–138

Janssen A, Willemse E, van der Hammen T (2003) Poor host plant quality causes omnivore to consume predator eggs. J Anim Ecol 72:478–483

Karban R (2001) Communication between sagebrush and wild tobacco in the field. Biochem Syst Ecol 29:995–1005

Karban R, English-Loeb G, Hougen-Eitzman D (1997) Mite vaccinations for sustainable management of spider mites in vineyards. Ecol App 7:183–193

Kessler A, Baldwin IT (2001) Defensive function of herbivore-induced plant volatile emissions in nature. Science 291:2141–2144

Kuijper LDJ, Kooi BW, Zonneveld C, Kooijman S (2003) Omnivory and food web dynamics. Ecol Mod 163:19–32

Lima SL (1998) Nonlethal effects in the ecology of predator–prey interactions – what are the ecological effects of anti-predator decision-making? Bioscience 48:25–34

Lima SL, Dill LM (1990) Behavioral decisions made under the risk of predation – a review and prospectus. Can J Zool-Rev Can Zool 68:619–640

Loreau M (2000) Biodiversity and ecosystem functioning: recent theoretical advances. Oikos 91:3–17

Loreau M, Naeem P, Inchausti J, Bengtsson J, Grime JP, Hector A, Hooper DU, Huston MA, Raffaelli D, Schmid B, Tilman D, Wardle DA (2001) Ecology – biodiversity and ecosystem functioning: current knowledge and future challenges. Science 294:804–808

McGrady-Steed J, Morin PJ (2000) Biodiversity, density compensation, and the dynamics of populations and functional groups. Ecology 81:361–373

Morin PJ, Lawler SP (1995) Food-web architecture and population dynamics – theory and empirical evidence. Annu Rev Ecol Syst 26:505–529

Mulder CPH, Koricheva J, Huss-Danell K, Hogberg P, Joshi J (1999) Insects affect relationships between plant species richness and ecosystem processes. Ecol Lett 2:237–246

Mylius S, Klumpers K, de Roos AM, Persson L (2001) Impact of omnivory and stage structure on food web composition along a productivity gradient. Am Nat 158:259–276

Naeem S, Li SB (1997) Biodiversity enhances ecosystem reliability. Nature 390:507–509

Naeem S, Wright JP (2003) Disentangling biodiversity effects on ecosystem functioning: deriving solutions to a seemingly insurmountable problem. Ecol Lett 6:567–579

Norton AP, English-Loeb G, Gadoury D, Seem RC (2000) Mycophagous mites and foliar pathogens: leaf domatia mediate tritrophic interactions in grapes. Ecology 81:490–499

O'Dowd DJ, Willson MF (1997) Leaf domatia and the distribution and abundance of foliar mites in broadleaf deciduous forest in Wisconsin. Am Midl Nat 137:337–348

Oksanen L, Fretwell S, Arruda J, Niemelä P (1981) Exploitation ecosystems in gradients of primary productivity. Am Nat 118:240–261

Pallini A, Janssen A, Sabelis MW (1997) Odour-mediated responses of phytophagous mites to conspecific and heterospecific competitors. Oecologia 110:179–185

Pemberton RW, Lee JH (1996) The influence of extrafloral nectaries on parasitism of an insect herbivore. Am J Bot 83:1187–1194

Polis GA, Strong DR (1996) Food web complexity and community dynamics. Am Nat 147:813–846

Polis GA, Winemiller KO (1996) Food webs. Integration of patterns and dynamics. Chapman & Hall, New York

Polis GA, Myers CA, Holt RD (1989) The ecology and evolution of intraguild predation – potential competitors that eat each other. Annu Rev Ecol Syst 20:297–330

Price PW, Bouton CE, Gross P, McPheron BA, Thompson JN, Weis AE (1980) Interactions among three trophic levels. Annu Rev Ecol Syst 11:41–65

Raffaelli D, van der Putten WH, Persson L, Wardle DA, Petchey OL, Koricheva J, van der Heijden M, Mikola J, Kennedy T (2002) Multitrophic dynamics and ecosystem processes. In: Loreau M, Naeem S, Inchausti P (eds) Biodiversity and ecosystem functioning: synthesis and perspectives. Oxford University Press, Oxford, pp 147–154

Rosenheim JA, Kaya HK, Ehler LE, Marois JJ, Jaffee BA (1995) Intraguild predation among biological control agents – theory and evidence. Biol Control 5:303–335

Ruesink JL, Srivastava SD (2001) Numerical and per capita responses to species loss: mechanisms maintaining ecosystem function in a community of stream insect detrivores. Oikos 93:221–234

Sabelis MW, Janssen A, Bruin J, Bakker FM, Drukker B, Scutareanu P, van Rijn PCJ (1999a) Interactions between arthropod predators and plants: a conspiracy against herbivorous arthropods? In: Bruin J, van der Geest LPS, Sabelis MW (eds) Ecology and evolution of the Acari. Kluwer, Dordrecht, pp 207–229

Sabelis MW, Janssen A, Pallini A, Venzon M, Bruin J, Drukker B, Scutareanu P (1999b) Behavioural responses of predatory and herbivorous arthropods to induced plant volatiles: from evolutionary ecology to agricultural applications. In: Agrawal AA, Tuzun S, Bent E (eds) Induced plant defenses against pathogens and herbivores. Am Phytopathol Soc Press, St. Paul, Minnesota, pp 269–296

Sabelis MW, van Baalen M, Pels B, Egas M, Janssen A (2002) Evolution of exploitation and defense in tritrophic interactions. In: Dieckmann U, Metz JA, Sabelis MW, Sigmund K (eds) Adaptive dynamics of infectious diseases: in pursuit of virulence management. Cambridge University Press, Cambridge, pp 297–321

Scheffer M, Carpenter S, Foley JA, Folke C, Walker B (2001) Catastrophic shifts in ecosystems. Nature 413:591–596

Schmid B, Hector A, Huston MA, Inchausti P, Nijs I, Leadley PW, Tilman D (2002) The design and analysis of biodiversity experiments. In: Loreau M, Naeem S, Inchausti P (eds) Biodiversity and ecosystem functioning: synthesis and perspectives. Oxford University Press, Oxford, pp 61–75

Schmitz OJ, Beckerman AP, O'Brien KM (1997) Behaviorally mediated trophic cascades: effects of predation risk on food web interactions. Ecology 78:1388–1399

Schmitz OJ, Hamback PA, Beckerman AP (2000) Trophic cascades in terrestrial systems: a review of the effects of carnivore removals on plants. Am Nat 155:141–153

Sessions L, Kelly D (2002) Predator-mediated apparent competition between an introduced grass, *Agrostis capillaris*, and a native fern, *Botrychium australe* (Ophioglossaceae), in New Zealand. Oikos 96:102–109

Shimoda T, Takabayashi J, Ashihara W, Takafuji A (1997) Response of predatory insect *Scolothrips takahashii* toward herbivore-induced plant volatiles under laboratory and field conditions. J Chem Ecol 23:2033–2048

Siemann E, Tilman D, Haarstad J, Ritchie M (1998) Experimental tests of the dependence of arthropod diversity on plant diversity. Am Nat 152:738–750

Sih A (1980) Optimal behavior: can foragers balance two conflicting needs? Science 210:1041–1043

Thaler JS (1999) Jasmonate-inducible plant defenses cause increased parasitism of herbivores. Nature 399:686–688

Thies C, Tscharntke T (1999) Landscape structure and biological control in agroecosystems. Science 285:893–895

Tilman D, Wedin D, Knops J (1996) Productivity and sustainability influenced by biodiversity in grassland ecosystems. Nature 379:718–720

Tilman D, Knops J, Wedin D, Reich P (2002) Plant diversity and composition: effects on productivity and nutrient dynamics of experimental grasslands. In: Loreau M, Naeem S, Inchausti P (eds) Biodiversity and ecosystem functioning: synthesis and perspectives. Oxford University Press, Oxford, pp 21–35

Tumlinson JH, Lewis WJ, Vet LEM (1993) How parasitic wasps find their hosts. Sci Am 268:100–106

Tuomi J, Augner M, Nilsson J (1994) A dilemma of plant defenses: is it really worth killing the herbivore? J Theor Biol 170:427–430

Turlings TCJ, Fritzsche ME (1999) Attraction of parasitic wasps by caterpillar-damaged plants. In: Proc Novartis Foundation Symp 223, Insect–plant interactions and induced plant defence. Wiley, Chichester, pp 21–38

Turlings TCJ, Tumlinson JH, Lewis WJ (1990) Exploitation of herbivore-induced plant odors by host-seeking parasitic wasps. Science 250:1251–1253

Turner CE, Pemberton RW (1989) Leaf domatia and mites: a plant protection–mutualism hypothesis. In: Bock JH, Linhart YB (eds) The evolutionary ecology of plants. Westview Press, Boulder, pp 341–359

van Dam NM, Hadwich K, Baldwin IT (2000) Induced responses in *Nicotiana attenuata* affect behavior and growth of the specialist herbivore *Manduca sexta*. Oecologia 122:371–379

van der Meijden E, Klinkhamer PGL (2000) Conflicting interests of plants and the natural enemies of herbivores. Oikos 89:202–208

van der Putten WH, Vet LEM, Harvey JA, Wackers FL (2001) Linking above-ground and belowground multitrophic interactions of plants, herbivores, pathogens, and their antagonists. Trends Ecol Evol 16:547–554

van Loon JJA, de Boer JG, Dicke M (2000) Parasitoid–plant mutualism: parasitoid attack of herbivore increases plant reproduction. Entomol Exp Appl 97:219–227 van Rijn PCJ, Van Houten YM, Sabelis MW (2002) How plants benefit from providing food to predators even when it is also edible to herbivores. Ecology 83:2664–2679

Walter DE (1996) Living on leaves: mites, tomenta, and leaf domatia. Annu Rev Entomol 41:101–114

Walter DE, O'Dowd DJ (1992) Leaves with domatia have more mites. Ecology 73:1514–1518

Wardle DA, van der Putten WH (2002) Biodiversity, ecosystem functioning and above-ground–below-ground linkages. In: Loreau M, Naeem S, Inchausti P (eds) Biodiversity and ecosystem functioning. Oxford University Press, Oxford, pp 155–168

10 A General Rule for Predicting When Insects Will Have Strong Top-Down Effects on Plant Communities: On the Relationship Between Insect Outbreaks and Host Concentration

W.P. CARSON, J. PATRICK CRONIN and Z.T. LONG

10.1 Summary

We provide a new and general rule that predicts when native insect herbivores will have a major influence on dominant native plant species in communities and ecosystems worldwide. We argue that native insect herbivores will function as classic keystone species whenever their hosts become abundant and form large, persistent, dense stands. Specifically, our Host Concentration Model predicts that the impact of specialist insect herbivores will be more severe on a per individual basis as host species build up to form large and dense stands. The impact of these native insect herbivores, while important at non-outbreak levels, will be most important during major bouts of defoliation that occur during periodic insect outbreaks. Our review of the literature suggests that such outbreaks are common from a phytocentric perspective. Consequently, these insect outbreaks will have a major influence on ecosystem function via their ability to regulate and reduce the abundance of host species that may typically be the superior competitor across the landscape. Finally, we believe that this Host Concentration Model will predict when specialist insects will regulate plant communities and ecosystems better than resource supply models that rely on gradients in fertility or productivity.

10.2 Introduction

"The role of rare events, such as outbreaks ... cannot be ignored as a force structuring ecological communities just because they are rare"

(Strong et al. 1984).

We propose a general rule that predicts when native insect herbivores will have very strong top-down effects on plant communities and ecosystems in

Ecological Studies, Vol. 173
W.W. Weisser and E. Siemann (Eds.)
Insects and Ecosystem Function

any terrestrial or wetland habitat worldwide. Specifically, native specialist insect herbivores will function as a keystone species (sensu Paine 1969) whenever their hosts become abundant and form large, persistent, dense stands (see also Carson and Root 2000; Long et al. 2003). This Host Concentration Model results from a modification and extension of Root's resource concentration hypothesis (Root 1973), which states "...herbivores are more likely to find and remain on hosts that are growing in dense or nearly pure stands". There is ample empirical evidence to support this hypothesis for plant populations (see a review by Andow 1991; Joshi et al., Chap. 6, this Vol.), but its importance has only been recently extended to predict when insects will control vegetation dynamics (Carson and Root 2000; Long et al. 2003). This rule will operate whenever insect herbivores can track host plant abundance and should work in nearly any ecosystem at spatial scales from a few tens of meters to thousands of square kilometers. While outbreaks of native insects are not required for the rule to operate (see Long et al. 2003) or for insect herbivory to alter community dynamics (e.g. Bach 1994; Carson and Root 1999), the influence of insects will be most pronounced and important during outbreak events at both the community and ecosystem levels (e.g. Carson and Root 2000; Lovett et al. 2002). This rule leads to the conclusion that species-level trophic cascades will commonly lead to community-level trophic cascades in terrestrial communities whenever hosts are abundant, long-lived perennial plant species (cf. Polis and Strong 1996; Polis 1999). Furthermore, this rule means that as a superior plant competitor increases in density, becomes widespread and suppresses the abundance of subordinate plant species, it will not be long before it is found and attacked by its enemies. These attacks will decrease the abundance of the dominant plant so as to promote plant species coexistence, increase plant species diversity, lower standing crop biomass for potentially long periods of time and likely alter successional trajectories (e.g. Carson and Root 2000; Bach 2001).

10.3 The Significance of Insect Outbreaks

This general rule runs counter to much conventional wisdom and several theoretical lineages in ecology. First, outbreaks of native insects have often been considered rare and unusual events (Hairston et al. 1960; Strong et al. 1984; Faeth 1987; Owen 1987; Hartley and Jones 1997; Lowman 1997; Weisser and Siemann, Chap. 1, this Vol.; but see Matson and Addy 1975). Consequently, a prevailing view is that outbreaks are too infrequent and their effect too ephemeral to cause substantial and enduring top-down effects on plant communities. Consistent with this view, theories or discussions of trophic cascades rarely give outbreaks a place of importance (e.g. Polis 1999). We disagree with this perspective. Outbreaks of native insects appear to be nearly ubiquitous in ecosystems worldwide (Table 10.1). In addition, it is likely that these outbreaks result in strong and lasting top-down effects on these communities (e.g. Carson and Root 2000; Bach 2001). Also, these outbreaks appear to be much more common and more devastating in dense host stands (see below). Finally, outbreaks tend to occur repeatedly during the life span of long-lived perennial plant species; thus these events are not uncommon from a phytocentric perspective.

Second, outbreaks by their very nature are instances where plant defences, enemies, intra- or interspecific interactions among insects, and other processes (see Polis 1999) fail to prevent runaway consumption which results in major food depletion for the outbreaking insect (Berryman 1987). These bottom-up and top-down forces may 'kick in' during or following the outbreak and reduce herbivore numbers and their impact. By then, however, insect damage already will have had a strong top-down effect on the plant community. Consequently, theories regarding top-down control of insect herbivores by their enemies and bottom-up theories regarding plant defences or nutrition do not apply or only ameliorate or modify the effect of the outbreak. Below, we provide evidence for and explore the community- and ecosystem-level consequences of the following observations: (1) insect outbreaks occur in the vast majority of ecosystems and community types worldwide; (2) insect outbreaks are more common and more devastating per host in large, dense and continuous host stands; (3) outbreaking insects function as keystone species by reducing the abundance of the dominant species and increasing diversity; (4) insect outbreaks are common relative to host life span yet may often go unnoticed; (5) chrysomelid beetles and lepidoptera seem to be responsible for the majority of outbreaks. Finally, we briefly discuss and compare the Host Concentration Model with resource supply models.

10.3.1 Insect Outbreaks Are Common in Numerous Community Types Worldwide

Very few terrestrial communities and ecosystems are spared from outbreaks by native phytophagous insects (Carson and Root 2000; Table10. 1). Insect outbreaks appear to be common in numerous ecosystems including grasslands, mangroves, old-fields, heathlands, dry tropical forests, *Eucalyptus* forest and even in some wetland communities. Overall, we have identified more than 30 ecosystem and community types worldwide where outbreaks occur and cause major defoliation, often over vast areas, of the dominant plant species. Even relatively abundant species in diverse tropical forests may suffer from repeated insect outbreaks (Wolda 1978; Wolda and Foster 1978; Janzen 1981). Insect outbreaks appear to be particularly devastating in forest ecosystems. For example, numerous papers describe these outbreaks and their impact in northern and boreal forests (Table 10.1; see also Berryman 1987;

Table 10.1. Occurrence of insect outbreaks that caused major defoliation to dominant or very abundant species in communities and biomes throughout the world. We define an outbreak as an explosive increase in the abundance of a particular insect species that occurs over a relatively short period of time (Berryman 1987) and causes major defoliation on host species. This list, although extensive, is not meant to be comprehensive but rather to demonstrate the potential of these outbreaks to strongly influence a diverse array of community types. With a few exceptions, the vast majority of these insect and plant species are native and most of the insects are specialized on their associated host plants or cause the majority of their damage on the dominant species in the community with little damage to subordinate species. This table omits (with one Chrysomelidae exception) the 24 species of lowland dry forest trees that suffered total defoliation by insects (mostly Lepidoptera) described by Janzen (1981) and the numerous grasshopper defoliations occurring in US grasslands described by Joern (1989). Berryman (1987) and Schultz (1987) also describe some additional outbreaks in forested systems

Insect taxon	Dominant plant species attacked	Community type and geographic region	Source
Coleoptera			
Trirhabda nitidicollis (Chrysomelidaeae)	*Chrysothamnus nauseosus*	Rangeland (USA)	Massey and Pierce (1960); Dalen et al. (1986)
Trirhabda pilosa; *T. attenuata* (Chrysomelidae)	*Artemisia tridentata*, *A. tripartita*	Sagebrush-dominated rangeland (Canada and USA)	Pringle (1960); Banham (1961)
Trirhabda virgata, *Trirhabda canadensis* (Chrysomelidae)	*Solidago altissima*, *S canadensis*	Early successional old-field (USA)	McBrien et al. (1983); Root (1996); Morrow and Olfelt (2003)
Microrhopala vitata (Chrysomelidae)	*Solidago altissima*	Early successional old-field (USA)	Carson and Root (2000)
Pyrrhalta nymphaeae (Chrysomelidae)	*Nuphar luteum*	Wetlands (USA)	Wallace and O'Hop (1985)
Lochmaea suturalis (Chrysomelidae)	*Calluna*	Heathlands (Europe and Scotland)	Berdowski and Zeilinga (1987); Scandrett and Gimingham (1991)
Altica cordata[a] (Chrysomelidae)	*Salix cordata*	*Salix*-dominated early successional sand dunes (USA)	Bach (1994, 2001)
Chrysophtharta bimaculata (Chrysomelidae)	*Eucalyptus* spp.	*Eucalyptus* forest (Australia)	Ohmart (1996)
Chrysomela scripta (Chrysomelidae)	*Populus fremontii*	Cottonwood gallery forest (USA)	Anderson and Nelson (2002)
Chrysomela lapponica (Chrysomelidae)	*Salix borealis*	Not reported (Russia)	Zvereva et al. (1997); Zvereva (2002)
Altica spp.[b] (Chrysomelidae)	*Ludwigia leptocarpa*, L. *decurrens*	Wetlands (USA)	Scott and Haskins (1987)

Dorynota sp. (Chrysomelidae)	*Tabebuia impetiginosa*	Tropical lowland dry forest (Central America)	Janzen (1981)
Dendroctonus rufipennis (Scolytidae)	*Picea engelmannii, Abies lasiocarpa*	Subalpine spruce-fir forests (USA including Alaska and Canada)	Veblen et al. (1991); Parish et al. (1999); Matouska et al. (2001)
Dendroctonus ponderosae (Scolytidae)	*Pinus contorta*	Lodgepole pine forests (USA)	Romme et al. (1986)
Hadramphus spinipennis (Curculionidae)	*Aciphylla dieffenbachii*	Pasture (New Zealand)	Schops (2002)
Elaphidinoides spp (Cerambycidae)	*Rhizophora mangle*	Mangroves (Belize)	Feller (2002)
Lepidoptera			
Choristoneura pinus pinus (Tortricidae)	*Pinus banksiana*	Suboreal forests (USA and Canada	McCullough (2000)
Choristoneura fumiferana, C. biennis C. occidentalis among other *Choristoneura* species (Tortricidae)	*Abies balsamea, A. lasiocarpa, Picea glauca, P. engelmannii*	Spruce-fir forests (USA and Canada)	Filion et al. (1998); McCullough et al. (1998); Parish and Antos (2002)
Acleris variana (Tortricidae)	*Abies balsamea*	Balsam fir forests	Miller (1966)
Hypsidae[c]	*Shorea albida*	*Shorea albida* forest/peat swamps (Sarawak, Malaysia)	Anderson (1961)
Perthida glyphopha (Incurvariidae)	*Eucalyptus marginata*	*Eucalyptus* forest (Australia)	Abbott (1990)
Epirrita autumnata also called *Oporinia autumnata* (Geometridae)	*Betula pubescens*	Mountain birch forests (Europe)	Tenow (1972); Laine and Nicmela (1980); Ruohomaki et al. (2000)
Voracia casuariniphaga	*Casuarina junghuhniana*	Ridge-top stands of *Voracia casuariniphaga* (Indonesia)	Nair (2000)
Eulepidotis phrygionia (Noctuidae)	*Peltogyne gracilipes*	Monodominant Amazonian rainforest (Brazil)	Nascimento and Proctor (1994)
Spodoptera eridania (Noctuidae)	*Phytolacca rivinoides, Impatiens wallerana, Ipomoea tiliacea, Cestrum macrophylum*	Early successional tropical forest[d] (Puerto Rico)	Torres (1992)
Spodoptera litura (Noctuidae)	*Pisonia grandis*	*Pisonia grandis* subtropical forest (Polynesia)	Brooke et al. (1996)
Cerapteryx graminis (Noctuidae)	Grasses including *Poa pratensis* and *Deschampsia cespitosa*	Meadow vegetation (Europe)	Danell and Ericson (1990)
Operophtera brumata (Geometridae)	*Calluna vulgaris, Prunus padus*	Heathlands (Scotland), *Prunus*-dominated secondary forest (Fennoscandia)	Kerslake et al. (1996); Tikkanen and Roininen (2001)
Malacosoma disstria (Lasiocampidae)	*Populus tremuloides, Liquidambar styraciflua*	Aspen stands (Canada), Bottomland hardwoods (USA)	Hildahl and Reeks (1960) Siemann and Rogers (2003)

Table 10.1. (*Continued*)

Insect taxon	Dominant plant species attacked	Community type and geographic region	Source
Eulepidotis sp. (Noctuidae)	*Quararibea asterolepis*	Seasonally moist tropical forest (Central America)	Wong et al. (1990)
Zunacetha annulata (Dioptidae)	*Hybanthus prunifolius*	Seasonally moist tropical forest (Central America)	Wolda and Foster (1978)
Ammalo sp. (Arctiidae)	*Ficus* sp.	Seasonally moist tropical forest (Central America)	Wolda and Foster (1978)
Orgyia vetusta (Lymantriidae)	*Lupinus arboreus*	Coastal grassland (USA)	Harrison and Maron (1995)
Hepialus californicus (Hepialidae)	*Lupinus arboreus*	Coastal grassland (USA)	Strong et al. (1995)
Aroga websteri (Gelechiidae)	*Artemisia tridentata*	Sagebrush-dominated rangeland (USA)	Andres (1971)
Cleora injectalia (Geometridae), *Ophiusa melicerta* (Noctuidae), *Nephopterix syntaractis* (Pyralidae), *Junonia evarete* (Nymphalidae), *Hyblaea purea* (Hyblaeidae) *Planotortrix avicenniae*, (Tortricidae), *Phocides pigmalion* (Hersperiidae)	*Avicennia alba*, *A marina*, *A. germinans*, *Rhizophora* spp., *Excoecaria agallocha*	Mangroves (Central America, Asia, Australia, New Zealand and Indonesia)	Piyakarnchana (1981); West and Thorogood (1985); Whitten and Damanik (1986); Anderson and Lee (1995); Ellison and Farnsworth (1996); Burns (1998); Saur et al. (1999)
Homoptera			
Magicicada spp. (Cicadidae)	*Cornus drummondii*	*Cornus*-dominated early successional shrubland (USA)	Cook et al. (2001); Cook and Holt (2002)
Magicicada septendecim (Cicadidae)	*Quercus ilicifolia*	Scrub oak communities (USA)	Karban (1980)
Orthoptera			
Didymuria violescens (Phasmatidae)	*Eucalyptus delegatensis*, *E. robertsoni*	Readshaw (1965); Mazanec (1968) *Eucalyptus* forest (Australia)	

[a] Insects may not have been at outbreak levels but caused heaving defoliation.
[b] Insects did not reach outbreak levels but were very abundant.
[c] The insect species responsible remains unclear from the published report.
[d] These species became 'very abundant' following Hurricane Hugo in 1989.

Carson and Root 2000). Lowman (1997) has suggested that outbreaks in forests occur frequently but are overlooked because they are patchy and go unnoticed high in the canopy (see also Wolda and Foster 1978).

10.3.2 Insect Outbreaks Are More Common and More Devastating Per Host in Large, Dense and Continuous Host Stands

The Host Concentration Model provides a viable insect–plant feedback mechanism whereby insect impact and damage increases as dominant plant species or superior competitors become more and more dense or concentrated (e.g. Long et al. 2003). If insect damage and subsequent impact are greater (per capita) and insect outbreaks more common in larger and denser stands of hosts, then insect herbivory may commonly exert strong top-down effects on plant communities whenever superior competitors increase sufficiently in density and area (cf. Strong 1992; Polis and Strong 1996; Polis 1999). Indeed, there seems to be a growing consensus that outbreaks are more likely and insect impact greater in natural vegetation where one or a few plant species are dominant, abundant or aggregated, thereby forming large, continuous host stands (Whitmore 1975; Morrow and Fox 1989; Bach 1994; Morris et al. 1996; Schowalter 1996a; Bergeron and Leduc 1998; Cappucino et al. 1998; Carson and Root 2000; Dymerski et al. 2001; Nair 2000, 2001; Schops 2002; Long et al. 2003).

That insect outbreaks occur more frequently and do more damage per host plant in large, continuous host stands comes as no surprise to ecologists who study northern, boreal and subalpine ecosystems (see a review by McCullough et al. 1998; Table 10.1). Indeed, Bergeron and Leduc (1998) concluded that in balsam fir forests (*Abies balsamea*) "...mortality due to outbreaks appears to be more important in regions where *Abies balsamea* is dominant". They further concluded that there was "...a direct correlation between mortality and the abundance of the host species". This relationship between host concentration and insect outbreaks appears to hold for numerous community types. For example, insect outbreaks have defoliated hundreds of square kilometers of mangroves in locations worldwide (Table 10.1; Anderson 1961; Whitten and Damanik 1986). Mangroves are typically characterized by dense stands of one or a few dominant species. Similarly, Australian *Eucalyptus* forests are low-diversity forests dominated by the genus *Eucalyptus*. These forests suffer heavy insect damage and outbreaks are common relative to the life span of the hosts (Morrow and Fox 1989; Fox and Morrow 1992; Table 10.1). Morrow and Fox (1989) concluded that the primary factor that caused these forests to suffer heavy damage was host concentration. Likewise, Nair (2000) concluded that in tropical Indonesian forests "...most outbreaks have been recorded in tree species that occur gregariously....A high host density appears to be a key factor promoting pest outbreaks" (see also Nair 2001).

Indeed, Nascimento and Proctor (1994) documented an outbreak event that severely damaged a monodominant Amazonian rainforest and Wolda and Foster (1978) documented an outbreak of a lepidopteran that attacked what is by far the most abundant woody species – *Hybanthus prunifolius* – on Barro Colorado Island in Panama. There are now numerous examples of insect outbreaks attacking dominant or abundant species in natural vegetation (Table 10.1).

10.3.3 Native Outbreaking Insects Function as Keystone Species by Reducing the Abundance of the Dominant Species and Increasing Diversity

Insect outbreaks typically reduce the abundance or growth of the dominant species within the community, subsequently freeing resources for subordinate species and thereby increasing diversity (MacLean 1988; Bach 1994; De Grandpré and Bergeron 1997; Carson and Root 2000; Dery et al. 2000; Matouska et al. 2001; Parish and Antos 2002; Long et al. 2003). In this way, outbreaking insects function as classic keystone species (e.g. Carson and Root 2000). Matouska et al. (2001) described the strong top-down impact of the spruce beetle (*Dendroctonus rufipennis*) on boreal forest systems:

"Outbreaks varying in size and severity have occurred in the region at approximately 30- to 50-year intervals since 1920. During large outbreaks, spruce beetles can cause widespread changes to the structure of forests by selectively attacking and killing the large diameter spruce (*Picea* sp.)....After the canopy forming trees have died, understorey vegetation may be released from competition for light, nutrients and water; this may result in the rapid growth of the smaller, subordinate trees, shade-intolerant herbs and grasses. As a result, plant communities following a large outbreak may be quite different in structure and composition from those that dominated before the irruption".

Unfortunately, with the exception of northern and boreal forests (e.g. MacLean 1988; Veblen et al. 1991; De Grandpré and Bergeron 1997; Dery et al. 2000; Matouska et al. 2001; Parish and Antos 2002) and sand dune and old-field successional communities in the USA and Canada (e.g. McBrien et al. 1983; Bach 1994; Carson and Root 2000), most studies have not considered the consequences of outbreaks for non-host species, for plant species composition or for diversity. Nonetheless, large-scale outbreaks of native insect herbivores often dramatically reduce the growth of their hosts and cause substantial mortality of the dominant plant species (e.g. Carson and Root 2000). If these outbreaks are common relative to host life span, then outbreaking insects may typically function as keystone species, causing strong top-down effects on plant communities. Indeed, Schowalter (1996a) concluded that when hosts become abundant, "...high intensities of herbivory represent a major mechanism for reversing site dominance ...and increasing diversity" in forest ecosystems.

10.3.4 Insect Outbreaks Are Common Relative to Host Life Span Yet May Often Go Unnoticed

There is a general perception that outbreaks of native insect species defoliating native plant species are uncommon or rare events (Hairston et al. 1960; Strong et al. 1984; Hartley and Jones 1997; Lowman 1997). Nonetheless, it is clear that in many cases outbreaks will occur repeatedly over the life span of long-lived hosts. For example, Root (1996) concluded that devastating outbreaks would occur on goldenrods in goldenrod-dominated fields in central New York, USA, every 5–15 years and goldenrod life span and period of dominance is 10–30+ years. These outbreaks dramatically reduce community biomass, increase plant diversity and alter successional trajectories for years following the outbreak (Carson and Root 2000). Outbreaks of insects in aspen stands in Canada, mountain birch forests in Europe, *Eucalyptus* forest in Australia, spruce-fir forests in North America and heathlands in Europe all experience outbreaks that occur at least once, if not repeatedly, during the life span of the dominant species in the stand (Table 10.1; Hildahl and Reeks 1960; Berdowski and Zeilinga 1987; Morrow and Fox 1989; Veblen et al. 1991; Fox and Morrow 1992; Carson and Root 2000; Ruohomaki et al. 2000; Matouska et al. 2001). For many community types, however, the frequency and scale of outbreaks remain unknown. If the probability of an outbreak is high for dominant or abundant plant species in a stand in any community type, then outbreaks may commonly have a strong top-down effect on plant communities. Unfortunately, in many community types, outbreaks are fairly short-term events that appear and then disappear relatively suddenly. In addition, in forests, outbreaks may occur high in the canopy and go unnoticed (e.g. Wolda and Foster 1978; Lowman 1997). Consequently, outbreaks may be poorly documented, dismissed or overlooked.

10.3.5 Chrysomelid Beetles and Lepidoptera Seem to Be Responsible for the Majority of Outbreaks

We found that Lepidoptera and Coleoptera accounted for approximately 58 and 36 % of outbreaking taxa respectively (Table 10.1). Thus, these two taxa accounted for more than 90 % of the outbreaks categorized in Table 10.1. Similarly, of the major defoliations described by Janzen (1981), Lepidoptera and Coleoptera accounted for 65 and 30 % respectively. These percentages among community types (Table 10.1) and within a community type (tropical dry forest, Janzen 1981) are strikingly similar and this may in part reflect the prominence of these two taxa among phytophagous insects (Strong et al. 1984) as well as reflecting the characteristics of these groups that make them more likely to outbreak (e.g. Schultz 1987).

Within the Lepidoptera, we found that 50 % of the outbreaks were caused by noctuids, tortricids and geometrids (Table 10.1). Janzen (1981) found that geometrids and noctuids accounted for 33 % of outbreaks in a dry tropical forest, but he did not describe any outbreaks by tortricids, which seem to be most important in northern and boreal forests (Table 10.1). Schultz (1987) concluded that geometrids were prone to being pests. We found that among the Coleoptera, a surprising 75 % of outbreaks were caused by chrysomelids. We suggest that chrysomelid beetles may frequently function as keystone species in plant communities (1) because they are a species-rich group, (2) owing to their tendency to break out (Table 10.1; Root 1996; White 1996; Carson and Root 2000), (3) owing to their tendency to aggregate on host plants or dense host patches (Bach and Carr 1990; Herzig 1995; Herzig and Root 1996; Morris et al. 1996; Long et al. 2003) and (4) because they cause strong top-down effects on biomass in plant communities (e.g. Carson and Root 2000). Thus, we suggest that it may be more than pure coincidence that three most 'spectacular cases' of biological control have used chrysomelid beetles to control abundant exotic perennial plants (*Hypericum perforatum*, *Senecio jacobaea* and *Alternanthera philoxeroids*; Huffaker 1964; Maddox et al. 1971; Crawley 1989; McEvoy et al. 1993; Buckingham 1996; Zwölfer and Zimmermann, Chap. 7, this Vol.). Overall, our findings lead us to the conclusion that chrysomelid beetles along with outbreaking species of Lepidoptera may commonly exert strong top-down effects on numerous community types worldwide. Lepidoptera appear to have their greatest impact on forest systems whereas chrysomelids appear to have their greatest impact on herbaceous or early successional systems, though exceptions to this generalization occur (Table 10.1).

10.4 The Host Concentration Model May Predict Insect Impact on Plant Communities at Multiple Spatial Scales Better Than Resource Supply Theory

10.4.1 Resource Supply Theory

The bottom-up, top-down debate has been dominated by a single and very important theoretical framework, called here Resource Supply Theory (RST; also called Resource Dependent Foodwebs by Grover 1997). RST encompasses a variety of models that are all fundamentally based on the premise that variation in the strength of top-down control by herbivores can be explained by changes in resource supply to primary producers (e.g. Oksanen et al. 1981, 1995; Leibold 1989, 1996; Schmitz 1992, 1993, Holt et al. 1994; Grover 1997). Much empirical work investigating top-down effects has taken its lead from these theories and has investigated how resource supply influences herbivore abundance (e.g. Siemann 1998; Ritchie 2000) and herbivore impact on communities (e.g. Fraser and Grime 1998). There is now a whole array of these models (see Chase et al. 2000) with varying degrees of empirical support.

We suggest that this theoretical framework may not accurately predict when *specialist insect herbivores* will have a strong top-down impact on a plant community. Specifically, RST models focus on how resource supply affects plant defences or net primary productivity or both, which then directly or indirectly (via predators) regulate herbivore control over primary producers. Alternatively, we suggest that insects will regulate vegetation dynamics whenever dominant host plants become concentrated and abundant and that this regulation will be most pronounced during outbreaks (but see Long et al. 2003 for a non-outbreak example). This means that to predict when specialist insect herbivores will have a strong top-down effect it may be more important to know the distribution and abundance of their hosts than to know how resource supply affects their enemies or plant traits (e.g. defences). Because the vast majority of insect herbivores are specialists (e.g. Fox and Morrow 1981; Thompson 1994; Janz et al. 2001; Nosil 2002), the Host Concentration Model may typically predict when insect herbivores in general will impact plant communities. We do not dispute the fact that resource supply and its relationship to productivity, plant defences and enemies have important consequences for insect herbivores. All we are suggesting is that host concentration may more accurately predict and determine when insect herbivores will have a strong top-down effect on a plant community.

10.4.2 The Host Concentration Model (HCM)

HCM predicts that specialist insect herbivores will have strong top-down effects on plant communities whenever their hosts form large, persistent dense stands (see also Long et al. 2003; Carson and Root 2000). This impact will be most pronounced and enduring following outbreaks. The host concentration model is more parsimonious than resource supply models (also see Polis 1999). It makes assumptions neither about predators, plant defences nor changes in resource supply. The only assumption of HCM is that the per stem impact of herbivores increases with host plant density or area sufficiently enough to have a top-down impact on plant communities. Furthermore, if native plants that are superior competitors are also widespread, which is typical, then these plants will have a substantially richer herbivorous insect fauna than inferior competitors that are more sparsely distributed

(Strong et al. 1984). Consequently, widespread, superior competitors may commonly have specialist insects that not only track their abundance but also periodically outbreak (Table 10.1), thereby reducing the impact of these dominant plant species on subordinate plant species within the community. Finally, insect herbivores appear to be able to track host plant abundance from spatial scales of a few tens of meters (e.g. Morrow et al. 1989; Herzig and Root 1996; Carson and Root 2000; Long et al. 2003) to tens of thousands of hectares (e.g. Bergeron and Leduc 1998; Cappucino et al. 1998; McCullough et al. 1998; Kruess et al., Chap. 8, this Vol.). Consequently, the host concentration model should apply across huge spatial scales depending upon the host finding behaviour of the specialist insect.

10.4.3 Distinguishing Between the Two Models

It is important to note that RST and the HCM can make very similar predictions about increasing regulation of plant communities with increasing resource supply when both the density and productivity of the dominant host plant increases with community productivity. If insect herbivores track host-plant abundance but host-plant abundance and productivity are positively correlated, then RST can make the correct prediction *based on the wrong mechanism*. Alternatively, if insect herbivores track host-plant abundance but host-plant abundance and resource supply are not correlated, RST *can make the wrong prediction*. It is only when measures of both host plant density, host plant productivity and host quality at both the population and community level are available that we can distinguish among the models.

10.5 Relationship to Other Related Processes Proposed to Promote Diversity

Our goal in this chapter was to provide a general rule to predict when *specialist insect herbivores* will have strong top-down effects on plant communities across multiple spatial scales and to argue that such effects may be pervasive and enduring from a phytocentric perspective. The current prevailing view is that top-down (e.g. predation and parasitism) and bottom-up forces (e.g. plant defences) in terrestrial food webs severely constrain the impact of insect herbivores on plant communities. We believe this focus may miss the brief but critical windows when insect herbivory will regulate plant communities: specifically when hosts form large dense stands and when insects outbreak in these stands. We recognize that the rule has many similarities to other general processes where enemies respond and suppress their prey or hosts in a density-dependent fashion. For example, our rule has many parallels to the

Janzen–Connell hypothesis that was originally proposed to explain the maintenance of tree diversity in tropical forests (Connell 1971; Janzen 1970). Our hypothesis, however, is focused specifically on the impact of specialist outbreaking phytophagous insects that attack large, dense concentrations of hosts. It is not focused on seed predation, the impact of predators on juveniles or the relationship between juvenile survivorship and distance to putative adult parents. Nonetheless, inherent within our rule and the Janzen-Connell hypothesis is that as host density increases the per host impact of predation (or herbivory) increases.

10.5.1 Does Pathogen Impact Increase with Host Concentration?

There is growing evidence that pathogens, like specialist insects, can have a strong top-down effect on plant communities (Dickman 1992; Van der Putten et al. 1993; Van der Putten and Peters 1997; Olff et al. 2000; Packer and Clay 2000; Rizzo et al. 2000; Joshi et al., Chap. 6, this Vol.; Kruess et al., Chap. 8, this Vol.). If the per host impact of these pathogens increases with host density (e.g. reviewed in Burdon and Chilvers 1982; Augspurger 1988; Burdon et al. 1989; Alexander 1992), then we believe the Host Concentration Model will also apply to many plant pathogens. Consequently, abundant plant species may have to contend with not only periodic insect outbreaks but also periodic pathogen outbreaks. Because of lack of empirical work at the community level, it remains unknown which is more important or how the combined impact of host-tracking insects and pathogens may interact to structure plant communities.

Acknowledgements. We thank Dan Bunker, Tony Baumert, Evan Siemann, Alex Royo, John Paul, Henry Schumacher, Tom Pendergast, Gay Allison and Scott Stark for comments on drafts of this manuscript. This work was supported by NSF grants DEB-95 27729 and DEB-95 15184 to Walter Carson.

References

Abbott I (1990) Insect outbreaks in forests of western Australia. In: Watt AD, Leather SR, Hunter MD, Kidd NAC (eds) Population dynamics of forest insects. Intercept, Hampshire, pp 95–103

Alexander HM (1992) Fungal pathogens and the structure of plant populations and communities. In: Carroll GC, Wicklow DT (eds) The fungal community: its organization and role in the ecosystem, 2nd edn. Marcel Dekker, New York, pp 481–497

Anderson C, Lee SY (1995) Defoliation of the mangrove *Avicennia marina* in Hong Kong – causes and consequences. Biotropica 27:218–226

Andersen DC, Nelson, SM (2002) Effects of cottonwood leaf beetle *Chrysomela scripta* (Coleoptera: Chrysomelidaeae) on survival and growth of fremont cottonwood (*Populus fremontii*) in northwest Colorado. Am Midl Nat 147:189–203

Anderson JAR (1961) The destruction of *Shorea albida* forest by an unidentified insect. Emp For Rev 40:19–29

Andow DA (1991) Vegetational diversity and arthropod population response. Annu Rev Entomol 36:561–586

Andres LA (1971) The suppression of weeds with insects. Proc Tall Timbers Conf Ecol Anim Control Habitat Manage 3:185–195

Augspurger CK (1988) Impact of pathogens on natural plant populations. In: Davey AJ, Hutchings MJ, Watkinson AR (eds) Proc 28th Symp of the British Ecological Society, Plant Population Ecology , Sussex, 1987. Blackwell, Oxford, pp 413–432

Bach CE (1994) Effects of a specialist herbivore (*Altica subplicata*) on *Salix cordata* and sand dune succession. Ecol Monogr 64:423–445

Bach CE (2001) Long-term effects of insect herbivory and sand accretion on plant succession on sand dunes. Ecology 82:1401–1416

Bach CE, Carr DS (1990) Aggregation behavior of a willow flea beetle, *Altica subplicata* (Coleoptera: Chrysomelidae). Great Lakes Entomol 23:65–76

Banham FL (1961) Distribution of *Trirhabda pilosa* Blake (Coleoptera: Chrysomelidaeae), attacking big sagebrush in the interior of British Columbia. Proc Entomol Soc BC 58:38–40

Berdowski JJM, Zeilinga R (1987) Transition from heathland to grassland: damaging effects of the heather beetle. J Ecol 75:159–175

Bergeron Y, Leduc A (1998) Relationships between change in fire frequency and mortality due to spruce budworm outbreak in the southeastern Canadian boreal forest. J Veg Sci 9:492–500

Berryman AA (1987) The theory and classification of outbreaks. In: Barbosa P, Schultz JC (eds) Insect outbreaks. Academic Press, San Diego, pp 3–30

Brooke M de L, Jones PJ, Vickery JA, Waldren S (1996) Seasonal patterns of leaf growth and loss, flowering and fruiting on a subtropical central Pacific island. Biotropica 28(2):164–179

Buckingham GR (1996) Biological control of alligatorweed, *Alternanthera philoxeroides*, the world's first aquatic weed success story. Castanea 61:232–243

Burdon JJ, Chilvers GA (1982) Host density as a factor in plant disease ecology. Annu Rev Phytopathol 20:143–166

Burdon JJ, Jarosz AM, Kirby GC (1989) Pattern and patchiness in plant–pathogen interactions – causes and consequences. Annu Rev Ecol Syst 20:119–136

Burns B (1998) Mangrove dieback at Waitakaruru. N Z Ecol Soc Newsl 88:6

Cappuccino N, Lavertu D, Bergeron Y, Regniere J (1998) Spruce budworm impact, abundance and parasitism rate in a patchy landscape. Oecologia 114:236–242

Carson WP, Root RB (1999) Top-down effects of insect herbivores during early succession: influence on biomass and plant dominance. Oecologia 121:260–272

Carson WP, Root RB (2000) Herbivory and plant species coexistence: community regulation by an outbreaking phytophagous insect. Ecol Monogr 70(1):73–99

Chase JM, Leibold MA, Downing AL, Shurin JB (2000) The effects of productivity, herbivory, and plant species turnover in grassland food webs. Ecology 81:2485–2497

Connell JH (1971) On the role of natural enemies in preventing competitive exclusion in some marine animals and in rain forest trees. In: den Boer PJ, Gradwell GR (eds) Dynamics of numbers in populations. Center for Agricultural Publishing and Documentation, Wageningen, The Netherlands, pp 298–312

Cook WM, Holt RD (2002) Periodical cicada (*Magicicada cassini*) oviposition damage: visually impressive yet dynamically irrelevant. Am Midl Nat 147:214–224

Cook WM, Holt RD, Yao J (2001) Spatial variability in oviposition damage by periodical cicadas in a fragmented landscape. Oecologia 127:51–61

Crawley MJ (1989) Insect herbivores and plant population dynamics. Annu Rev Entomol 34:531–564

Dalen RS, Fletcher RA, Winter FA (1986) Rabbitbrush (*Chysothamnus nauseosus* ssp. *consimilis*) mortality associated with defoliation by a leaf-feeding beetle, *Trirhabda nitidicollis*. In: Proc Symp on the Biology of *Artemisia* and *Chysothamnus*, USDA Forest Service General Tech Rep INT-200, pp 199–204

Danell K, Ericson L (1990) Dynamic relations between the antler moth and meadow vegetation in northern Sweden. Ecology 71:1068–1077

De Grandpré L, Bergeron Y (1997) Diversity and stability of understory communities following disturbance in the southern boreal forest. J Ecol 85:777–784

Dery S, Belanger L, Marchand S, Cote S (2000) Succession apres epidemie de la tordeuse des bourgeons de l'epinette (*Choristoneura fuminiferana*) dans des sapinieres boreales pluviales de seconde venue. Can J For Res 30:801–816

Dickman A (1992) Plant pathogens and long-term ecosystem changes. In: Carroll GC, Wicklow DT (eds) The fungal community: its organization and role in the ecosystem, 2nd edn. Marcel Dekker, New York, pp 499–520

Dymerski AD, Anhold JA, Munson A (2001) Spruce beetle (*Dendroctonus engelmannii*) outbreak in Engelmann spruce (*Picea engelmannii*) in central Utah, 1986–1998. West N Am Nat 61:19–24

Ellison AM, Farnsworth EJ (1996) Spatial and temporal variability in growth of *Rhizophora mangle* saplings on coral cays: links with variation in isolation, herbivory, and local sedimentation rate. J Ecol 84:717–731

Faeth SH (1987) Community structure and foliverous insect outbreaks: the roles of vertical and horizontal interactions. In: Barbosa P, Schultz JC (eds) Insect outbreaks. Academic Press, San Diego, pp 135–171

Feller IC (2002) The role of herbivory by wood-boring insects in mangrove ecosystems in Belize. Oikos 97:167–176

Filion L, Payette S, Delwaide A, Bhiry N (1998) Insect defoliators as major disturbance factors in the high-altitude balsam fir forest of Mount Megantic, southern Quebec. Can J For Res 28:1832–1842

Fox LR, Morrow PA (1981) Specialization: species property or local phenomenon? Science 211:887–893

Fox LR, Morrow PA (1992) Eucalypt responses to fertilization and reduced herbivory. Oecologia 89:214–222

Fraser LH, Grime JP (1998) Top-down control and its effect on the biomass and composition of three grasses at high and low soil fertility in outdoor microcosms. Oecologia 113:239–246

Grover JP (1997) Resource competition. Chapman and Hall, New York

Hairston NG, Smith FE, Slobodkin LB (1960) Community structure, population control and competition. Am Nat 94:421–425

Harrison S, Maron JL (1995) Impacts of defoliation by tussock moths (*Orgyia vetusta*) on the growth and reproduction of bush lupine (*Lupinus arboreus*). Ecol Entomol 20:223–229

Hartley SE, Jones CG (1997) Plant chemistry and herbivory, or why the world is green. In: Crawley M (ed) Plant ecology. Blackwell, Cambridge, Massachusetts, pp 284–324

Herzig AL (1995) Effects of population density on long-distance dispersal in the goldenrod beetle *Trirhabda virgata*. Ecology 76:2044–2054

Herzig A, Root RB (1996) Colonization of host patches following long-distance dispersal by a goldenrod beetle, *Trirhabda virgata*. Ecol Entomol 21:344–351

Hildahl V, Reeks WA (1960) Outbreaks of the forest tent caterpillar, *Malacosoma disstria* Hbn., and their effects on stands of trembling aspen in Manitoba and Saskatchewan. Can Entomol 92:199–209

Holt RD, Grover J, Tilman D (1994) Simple rules for interspecific dominance in systems with exploitative and apparent competition. Am Nat 144:741–771

Huffaker C (1964) Fundamentals of biological weed control. In: Debach, Schlinger E (eds) Control of insect pests and weeds. Chapman and Hall, London, pp 74–117

Janz N, Nyblom K, Nylin S (2001) Evolutionary dynamics of host-plant specialization: a case study of the tribe Nymphalini. Evolution 55:783–796

Janzen DH (1970) Herbivores and the number of tree species in tropical forests. Am Nat 104: 501–529

Janzen DH (1981) Patterns of herbivory in a tropical deciduous forest. Biotropica 13:271–282

Joern A (1989) Insect herbivory in the transition to California annual grasslands: did grasshoppers deliver the coup de grass? In: Huenneke L, Mooney H (eds) Grassland structure and function: California annual grassland. Kluwer, Boston, pp 117–134

Karban R (1980) Periodical cicada nymphs impose periodical oak tree wood accumulation. Nature 287:326–327

Kerslake JE, Kruuk LEB, Hartley SE, Woodin SJ (1996) Winter moth (*Operophtera brumata* (Lepidoptera: Geometridae)) outbreaks on Scottish heather moorlands: effects of host plant and parasitoids on larval survival and development. Bull Entomol Res 86:155–164

Laine KJ, Niemela P (1980) The influence of ants on the survival of mountain birches during an *Oporinia autumnata* (Lep., Geometridae) outbreak. Oecologia 47:39–42

Leibold MA (1989) Resource edibility and the effects of predators and productivity on the outcome of trophic interactions. Am Nat 134:922–949

Leibold MA (1996) A graphical model of keystone predators in food webs: trophic regulation of abundance, incidence, and diversity patterns in communities. Am Nat 147:784–812

Long ZT, Mohler CL, Carson WP (2003) Extending the resource concentration hypothesis to plant communities: effects of litter and herbivores. Ecology 84(3):652–665

Lovett GM, Christenson LM, Groffman PM, Jones CG, Hart JE, Mitchell NJ (2002) Insect defoliation and nitrogen cycling in forests. BioScience 52:335–341

Lowman MD (1997) Herbivory in forests – from centimeters to megameters. In: Watt AD, Stork NE, Hunter MD (eds) Forests and insects. Chapman and Hall, New York, pp 135–149

MacLean D (1988) Effects of spruce budworm outbreaks on vegetation, structure, and succession of balsam fir forests on Cape Breton Island, Canada. In: Werger MJA, van der Aart PJM, During HJ, Verhoeven JTA (eds) Plant form and vegetation structure. SPB Academic Publishing, The Hague, pp 253–261

Maddox DM, Andres LA, Hennessey RD, Blackburn RD, Spencer NR (1971) Insects to control alligatorweed: an invader of aquatic ecosystems in the United States. BioScience 21:985-991

Massey CL, Pierce DA (1960) *Trirhabda nitidicollis*, a pest of rabbitbrush in New Mexico. J Range Manage 13:216–217

Matson WJ, Addy ND (1975) Phytophagous insects as regulators of forest primary production. Science 190:515–522

Matouska SM, Handel CM, Ruthrauff DR (2001) Densities of breeding birds and changes in vegetation in an Alaskan boreal forest following a massive disturbance by spruce beetles. Can J Zool 79:1678–1690

Mazanec Z (1968) Influence on defoliation by the phasmatid *Didymuria violescens* on seasonal diameter growth and the pattern of growth rings in alpine ash. Aust For 32:3–14

McBrien H, Harmson R, Crowder A (1983) A case of insect grazing affecting plant succession. Ecology 64:1035–1039

McCullough DG (2000) A review of factors affecting the population dynamics of jack pine budworm (*Choristoneura pinus pinus* Freeman). Popul Ecol 42:243–256

McCullough DG, Werner RA, Neumann D (1998) Fire and insects in northern and boreal forest ecosystems of North America. Annu Rev Entomol 43:107–27

McEvoy PB, Rudd NT, Cox CS, Huso M (1993) Disturbance, competition, and herbivory effects on ragwort *Senecio jacobaea* populations. Ecol Monogr 63:55–75

Miller CA (1966) The black-headed budworm in eastern Canada. Can Entomol 98:592–613

Morris W, Grevstad F, Herzig A (1996) Mechanisms and ecological functions of spatial aggregation in Chrysomelidae beetles. In: Jolivet PHA, Cox ML (eds) Chrysomelidaeae biology, vol 2. Ecological studies. SPB Academic Publishing, Amsterdam, pp 303–322

Morrow PA, Fox LR (1989) Estimates of pre-settlement insect damage in Australian and North American forests. Ecology 70:1055–1060

Morrow PA, Olfelt JP (2003) Phoenix clones: recovery after long-term defoliation induced dormancy. Ecol Lett 6:119–125

Morrow PA, Tonkyn D, Goldburg R (1989) Patch colonization by *Trirhabda canadensis* (Coleoptera: Chrysomelidae): effects of plant species composition and wind. Oecologia 81:43–50

Nair KSS (2000) Insect pests and diseases in Indonesian forests: an assessment of the major threats, research efforts and literature. Center for International Forestry Research, Bogor, Indonesia

Nair KSS (2001) Pest outbreaks in tropical forest plantations: is there a greater risk for exotic tree species? Center for International Forestry Research, Bogor, Indonesia

Nascimento MT, Proctor J (1994) Insect defoliation of a monodominant Amazonian rainforest. J Trop Ecol 10:633–636

Nosil P (2002) Transition rates between specialization and generalization in phytophagous insects. Evolution 56:1701–1706

Ohmart CP (1996) Population dynamics of Chrysomelidae beetles feeding on *Eucalyptus*. In: Jolivet PHA, Cox ML (eds) Chrysomelidaeae biology, vol 2. Ecological studies. SPB Academic Publishing, Amsterdam, pp 263–269

Oksanen L, Fretwell SD, Arruda J, Niemela P (1981) Exploitation ecosystems in gradients of primary productivity. Am Nat 118:240–261

Oksanen T, Power ME, Oksanen L (1995) Ideal free habitat selection and consumer-resource dynamics. Am Nat 146:565–585

Olff H, Hoorens B, de Goede RGM, Van der Putten WH, Gleichman JM (2000) Small-scale shifting mosaics of two dominant grassland species: the possible role of soil-borne pathogens. Oecologia 125:45–54

Owen DF (1987) Leaf eating as mutualism. In: Barbosa P, Schultz JC (eds) Insect outbreaks. Academic Press, San Diego, pp 81–95

Packer A, Clay K (2000) Soil pathogens and spatial patterns of seedling mortality in a temperate tree. Nature 404:278–281

Paine RT (1969) A note on trophic complexity and community stability. Am Nat 103:91–93

Parish R, Antos JA (2002) Dynamics of an old-growth, fire-initiated, subalpine forest in southern interior British Columbia: tree-ring reconstruction of 2 year cycle spruce budworm outbreaks. Can J For Res 32:1947–1960

Parish R, Antos JA, Fortin M-J (1999) Stand development in an old-growth subalpine forest in southern interior British Columbia. Can J For Res 29:1347–1356

Piyakarnchana T (1981) Severe defoliation of *Avicennia alba* Bl. by larvae of *Cleora injectaria* walker. J Sci Soc Thailand 7:33–36
Polis GA (1999) Why are parts of the world green? Multiple factors control productivity and the distribution of biomass. Oikos 86:3–15
Polis GA, Strong DR (1996) Food web complexity and community dynamics. Am Nat 147:813–846
Pringle WL (1960) The effect of a leaf feeding beetle on big sagebrush in British Columbia. J Range Manage 13:139–142
Readshaw JL (1965) A theory of phasmatid outbreak release. Aust J Zool 13:475–490
Ritchie ME (2000) Nitrogen limitation and trophic vs. abiotic influences on insect herbivores in a temperate grassland. Ecology 81:1601–1612
Rizzo DM, Slaughter GW, Parmeter JR Jr (2000) Enlargement of canopy gaps associated with a fungal pathogen in Yosemite Valley, California. Can J For Res 30:1501–1510
Romme WH, Knight DH, Yavitt JB (1986) Mountain pine beetle outbreaks in the Rocky Mountains: regulators of primary productivity? Am Nat 127:484–494
Root RB (1973) Organization of a plant–arthropod association in simple and diverse habitats: the fauna of collards (*Brassica oleracea*). Ecol Monogr 43:95–124
Root RB (1996) Herbivore pressure on goldenrods (*Solidago altissima*): its variation and cumulative effects. Ecology 77:1074–1087
Ruohomaki K, Tanhuanpaa M, Ayres MP, Kaitaniemi P, Tammaru T, Haukioja E (2000) Causes of cyclicity of *Epirrita autumnata* (Lepidoptera, Geometridae): grandiose theory and tedious practice. Popul Ecol 42:211–223
Saur E, Imbert D, Etienne J, Mian D (1999) Insect herbivory on mangrove leaves in Guadeloupe: effects on biomass and mineral content. Hydrobiology 413:89–93
Scandrett E, Gimingham CH (1991) The effect of heather beetle *Lochmaea suturalis* on vegetation in a wet heath in NE Scotland. Holarct Ecol 14:24–30
Schmitz OJ (1992) Exploitation in model food chains with mechanistic consumer-resource dynamics. Theor Popul Biol 41:161–183
Schmitz OJ (1993) Trophic exploitation in grassland food chains: simple models and a field experiment. Oecologia 93:327–335
Schops K (2002) Local and regional dynamics of a specialist herbivore: overexploitation of a patchily distributed host plant. Oecologia 132:256–263
Schowalter TD (1996a) Insect ecology: an ecosystem approach. Academic Press, San Diego
Schultz JC (1987) What is a forest pest? In: Barbosa P, Schultz JC (eds) Insect outbreaks. Academic Press, San Diego, pp 59–80
Scott ML, Haskins JL (1987) Effects of grazing by Chrysomelidae beetles on two wetland herbaceous species. Bull Torr Bot Club 114:13–17
Siemann E (1998) Experimental tests of effects of plant productivity and diversity on grassland arthropod diversity. Ecology 79(6):2057–2070
Siemann E, Rogers W (2003) Herbivory, disease, recruitment limitation and the success of an alien tree species. Ecology 84:1489–1505
Strong DR (1992) Are trophic cascades all wet? Differentiation and donor control in speciose ecosystems. Ecology 73:747–754
Strong DR, Lawton JH, Southwood TRE (1984) Insects on plants: community patterns and mechanisms. Harvard University Press. Cambridge, Massachusetts
Strong DR, Maron J, Connors P, Whipple A, Harrison S, Jefferies R (1995) High mortality, fluctuation in numbers, and heavy subterranean insect herbivory in bush lupine, *Lupinus arboreus*. Oecologia 104:85–92
Tenow O (1972) The outbreaks of *Oporinia autumnata* Bkh. and *Operophthera* spp. (Lep., Geometridae) in the Scandinavian mountain chain and northern Finland 1862–1968. Zool Bidrag Uppsal S 2:1–107

Thompson JN (1994) The co-evolutionary process. University of Chicago Press, Chicago
Tikkanen O-P, Roininen H (2001) Spatial pattern of outbreaks of *Operophtera brumata* in eastern Fennoscandia and their effects on radial growth of trees. For Ecol Manage 146:45–54
Torres JA (1992) Lepidoptera outbreaks in response to successional changes after the passage of Hurricane Hugo in Puerto Rico. J Trop Ecol 8:285–298
Van der Putten WH, Peters BAM (1997) How soil-borne pathogens may affect plant competition. Ecology 78:1785–1795
Van der Putten WH, Van Dijk C, Peters BAM (1993) Plant specific soil-borne diseases contribute to succession in foredune vegetation. Nature 362:53–56
Veblen T, Hadley K, Reid M, Rebertus AJ (1991) The response of subalpine forests to spruce beetle outbreak in Colorado. Ecology 72:213–231
Wallace JB, O'Hop J (1985) Life on a fast pad: waterlily leaf beetle impact on water lilies. Ecology 66:1534–1544
West RJ, Thorogood CA (1985) Mangrove dieback in Hunter River caused by caterpillars. Aust Fish 44:27–28
White RE (1996) Leaf beetles as biological control agents against injurious plants in North America. In: Jolivet PHA, Cox ML (eds) Chrysomelidaeae biology, vol 2. Ecological studies. SPB Academic Publishing, Amsterdam, pp 373–399
Whitmore TC (1975) Tropical rain forests of the Far East. Clarendon Press, Oxford
Whitten AJ, Damanik SJ (1986) Mass defoliation of mangroves in Sumatra, Indonesia. Biotropica 18:176
Wolda H (1978) Fluctuations in abundance of tropical insects. Am Nat 112:1017–1045
Wolda H, Foster R (1978) *Zunacetha annulata* (Lepidoptera; Dioptidae), an outbreak insect in a neotropical forest. Geo-Eco-Trop 2(4):443–454
Wong M, Wright SJ, Hubbell SP, Foster RB (1990) The spatial pattern and reproductive consequences of outbreak defoliation in *Quararibea asterolepis*, a tropical tree. J Ecol 78:579–588
Zvereva EL (2002) Effects of host plant quality on overwintering success of the leaf beetle *Chrysomela lapponica* (Coleoptera: Chrysomelidaeae). Eur J Entomol 99:189–195
Zvereva EL, Kozlov MV, Niemela P, Haukioja E (1997) Delayed induced resistance and increase in leaf fluctuating asymmetry as response of *Salix borealis* to insect herbivory. Oecologia 109:368–373

11 The Ecology Driving Nutrient Fluxes in Forests

B. Stadler, E. Mühlenberg and B. Michalzik

11.1 Summary

Phytophagous insects in the canopies of forest trees play a considerable role in the cycling of nutrients and energy not only in outbreak situations, but also at endemic density levels. However, nutrient fluxes through ecosystems are often studied without a detailed knowledge of the biology of the organisms that affect them. Here, we will address the key features of aphids, adelgids and lepidopterous larvae, which affect ecosystem processes via specific life-history characteristics, fluctuations in population size and trophic relationships with other canopy organisms. For example, aphids and adelgids produce large quantities of sugary excreta and wax wool respectively, which are a source of organic carbon in the canopy. Aphids show erratic population fluctuations, while an introduced pest species such as the hemlock woolly adelgid kills its host within 10–15 years. The winter moth often shows cyclic population fluctuations spanning several years without killing the various host species. These different features in the ecology of canopy insects are expected to influence the availability of energy within the canopies of trees and subsequent processes in nutrient cycling, which eventually affect the forest floor. The availability of energy-rich excreta of canopy herbivores significantly increased the growth of epiphytic micro-organisms, the organic carbon concentrations in throughfall and decreased the nitrogen concentrations beneath trees infested by aphids and lepidoptera. Beneath adelgid-infested hemlock trees, however, significantly higher concentrations of nitrogen were found in the throughfall, which is due to a significant increase in needle N content of infested trees. Therefore, we suggest that the many facets in the biology of the herbivores need to be known to understand the direction of change in flows of nutrient beneath infested trees. Results on vertical nutrient and energy flows are reviewed from different temperate forest ecosystems, and new areas of research linking biotic processes and ecosystem functions are identified.

Ecological Studies, Vol. 173
W.W. Weisser and E. Siemann (Eds.)
Insects and Ecosystem Function

11.2 Introduction

It is increasingly being appreciated that phytophagous insects living in the canopies of forest trees play an important role in the cycling of nutrients and energy, and not only in outbreak situations (Stadler et al. 2001b; Hunter et al. 2003). However, it is not easy to quantify their short-term and long-term effects on ecosystem processes, and insect biomass or leaf area consumed inadequately represent their role and function in ecosystems. As a consequence of these difficulties, nutrient fluxes through ecosystems are often studied without a detailed knowledge of the biology of the organisms that affect them. Such a simplified approach has nevertheless revealed several interesting patterns in energy, nutrient and ion fluxes common to different forest ecosystems (Qualls and Haines 1991; Currie et al. 1996; Michalzik and Matzner 1999; Aber et al. 2002). For example, the forest floor and the canopy are the main net sources of dissolved organic matter in forested ecosystems and nitrogen losses are negatively correlated with the C/N ratio of the forest soil. Finding patterns is at the heart of good science, but understanding the mechanisms that produce them is equally important and valuable, especially if the aim is the sustainable use of natural resources or maintenance of ecosystem services. The mechanistic and flux-oriented perception of the natural environment has led to the well-known dichotomy in the approach to the study of ecological processes (Jones and Lawton 1995; Weisser and Siemann, Chap. 1, this Vol.).

The title of our chapter reflects this dichotomy. Currently, there are few links between the ecology of insects and nutrient fluxes, but even so organisms and their biological characters have a strong imprint on fluxes in every compartment of an ecosystem. For example, a search for papers sharing the terms 'population ecology/community ecology' and 'nutrient fluxes' in the BIOSIS database did not yield a single publication. This does not imply that there are no studies on the effects of insects on ecosystem processes and ecosystem functions (for recent reviews see Schowalter 2000; Hunter 2001), but it does indicate that there is no thorough understanding of the temporal and spatial processes linking insect ecology (life history aspects and population biology) with ecosystem processes. Because of this limited understanding of the role of insects in ecosystem processes, we opted for a rather simple approach. For example, we are not concerned with whether insects *regulate* nutrient fluxes, or their amount of biomass per square meter, but we are interested in their effects via specific life histories, population dynamics and trophic links with other organisms. Also, we do not address the effects of insect diversity on ecosystem functioning because the cascading effects of individual species within their trophic framework are not sufficiently quantified to warrant such an approach. Our specific objective is to identify the important life-history characters and population features of key phytophagous insects in forests that are likely to affect ecosystem processes; in particular, those of sap feeders, such as aphids and adelgids, the latter of which is an example of an exotic pest species introduced into North America, and leaf feeders such as lepidopterous larvae. Exotic pest species are included because they might not only pose threats to ecosystems, but also offer the opportunity to unravel hidden connections between different components/processes. We try to quantify the impacts of these species and offer suggestions on what research is needed for a better understanding of how insects affect ecosystem functioning.

11.3 Life Histories of Canopy Insects

11.3.1 Aphids

The life cycle of most aphids usually consists of several asexual and a sexual generation. Most species live on one or a few species of a particular genus of plant. Only about 10 % are heteroecious, spending their life on a primary host plant during winter and on a secondary host during summer (Dixon 1998). Aphids feed on the phloem sap of their host plants, which is a nutritionally poor diet. They may show an enormous degree of phenotypic plasticity in response to changes in environmental conditions such as food quality (Honek 1991; Stadler 1995; Stadler et al. 2002), amount and quality of excreted honeydew (Fischer and Shingleton 2001; Yao and Akimoto 2001, 2002) or predator pressure (Dixon and Agarwala 1999; Kunert and Weisser 2003). They can be either winged or wingless and have a range of associations with ants from close attendance to unattendance (Bristow 1991; Stadler et al. 2003). Ants may collect their sugary excreta (honeydew), but aphid–ant associations are highly variable in space and time (Cushman 1991; Cushman and Whitham 1991), which affects the amount of honeydew consumed. Honeydew may be excreted in large droplets crystallizing on needle and leaf surfaces, or flicked away resulting in a shiny coating to the leaves. The amount of honeydew excreted per hour might be equivalent to the body mass of an aphid. The absolute amount of honeydew excreted increases with aphid development and declining plant quality (Stadler and Müller 2000) and may reach up to 700 kg fresh mass ha^{-1} $year^{-1}$ on 60- to 80-year-old spruce (Zoebelein 1954).

11.3.2 Scale Insects

In contrast to leaf feeders, but similar to aphids, the damage done to forest trees by native scale insects is often not apparent because of their feeding habit. Immature scales, upon hatching from eggs, are soft-bodied and mobile (crawlers). These crawlers seek suitable feeding sites on a tree, secrete a protective shell, and develop to maturity. The adult females are immobile, whereas the males are small, fly-like, mobile and relatively rare. Scale insects usually produce one to two generations per year. Soft scales generally secrete a thin waxy layer over themselves. They typically move between branches and leaves during their life cycle. They also produce honeydew, but in smaller quantities than aphids although remarkable exceptions are known (Gaze et al. 1983; Beggs 2001).

Some scale insects damage their hosts by removing cell fluids. Leaf and needle stunting and yellowing, needle, shoot and branch dieback as well as plant death can occur depending on population size. In some instances, scales weaken plants, making them susceptible to damage from secondary pests such as borers or environmental extremes, which may ultimately kill the plant. Their population dynamics is often characterized by a gradual increase in infestation rates followed by a sudden crash in numbers (Franz 1956; Karafiat and Franz 1956). Several scale insects, like the balsam woolly adelgid (*Adelges piceae* Ratz.) and the hemlock woolly adelgid (HWA, *Adelges tsugae*, Annand), have become pests when introduced into new environments (Pschorn-Walcher and Zwölfer 1958; Eichhorn 1969). HWA was introduced from Asia into North America in the 1920s and is now a destructive pest of hemlock [*Tsuga canadensis* (L.) and *T. caroliniana* Engelm.; McClure 1991). It attacks trees of all age classes and may cause the trees to die within 10–15 years. There are two phenotypes, with the winged morph migrating to spruce, but they are unable to survive on these summer hosts in eastern North America (McClure 1989). They feed at the base of the needles by inserting their stylets into the parenchyma cells of the xylem rays (McClure et al. 2001).

11.3.3 Lepidopterous Larvae

Lepidoptera, like *Lymantria dispar* (L.), *L. monacha* (L.), *Operophtera brumata* (L.), *Zeiraphera diniana* (Gn.) and *Choristoneura fumiferana* (Clemens), sometimes become pests and have attracted attention because their feeding damage is immediately and easily recognized. There are good long-term population census data on these species (Royama 1984; Baltensweiler and Fischlin 1988; Klemola et al. 2003) and a detailed understanding of the mortality factors operating on the different developmental stages (Den Boer and Reddingius 1996). They usually produce a single generation per year and changes in abundance often follow regular cycles (Royama 1984). Much effort was invested in identifying potential regulatory mechanisms and their response to environmental or climate change (Varley and Gradwell 1970; Ludwig et al. 1978; Williams and Liebhold 1995, 2000; Hunter et al. 1997; Liebhold et al. 2000; Visser and Holleman 2001; Haukioja 2003). In spite of the focus on outbreaks, the effects of leaf feeders are visible even at nominal infestation levels (Seastedt et al. 1983).

11.4 Population Ecological Background of Nutrient Fluxes

The three insect groups mentioned above show three types of population fluctuations/cycles (Fig. 11.1), which could affect the carbon and nitrogen cycling in the canopies of forest trees. Due to their short life cycles and parthenogenetic reproduction, aphids quickly become abundant. While plants are expanding their leaves they usually are of high quality, allowing rapid multiplication. In addition, aphids associated with ants might be closely attended and sheltered from natural enemies. During mid summer, plant quality deteriorates and ants are often less interested in sugar resources (Sudd and Sudd 1985), which might lead to ants abandoning aphids. Subsequently, aphids disperse or are killed by natural enemies, causing a crash in local aphid abundance (Fig. 11.1a). This pattern is repeated almost every year and observed for many species of aphids (Scheurer 1964; Kidd 1990; Kindlmann and Dixon 1996; Stadler et al. 2001b). Initially, in a year, there are indications that some trees in forests are heavily infested (hot spots) while others are uninfested. After dispersal, virtually all trees become infested. For example, in the Waldstein area in northern Bavaria, Germany, the infestation of Norway spruce by *Cinara pilicornis* (Hartig) shows the following pattern: in spring: 53.2 % of all trees are uninfested, 42.6 % are infested with up to 100 aphids m^{-2} and 4.2 % are heavily infested (hot spots, >100 aphids m^{-2}). After dispersal, almost all of the trees (97.5 %) became infested with aphids, however, only in low numbers. Thus, the infestation pattern is characterized by a clumped distribution in spring and an even distribution during summer. It is reasonable to assume that this change in infestation pattern is reflected in ecosystem processes (see below).

Pest species such as the hemlock woolly adelgid, which was introduced into North America, show a different infestation pattern. Because winged migrants are unable to find suitable secondary hosts in eastern North America (McClure 1983), dispersal is mainly by crawlers, moving within a tree or forest stand. Other means of dispersal are wind, birds and humans. Currently, the spread is about 15–30 km $year^{-1}$ (http://www.fs.fed.us/na/morgantown/fhp/hwa/hwa-site.html). HWA multiplies rapidly on newly infested trees, and due to its destructive feeding on all host age classes, hemlock becomes extinct in infested stands after 10–15 years (Fig. 11.1b). Thus, a climax species is eliminated and replaced by hardwood trees such as black birch and maple. Again, it is reasonable to assume that the decline in the vitality of the trees and changing trophic relationships between canopy organisms should affect ecosystem processes and become increasingly more pronounced as the infestation progresses.

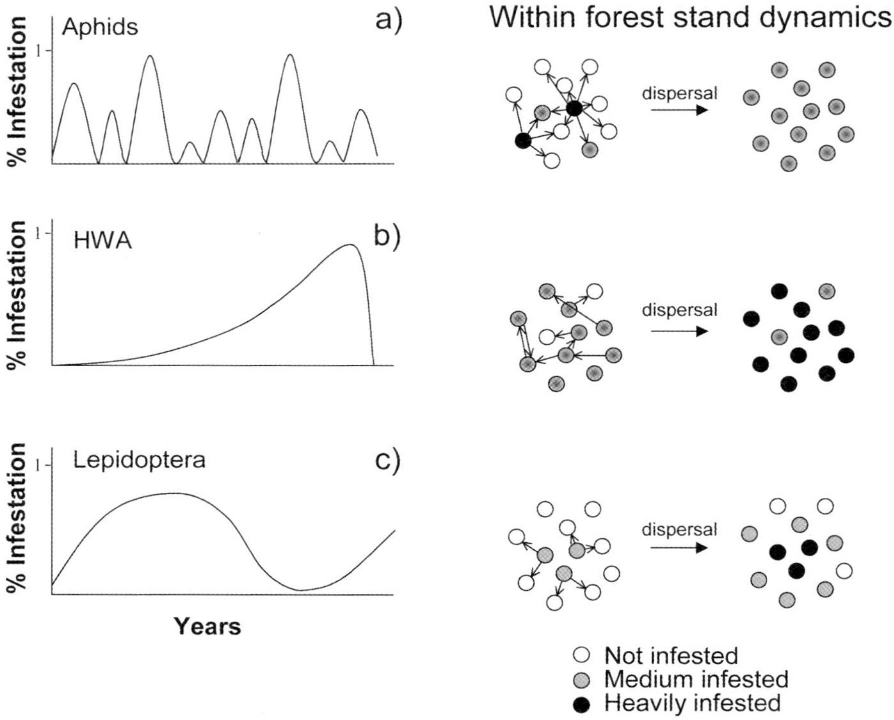

Fig. 11.1. Simplified conceptual model of changes in infestation patterns for **a** aphids on Norway spruce, **b** hemlock woolly adelgid (*HWA*) on hemlock and **c** lepidopterous larvae on beech and oak. Aphids usually show rather erratic changes in densities in successive years, while HWA increases in numbers and kills the host tree within 10–15 years. All age classes are attacked and eventually succumb. Lepidopterous larvae like *Operophtera brumata* or *Lymantria dispar* show a more cyclic infestation pattern spanning several years but do not necessarily reach outbreak densities. Within a forest stand aphids initially are very patchily distributed with some trees heavily infested, but after dispersal in June typically all trees become infested. In contrast, HWA gradually builds up in numbers within a stand until all hemlock trees are infested. Due to the relatively poor migratory ability of female lepidoptera, infested forest stands are patchily distributed and the infestation radiates out from the centres of infestation

Lepidopterous larvae, even those that do not achieve outbreak levels, often show a gradual increase and subsequent decline in numbers spanning several years (Fig. 11.1 c). The gradual increase and decline in abundance are consequences of their annual life cycle, and most likely a combination of different environmental conditions such as precipitation, natural enemies and pathogens, induced defence in the host plants, long-range dispersal or recruitment rates. The debate about which mechanism, or combination of mechanisms, is important in regulating local abundance is far from over (Royama 1996; Selås

1997; Zhang and Alfaro 2003). Females are often less likely to disperse, thus creating epicentres of infestations in favourable environments. Radiating out from these spots, neighbouring trees and stands suffer increasing herbivore pressure. Again, the assumption is that infested stands should differ from uninfested stands if leaf feeders affect the nutrient cycling in the canopies of trees.

In summary, the expectation is that aphids increase variability in ecosystem processes and the pattern is repeated each year. Especially early in a season, differences in energy and nutrient flows beneath infested and uninfested trees should be apparent on a local scale. Hemlock infested with HWA should show a progressive decline in vitality, which might have a pronounced effect on ecosystem processes at the stand level. Leaf feeders directly damage foliage and produce more solid faeces, which are less soluble in rain water. Their effects should be visible in successive years and uninfested and infested trees are likely to be close to one another. However, within stands effects on nutrient flow should largely depend on the feeding activity of the larvae.

We do not know of any long-term investigation in which the between-year and local to regional variability in herbivore-mediated changes in nutrient fluxes were measured. Therefore, we shall present the results of short-term investigations on these three types of phytophagous insects and try to deduce potential long-term trends. The experiments we report span the period from 1996–2002.

11.4.1 Sites and Experimental Setup

At our main experimental site in the Waldstein area (50° 09′N, 11° 52′E) throughfall was collected in 1996 by placing standardized throughfall samplers (5-l polyethylene bottles, 20 cm in diameter) beneath five 10-year-old Norway spruce [*Picea abies* (L.) Karst.] trees infested with *Cinara* spp. The same number of samplers were installed beneath uninfested control trees. Because it is difficult to predict which trees are going to be heavily infested in the field, we simulated heavy infestation during a cage experiment in 1998, in which 10-year-old spruce trees were artificially infested with *Cinara* spp. The cage prevented natural enemies from preying on the aphids. When aphid numbers peaked in the field, the cage was removed, allowing the winged aphids to disperse and natural enemies to attack the aphids. Three uninfested control trees were treated in the same way.

In 2002 the same method was used to collect throughfall beneath hemlock trees with heavy and medium infestations of HWA and beneath uninfested control trees. Sites were located on a south–north gradient from southern Connecticut (heavy and medium infestations) to central Massachusetts, Harvard Forest (uninfested control stands). In order to capture the spatial variability in throughfall concentrations beneath individual trees, three samplers were placed beneath a tree, one close to the periphery, one close to the trunk and one in between. Because these sites received different amounts of precipitation it is not sensible to compare fluxes directly; rather we present the concentrations of different energy and nutrient compounds collected beneath infested and uninfested trees.

The third experimental site was a 130-year-old mixed beech (*Fagus sylvatica* L.) oak [*Quercus petraea* (Matt.) Liebl.] stand situated in the Steigerwald area, northern Bavaria (49° 52′N, 10° 27′E). In some stands the beech and oak trees are attacked mainly by the winter moth *Operophtera brumata* L. which consumed between 25 and 40 % of the leaf area during the main feeding period from bud burst to mid July. In 1998, throughfall was collected at the periphery of the canopies of five trees where the number of damaged leaves and leaf area loss were monitored regularly (Stadler et al. 2001b). Only some 300 m away from this infested stand we collected throughfall from beneath uninfested trees. During a field experiment, it is not possible to exclude the activities of other leaf feeders. Therefore, our 'control trees' also experienced a small loss of leaf area, ranging between 1 and 5 %. However, we expected the effects of the winter moth to be large enough to be identified in ecosystem processes.

During the summer months throughfall was collected at biweekly intervals at all sites and immediately filtered through a 0.45-μm cellulose acetate membrane. Dissolved organic carbon was determined as CO_2 by infrared detection, after persulfate UV oxidation using a DOC analyzer (Liqui TOC, Foss Heraeus, Hanau, Germany). Ammonium nitrogen (NH_4-N) and nitrate-N (NO_3-N) were measured by ion chromatography (Dionex, Idstein). Dissolved organic nitrogen (DON) was calculated by subtracting NH_4-N plus NO_3-N from total dissolved nitrogen (N_{total}). Total dissolved nitrogen was measured as NO_x after thermo-oxidation at 700 °C (Abimed: TN-05).

11.4.2 Results

The DOC concentration in throughfall collected beneath heavily infested trees increased with increasing aphid numbers and declined after the cage was removed and the aphids dispersed in early July (Fig. 11.2a). Beneath uninfested trees DOC concentrations were always lower than 10 mg l^{-1}. The same trends were observed for DON. The large peak in DON beneath uninfested trees does not appear to be an artefact of caging because it was also observed in the field trial. Initially, there was a large difference in the inorganic nitrogen in throughfall concentrations collected beneath infested and uninfested trees, but with the decline in aphid numbers concentrations beneath infested trees returned to the levels in the throughfall collected beneath uninfested trees.

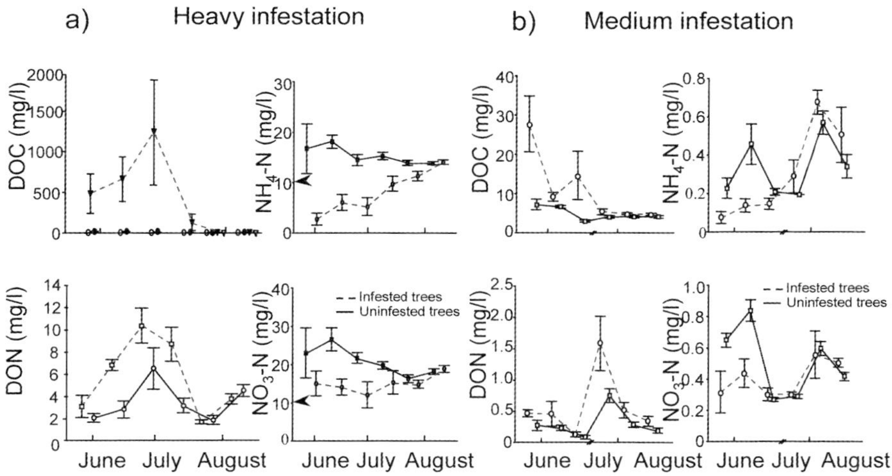

Fig. 11.2a, b. Concentrations of dissolved organic carbon (*DOC*), dissolved organic nitrogen (*DON*), ammonium-N (*NH_4-N*) and nitrate-N (*NO_3-N*) in throughfall collected beneath Norway spruce infested with aphids (*Cinara* species). **a** Cage experiment simulating heavy infestation; **b** field experiment at the Waldstein site (mean ± SE). *Arrowheads* indicate concentrations in the artificial rain water

Medium-infested trees in the field basically showed the same trends, with higher DOC concentrations in throughfall beneath infested trees (Fig. 11.2b). The correlation between aphid numbers above throughfall samplers and the DOC concentration in throughfall is highly significant (r=0.72, n=30, P <0.001). With declining infestation DOC concentrations declined while DON concentrations showed a marked peak beneath infested trees in mid July. Inorganic nitrogen species were lower in throughfall as long as aphids were present but similar to that collected beneath uninfested trees after aphid dispersal. Thus, depending on population size the effects of aphids on carbon and nitrogen cycling might be reflected in throughfall compounds during summer (e.g. hot spots) or decline quickly because honeydew is readily consumed by ants or other canopy organisms (see below).

Throughfall concentrations beneath hemlock are also strongly dependent on the abundance of HWA (Fig. 11.3). Compared to aphids, the differences in DOC concentrations in throughfall were less pronounced, with the overall concentrations beneath trees with medium and heavy infestations 1.6 times larger than beneath uninfested trees. It is likely that other processes blur a simple cause–effect relationship. First, uninfested control trees had much more foliage than infested trees, which show a progressive decline in vitality when needles are shed (Orwig and Foster 1998). Thus, DOC leaching from the foliage of uninfested trees is likely to be higher. Second, wax wool remains in

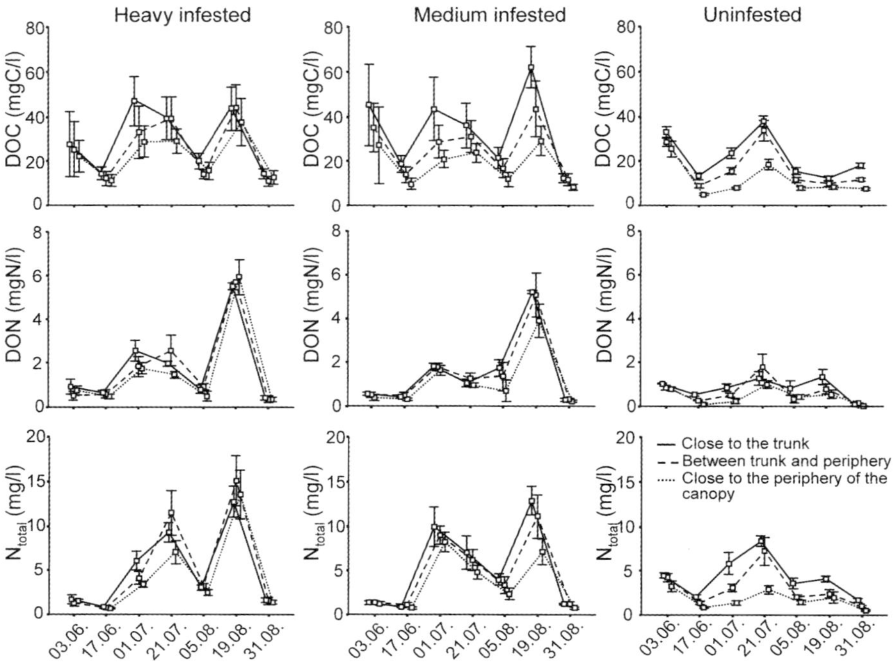

Fig. 11.3. Concentrations of dissolved organic carbon (*DOC*), dissolved organic nitrogen (*DON*) and total nitrogen (N_{total}) in throughfall collected beneath hemlock (*Tsuga canadensis*) at three sites in eastern North America. *Left* Heavily infested site in southern Connecticut; *middle* medium-infested site in northern Connecticut; *right* control site in central Massachusetts. Concentrations were separated according to position of the throughfall samplers (mean ± SE)

the canopy for a longer time because it is not readily soluble in precipitation like honeydew. These opposing effects only lead to a moderate increase in DOC concentrations beneath infested trees.

Concentrations of DON and total nitrogen were significantly higher in throughfall collected beneath HWA-infested trees compared to uninfested trees (Fig. 11.3), with average throughfall concentrations (in mg l^{-1}) for the control of DON 0.73 and N_{total} 3.00, for medium-infested trees of DON 1.26 and N_{total} 4.49 and for heavily infested trees of DON 1.51 and N_{total} 4.98. The total nitrogen was higher than observed in the aphid study. The explanation is that unlike aphids, HWA significantly increases the nitrogen content of newly developing and 1-year-old needles by an average of 29.0 % during summer (Stadler et al. 2004). This implies that the small amount of foliage on infested trees affects throughfall more strongly than the large mass of foliage of uninfested trees.

The spatial variability in throughfall volumes, between trunk and periphery of the canopy, was less beneath heavily infested than medium- and uninfested trees. This is a direct consequence of progressive needle loss due to the HWA infestation. In spite of their low needle biomass heavily infested hemlock shed significantly more needles than medium- and uninfested trees, which indicates the accelerating loss in vitality with duration of infestation (Stadler et al. 2004).

Lepidopterous larvae feeding on beech and oak affected throughfall by increasing DOC concentrations and decreasing inorganic nitrogen concentrations (Fig. 11.4). High DOC concentrations at the end of May were attributed to pine pollen, which cannot clearly be separated from the effects of herbivores. Beneath infested trees the decline in concentrations of most throughfall compounds were pronounced up until the end of July when feeding by caterpillars declined. Thus, the effect of leaf feeders was similar to that observed for aphids. The effect on DON concentrations in throughfall was less clear and higher beneath infested trees only in June.

In conclusion, differences in the energy content and the nutrient compounds in throughfall collected beneath infested and uninfested trees were pronounced and reflected the abundance and specific life history features of the canopy herbivores. Aphids produce honeydew, which is readily soluble and results in an immediate flush of energy when it rains, while the effects of lepidopterous larvae and HWA were detectable over a longer period of time.

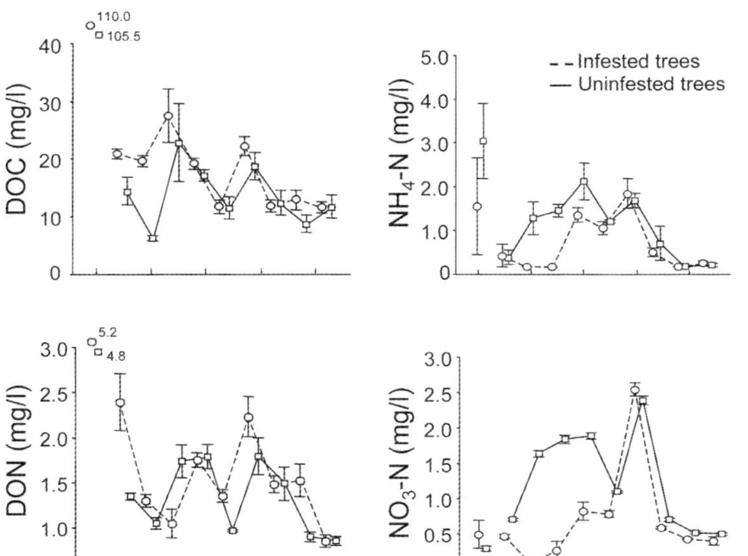

Fig. 11.4. Concentrations of dissolved organic carbon (*DOC*), dissolved organic nitrogen (*DON*), ammonium-N (*NH_4-N*) and nitrate-N (*NO_3-N*) in throughfall collected beneath mature beech/oak stands at the Steigerwald site in 1998 (mean ± SE)

These effects were even apparent in throughfall collected at the periphery of the canopy, where there are few needles and herbivores.

11.5 Trophic Effects and Organic Pathways

Dissolved organic nitrogen is an important constituent of the total nitrogen in throughfall in many forested ecosystems, accounting for 15–30 % in temperate forests (Matzner 1988; Currie et al. 1996; Michalzik and Matzner 1999) and up to 60 % in boreal forests (Piirainen et al. 1998). Within the forest floor DON as a proportion of the total N ranges from 20–60 % at temperate sites and is around 80 % at boreal sites. Therefore, it is reasonable to try to identify potential mechanisms that are likely to influence the concentrations of organic compounds in throughfall. An obvious starting point is the trophic interactions of canopy herbivores with epiphytic micro-organisms, especially of species that are abundant, fast growing and potential consumers/sources of organic compounds. We studied the links between aphids, HWA and lepidopterous larvae and the epiphytic bacteria on needle and leaf surfaces. At each of the three sites described above three shoots comprising three needle age classes were cut from the periphery of the lower canopy in July, prior to which it had not rained for at least 1 week.

For the microbial analyses 5 g of beech and oak leaves and spruce needles and 3 g of hemlock needles were cut off with sterile scissors and blended with 145 ml of distilled water for 2 min. To enumerate the numbers of aerobic heterotrophic bacteria in each sample these washings were logarithmically diluted in 1/4 strength Ringer's solution and spread-plated onto 1/10 strength tryptic soy agar (Merck; pH 7.2), supplemented with 0.4 g l^{-1} cycloheximide (Merck). All plates were incubated at 25 °C for 5 days. Results were expressed in terms of colony forming units (CFU) per gram of fresh matter. In addition, to follow the effect of aphid infestation on the composition of the bacterial community, direct cell counts and fluorescence in situ hybridization (FISH) of bacteria from needle washings were carried out for shoots unaffected by aphids and those contaminated with visible amounts of honeydew. At the Waldstein site two shoots per sample, with 1- to 2-year-old needles, were cut from the periphery of four infested and four uninfested spruce trees using sterile scissors and blended 1:10 (fresh mass:solution) with 0.9 % NaCl solution. After shaking (8 h at 5 °C) and sonication (2×1 min) cells were enriched on polycarbonate membranes, fixed, washed, hybridized with specific rRNA binding probes and stained with a DNA binding dye (DAPI: 4′,6-diamidino-2-phenylindoldihydrochlorid).

The number of CFU of epiphytic bacteria was always significantly higher on infested trees irrespective of the type of herbivore or tree species (Fig. 11.5). The differences were often of two to three orders of magnitude.

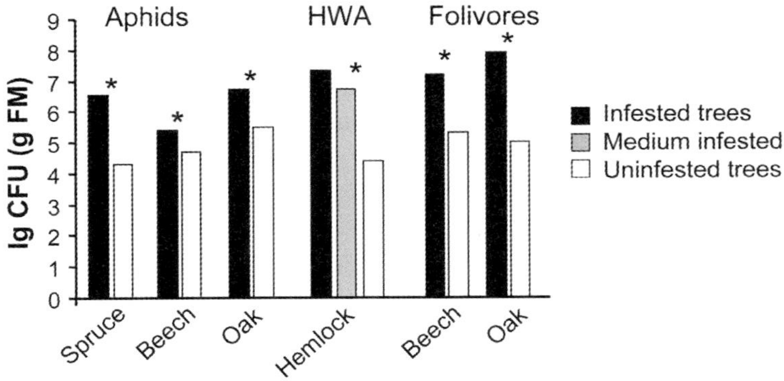

Fig. 11.5. Log number of colony forming units (*lgCFU*) of heterotrophic, epiphytic bacteria per gram of needle fresh mass (*FM*), collected from different tree species experiencing different infestation regimes. Samples were collected in July and only from the periphery of the canopy. *Columns separated with an asterisk* indicate significant differences (Mann and Whitney test, $P < 0.05$) in numbers of bacteria between infested and uninfested trees

The differences between medium and heavily infested hemlock were not significant. Thus, different functional groups of herbivores (sap feeders, leaf feeders, cell feeders), with different modes of excreting waste products and damaging leaves/needles, affect the epiphytic microbes similarly. This result is very robust and corroborated by all our studies carried out over several seasons and involving several groups of micro-organisms, such as yeasts and fungi (Stadler and Müller 1996). In addition, the foliovores on beech and oak affected not only the abundance of epiphytic bacteria but also their diversity (Müller et al. 2004). On spruce, FISH of bacteria from needle washings revealed fewer gamma-proteobacteria on infested needles, and a slightly higher proportion of active cells relative to all cells (Table 11.1). However, the differences were not statistically significant. Only active cells with their high rRNA content are detected by an rRNA binding probe. Honeydew seemed to enhance the growth of groups of bacteria that were not determined, since the ratio of alpha- and beta-proteobacteria based on total cell numbers remained unchanged. It is likely that changes in the composition of the bacterial community or changes in activity patterns will affect nutrient cycling in the canopy and in the forest floor, especially the organic fraction in throughfall (Figs. 11.2–11.4).

Table 11.1. Bacteria in the washings from needles collected from aphid infested and uninfested spruce branches at the Waldstein site in June 2003. Shown are means ± SD of the target cells expressed as percentages of all cells resulting from samples of four different trees. *P* values give the level of significance for differences in infestation (Mann-Whitney *U*-test). Target groups were identified using fluorescence in situ hybridization (FISH) in which the probes bind specifically to the rRNA of particular cells. Total bacterial cells were counted by staining with DAPI, which binds to the DNA of all cells

Target group	Probe	Percentage of all cells		
		Infested	Uninfested	*P*
Alpha-proteobacteria (except Rickettsiales)	ALF968-CY3	11.21±2.97	10.57±3.25	0.773
Beta-proteobacteria	BET42a-CY3/GAM42a	1.30±0.38	1.38±0.28	0.773
Gamma-proteobacteria	GAM42a-CY3/BET42a	1.54±0.33	2.08±0.16	0.083
All bacteria	EUB338-CY3/EUB338-II-CY3/EUB338-III-CY3	56.66±6.89	46.66±6.18	0.083

11.6 Herbivore-Mediated Changes in Quality and Quantity of Nutrient Fluxes

In all systems studied the herbivores in the canopy positively affect the organic portion in the throughfall. In addition, several studies indicate that canopy herbivores affect the inorganic compounds, ion fluxes and needle litter eventually reaching the forest floor (Table 11.2). Here, basic qualitative effects on soil processes, especially on inorganic N dynamics, become evident and appear to be quite diverse. While some authors observe increased N mobilization, attributable to enhanced litter and frass input, at both the plot and catchment scale (Swank et al. 1981; Hollinger 1986; Pedersen and Bille-Hansen 1995; Christensen et al. 2002), others report N immobilization, especially in response to elevated inputs of labile C compounds (Grier and Vogt 1990; Lovett and Ruesink 1995; Michalzik and Stadler 2000), which are mainly provided by sap feeders. The fact that there is no consistent trend in soil nutrient availability in herbivore-infested stands might be due to specific site properties such as nutrient status (e.g. N saturation), litter quality, herbivore species and microbial soil community. Nevertheless, the effects on the nutrient dynamics in the soil mediated by canopy herbivores are pronounced, and the majority of ecosystem studies focus on element and nutrient fluxes of the dissolved fraction, which is defined as those solutes that pass through filters with pores sizes $<0.45\ \mu m$.

It is reasonable to assume that leaf feeders contribute significantly to fluxes, of which the size classes are $>0.45\ \mu m$, due to their frass and green lit-

Table 11.2. Effects of phytophagous insects living in the canopy of trees in temperate forests on soil processes – results from field and laboratory experiments

Herbivory + host plants	Study site and method	Nutrient soil input	Effects on soil processes	Source
(a) Leaf feeders				
Fall cankerworm (*Alsophila pometaria*) in a mixed hardwood forest	Field observation in the southern Appalachians (Coweeta Hydrologic Laboratory)	33 % of the total leaf mass consumed; throughfall concentrations not measured	Increase in stream export of nitrate nitrogen (NO_3-N); no effects on ammonium nitrogen, phosphate and major cations	Swank et al. (1981)
Chewing insects not identified, on 4-year-old trees	Watershed study in the southern Appalachians (Coweeta Hydrologic Laboratory); insecticide treatment	Decreased P and increased K input beneath untreated trees		Seastedt et al. (1983)
California oak moth (*Phrygamidia californica*) on oak	Field observation, evergreen oak (*Quercus agrifolia*), deciduous oak (*Quercus lobata*), Palo Alto, California, USA	Increased P and N input with litterfall and frass	Altered litter quantity and quality (leachability), increased pulses of N and P from litter and frass into the soil	Hollinger (1986)
Gypsy moth (*Lymantia dispar*) on deciduous trees	Incubation experiment with caterpillar frass from oak forests, Millbrook, New York, USA	Enhanced input of labile C and extractable N	Increased microbial growth, N immobilization and C mineralization (CO_2 production)	Lovett and Ruesink (1995)

Table 11.2. (*Continued*)

Herbivory + host plants	Study site and method	Nutrient soil input	Effects on soil processes	Source
Frass from several moth species	Elevational gradient at Coweeta Hydrologic Laboratory	Addition of frass	Increased soil respiration	Reynolds and Hunter (2001)
Gypsy moth (*Lymantia dispar*) on deciduous trees	Field manipulation experiment using N^{15}-labelled caterpillar frass input on plots with red oak (*Q. rubra*) seedlings, Millbrook, New York, USA	Frass input; 50 % defoliation	Enhanced N mobilization by microbes and direct dissolution, reduced availability to plants	Christensen et al. (2002)
(b) Sap feeders				
Aphid (*Pterocallis alni*)	Field observation on a year-old red alder (*Alnus rubra*), Washington, Washington, USA	Enhanced input of honeydew (melezitose)	Reduced availability of soil N, N mineralization, aboveground NPP and N uptake by trees	Grier and Vogt (1990)
Spruce aphid (*Elatobium abietinum*) on spruce	Field observation on three 30-year-old sitka spruce (*Picea sitchensis*) sites, Denmark	Enhanced N input with litterfall due to aphid damage	Enhanced NO_3 leaching with soil solution	Pedersen and Bille-Hansen (1995)

Hemlock woolly adelgid (*Adelges tsugae*) on hemlock	Study sites in central and southern New England, litter bags, leaf fall measurements	Increased needle litter input	Increased N availability and nitrification rates	Jenkins et al. (1999),
Aphid (*Cinara pilicornis*)	Column irrigation experiment with soil from a spruce forest in Germany (high temporal resolution)	Addition of honeydew (simulation of different degrees of infestation)	Increased respiration rates; decreased inorganic N concentrations in leachates	Michalzik and Stadler (2000)
Spruce aphid (*Elatobium abietinum*) on spruce	Controlled field irrigation experiment in Scotland; application of pollutants	Increased needle litter input; increased DOC, reduced nitrogen input		Stadler et al. (2001a)

ter production; however, data on particulate organic matter (POM) (0.45 μm <POM <2 mm) fluxes are rare. The only study we are aware of is that of Carlisle et al. (1966) on oak, who found for the size class up to 200 μm (0.2 mm) an annual input of particulate organic carbon (POC) in throughfall of 227 kg C ha^{-1}, including a carbohydrate input of 89 kg ha^{-1}. Compared to DOC throughfall inputs of up to 130 kg C ha^{-1} $year^{-1}$ in hardwood forests (Qualls and Haines 1991), the additional amount of organic carbon transported via the particulate pathway is about 100 kg ha^{-1} $year^{-1}$. Nevertheless, organic matter fluxes covering the size class >0.45 μm and <2 mm (fine particulate matter) are not measured in routine sampling and used to budget matter and energy fluxes in forest ecosystems, nor is their origin studied.

The extent to which throughfall fluxes of POM (hexose-C) are affected by different groups of phytophagous insects (leaf feeders/sap feeders) is evident in the laboratory, where leachates from needle and leaf washings can be measured. Leaf feeders (winter moth larvae from the Steigerwald site) on beech produced frass larger than 0.45 μm; therefore, this particle size class is most obvious during the period of herbivore feeding. In the filtered leachates (leaf washings) from infestation and control treatments there was no difference in the amount of hexose-C, whereas it was significantly higher in unfiltered solutions for beech, in both the control (38 % increase, P=0.002, t-test) and infested treatments (65 % increase, P <0.001, t-test). Aphids, in contrast, excrete honeydew, which is readily soluble in water and therefore affects mainly the dissolved fraction. As expected, infestation with aphids resulted in significantly higher amounts of hexose-C (a factor of about 180) in filtered and particulate form in the leachates from infested compared to control spruce shoots (Michalzik and Stadler 2004).

11.7 Synthesis and Conclusions

Why should we want to understand fluxes from a life history and population biology perspective? The answer, we believe, should address four aspects.

11.7.1 Understanding the Temporal Dynamics of Energy and Nutrient Fluxes

There is now a good understanding of the annual concentrations and fluxes of nutrients and matter in the different compartments of forested ecosystem (Qualls and Haines 1991; Likens and Bormann 1995; Currie et al. 1996). For example, fluxes of DOC and DON in the forest floor are correlated with precipitation and with DOC and DON fluxes with throughfall (Michalzik et al. 2001), indicating a significant role of the forest canopy in soil processes

(Lovett et al. 1989; Prescott 2002). Organic material is a large component of the C and N cycle in forest ecosystems; however, its origin is often unclear. Annual averages of the concentrations and fluxes at local sites only offer a coarse-grained picture of the environment and little understanding of the mechanisms that produce these patterns. In most forested ecosystems there is a marked temporal variability in fluxes of energy and nutrients in throughfall, soil solutions or stream water export (McDowell and Likens 1988; Michalzik et al. 2001). For example, Guggenberger and Zech (1994) recorded peak DOC and carbohydrate C concentrations in throughfall solution around June in 2 years in two spruce stands. The interpretation of the results vary but usually favour abiotic mechanisms, like the amount of precipitation (Lovett et al. 1996) or plant-related features, such as leaching (Tukey 1970) and stand structure (Prescott 2002). This is understandable because trees dominate forests in terms of visible biomass. However, there is growing evidence that insects have a considerable effect on the nutrient dynamics of forests (Seastedt et al. 1983; Stadler et al. 2001b). The strength and duration of these effects vary according to the ecology of the insect, and all are seasonal (Figs. 11.2–11.4). DOC in throughfall beneath infested trees usually peaks with the maximum in aphid abundance or with the peak in the feeding activity of lepidopterous larvae. HWA, as a non-native species, must be considered separately, but the results corroborate the point that the ecology of canopy herbivores is identifiable in throughfall fluxes. For example, even though the medium-infested trees had a lower infestation density (number of egg sacks/shoot) the DOC concentrations were as high as beneath heavily infested trees. This is because recently infested trees still have a large foliage biomass and therefore the total HWA biomass on medium-infested trees exceeds that on heavily infested trees, which have lost more than 50 % of their foliage. In addition, HWA changes the chemistry of the needles, which possibly accounts for the more nitrogen in throughfall collected beneath infested hemlock. Producers of large quantities of faeces, like aphids and lepidopterous larvae, significantly decreased the inorganic nitrogen concentration in throughfall, possibly because their excreta fuel the rapid growth of epiphytic micro-organisms, which in turn immobilize the N compounds in the leachates. The effect of HWA on bacteria is also pronounced, resulting in a two to three orders of magnitude increase in their abundance (Fig. 11.5). In conclusion, these trophic interactions have a strong imprint on the temporal dynamics of nutrient cycling in forest canopies.

11.7.2 Understanding the Spatial Variability in Fluxes

The three insect taxa described in this chapter have very different population dynamics, with aphids able to travel large distances via their winged morphs. HWAs and lepidopterous larvae, in contrast, are likely to migrate much shorter distances, but are more persistent in infested patches (Fig. 11.1).

Therefore, the expectation is that the effect of aphids on within-stand variability in DOC/DON throughfall fluxes should be larger within and between years compared to that of HWA and lepidopterous larvae. At the end of July the aphid effect was largely invisible because the aphids are then sparsely and evenly distributed between trees. The honeydew they excrete is consumed entirely by other canopy organisms, and little reaches the forest floor at that time. During spring, however, heavily infested trees ('hot spots' of infestation) may be situated next to uninfested trees, leading to local differences in throughfall DOC concentrations of several orders of magnitude (Fig. 11.2a). We are not aware of any study of the local and regional population dynamics of canopy insects in non-outbreak years that correlate their life history strategies and throughfall fluxes. The biggest problem is gaining access to the canopy of mature trees to identify the sources of variability in throughfall fluxes. Our preliminary analysis of aphid abundance on mature Norway spruce strongly indicates that our understanding of variability in the spatial/temporal DOC dynamic in throughfall is considerably improved if the local abundance of aphids is considered (Kindlmann and Stadler 2004).

A good example of the spatial variability of throughfall concentrations within infested and uninfested trees is presented for HWA (Fig. 11.3). Heavily infested hemlock showed little variability in throughfall concentrations, irrespective of whether collected close to the trunk or at the periphery of the canopy. This is owing to the fact that HWA is evenly distributed in the canopy and the patchy nature of needle loss. Precipitation is relatively unaffected by the needles and passes through the canopy unimpeded. Medium-infested trees, in contrast, which still retain most of their needles, showed a much larger spatial variability in the energy and nutrient fluxes recorded at different locations beneath a tree, especially when the amount of precipitation is high. For example, close to the trunk 41.2 % of the total DOC flux passed through the canopy, while at the periphery only 26.7 % did so. Beneath heavily infested trees the differences in DOC fluxes between trunk and periphery declined to only 4.8 %. Identical results were obtained for the spatial variability in nitrogen fluxes. Therefore, in the case of HWA the result appears to be a progressive decrease in the spatial variability in nutrient fluxes within the canopy with duration of infestation. The demise of hemlock and the increased input of nitrogenous compounds into the forest floor eventually lead to a gradual development of extensive and relatively homogeneous stands of an uncommon forest type that is dominated by black birch, a nitrophilous species.

11.7.3 Understanding the Mechanisms That Regulate Fluxes

Ecosystem ecology is dominated by studies that seek to identify patterns and correlations in energy and nutrient cycling. This mass perspective approach assumes that the functioning of a system is controlled by those organisms with the largest biomass and by the abiotic environment. It emphasizes physical and chemical constraints and regulation and predictability at the system level. We appreciate that this approach has produced some interesting results, e.g. that water fluxes are one of the most important driving forces in DOC and DON mobilization, accumulation and transport between organic matter pools. However, dilution occurs when small pools are depleted, resulting in non-linear and therefore non-predictable matter concentrations and fluxes (Currie et al. 1996; Michalzik and Matzner 1999; Schulze 2000). Population and community ecology offers an historical perspective and treats variability as a necessary precondition for the development of biological systems. Such studies are more likely to identify mechanisms that lead to certain patterns on the small to medium scale. Therefore, we believe these approaches are complementary in their contribution to our understanding of the behaviour of ecological systems and support calls to merge these different perspectives of our natural environment. However, the identification of mechanisms that produce patterns at the ecosystem level is much more difficult, because they operate at a multitude of different temporal and spatial scales.

Our results on canopy herbivores indicate that it is important to appreciate the role of small herbivores, because, e.g., aphids can excrete daily an amount of honeydew equivalent to their own body mass and low numbers of adelgids can change the flow of water and nutrients in the canopy by affecting needle loss and needle nitrogen content. In addition, in order to identify ecosystem changes well before irrevocable damage is done, it is necessary to develop indicators that are more sensitive than annual averages in fluxes or throughfall concentrations. Potassium is one of the elements that repeatedly is found to indicate early stages of herbivore activity, well before macroscopic damage is visible (Seastedt and Crossley 1984; Stadler et al. 2001a). For successful management strategies early indicators of increases in herbivore abundance and a better understanding of the individual ecology of herbivores need to be developed.

11.7.4 Generating Testable Hypotheses

There is a long history in population/community ecology and life history theory of hypotheses development and testing. The relatively young discipline of ecosystem science often seems to be less concerned with the development and testing of theories or experimental manipulations because of the difficulty of manipulating whole ecosystems. A few notable exceptions, however, do exist. Thus, ecosystem ecology is largely based on observations, correlative and modelling approaches. An ecological approach to the study of energy and nutrient fluxes offers the additional advantage of combining patterns identified on a large scale with mechanisms operating at and below the population/community level. Both approaches are intertwined because the abiotic features of an ecosystem set the limits for the evolution of species traits, and ecosystem properties are constrained by the ecological and evolutionary history of the interacting species. For example, the adoption of a metapopulation approach to the study of throughfall fluxes might offer a theoretical framework for predicting the seasonal, temporal and spatial variability in throughfall concentrations and fluxes. The patch colonization/extinction perspective (Wiens et al. 1993; Wiens 1995; Wu and Loucks 1995) essentially captures many biological features that are relevant to ecosystem studies. For example, different species of herbivore in the canopies of trees show different abilities to migrate to adjacent patches, or periods of infestation in local patches (Fig. 11.1). The identification of the key players and their direct and indirect trophic effects within patches of infested stands should help to explain the variability in nutrient cycling.

An evolutionary perspective of ecosystem ecology (Loehle and Pechmann 1988; Loreau 2001) will enable the prediction and provide a mechanistic understanding of the effects of climate change on ecosystem processes and function. This does not mean that there is consensus on the direction of change, e.g. in insect abundance (Lawton 2000), phenological synchronicity between insects and host plants (Koricheva et al. 1988; Watt and McFarlane 2002; Dixon 2003) or frequency of pest species outbreaks (Buse et al. 1998; Cannon 1998; Bezemer et al. 1999), but a blend of approaches makes it more likely that the chain of mechanisms that link the different types and scales of ecological processes involved in ecosystem dynamics will be identified and understood. Opportunistic species are likely to benefit most from environmental change irrespective of whether this change comes in the form of increased temperature, introduction into new environments or larger fluctuations in weather conditions. A promising road to a better understanding of the patterns of change is to link ecosystem processes with food web interactions and evolutionary aspects of phenotypic adaptation to environmental change. A combination of approaches appears to be better suited to answer questions on how to best recognize changes in ecosystem function or directions of change in the ecology of nutrient fluxes.

Although we realize that the results and ideas presented in this chapter are far from comprehensive, it is likely to be the first paper found in a literature search when entering the terms 'population ecology/community ecology' and 'nutrient fluxes'.

Acknowledgements. We are indebted to Tony Dixon for a critical reading of an earlier draft of the manuscript. Financial support was given by the German Ministry for Research and Technology (Fördernummer: BMBF no. PT BEO 51-0339476D).

References

Aber JD, Ollinger SV, Driscoll CT, Likens GE, Holmes RT, Freuder RJ, Goodale CL (2002) Inorganic nitrogen losses from a forested ecosystem in response to physical, chemical, biotic, and climatic perturbations. Ecosystems 5:648–658

Baltensweiler W, Fischlin A (1988) The larch budmoth in the Alps. Dynamics of forest insect populations. Plenum Press, New York, pp 331–351

Beggs JR (2001) The ecological consequences of social wasps (*Vespula* spp.) invading an ecosystem that has an abundant carbohydrate resource. Biol Conserv 99:17–28

Bezemer TM, Knight KJ, Newington JE, Jones TH (1999) How general are aphid responses to elevated atmospheric CO_2? Ann Entomol Soc Am 92:724–730

Bristow CM (1991) Why are so few aphids ant-tended? In: Huxley CR, Cutler DF (eds) Ant–plant interactions. Oxford University Press, Oxford, pp 104–119

Buse A, Good JEG, Dury S, Perrins CM (1998) Effects of elevated temperature and carbon dioxide on the nutritional quality of leaves of oak (*Quercus robur* L.) as food for the winter moth (*Operophtera brumata* L.). Funct Ecol 12:742–749

Cannon RJC (1998) The implications of predicted climate change for insect pests in the UK, with emphasis on non-indigenous species. Global Change Biol 4:785–796

Carlisle A, Brown AHF, White EJ (1966) Organic matter and nutrient elements in precipitation beneath a sessile oak (*Quercus petraea*) canopy. J Ecol 54:87–98

Christensen LM, Lovett GM, Mitchell MJ, Groffman PM (2002) The fate of nitrogen in gypsy moth frass deposited to an oak forest floor. Oecologia 131:444–452

Currie WS, Aber JD, McDowell WH, Boone RD, Magill AH (1996) Vertical transport of dissolved organic C and N under long-term N amendments in pine and hardwood forests. Biogeochemistry 35:471–505

Cushman JH (1991) Host-plant mediation of insect mutualisms: variable outcomes in herbivore–ant interactions. Oikos 61:138–144

Cushman JH, Whitham TG (1991) Competition mediating the outcome of a mutualism – protective services of ants as a limiting resource for membracids. Am Nat 138:851–865

Den Boer PJ, Reddingius J (1996) Regulation and stabilisation paradigms in population ecology. Chapman and Hall, London

Dixon AFG (1998) Aphid ecology. Chapman and Hall, London

Dixon AFG (2003) Climate change and phenological asynchrony. Ecol Entomol 28:380–381

Dixon AFG, Agarwala BK (1999) Ladybird-induced life-history changes in aphids. Proc R Soc Lond B 266:1549–1553

Eichhorn O (1969) Natürliche Verbreitungsareale und Einschleppungsgebiete der Weißtannen-Wolläuse (Gattung *Dreyfusia*) und die Möglichkeiten ihrer biologischen Be - kämpfung. Z Ang Entomol 63:113–131

Fischer MK, Shingleton AW (2001) Host plant and ants influence the honeydew sugar composition of aphids. Funct Ecol 15:544–550

Franz JM (1956) The effectiveness of predators and food in limiting gradations of *Adelges* (*Dreyfusia*) *piceae* (Ratz.) in Europe. Proc 10th Int Congr Entomol 4:781–787

Gaze P, Grant D, Clout MN (1983) Honeydew and its importance to birds in beech forests of South Island, New Zealand. N Z J Ecol 22:131–142

Grier CC, Vogt DJ (1990) Effects of aphid honeydew on soil nitrogen availability and net primary production in an *Alnus rubra* plantation in western Washington. Oikos 57:114–118

Guggenberger G, Zech W (1994) Composition and dynamics of dissolved carbohydrates and lignin – degradation products in two coniferous forests, N.E. Bavaria, Germany. Soil Biol Biochem 26:19–27

Haukioja E (2003) Putting the insect into the birch–insect interaction. Oecologia 136:161–168

Hollinger DY (1986) Herbivory and the cycling of nitrogen and phosphorus in isolated California oak trees. Oecologia 70:291–297

Honek A (1991) Environment stress, plant quality and abundance of cereal aphids (Hom., Aphididae) on winter wheat. J Appl Entomol 112:65–70

Hunter MD (2001) Insect population dynamics meets ecosystem ecology: effects of herbivory on soil nutrient dynamics. Agric Forest Entomol 3:77–84

Hunter MD, Varley GC, Gradwell GR (1997) Estimating the relative roles of top-down and bottom-up forces on insect herbivore populations: a classic study revisited. Proc Natl Acad Sci Am 94:9176–9181

Hunter MD, Linnen CR, Reynolds BC (2003) Effects of endemic densities of canopy herbivores on nutrient dynamics along a gradient in elevation in the southern Appalachians. Pedobiologia 47:231–244

Jenkins JC, Aber JD, Canham CD (1999) Hemlock woolly adelgid impacts on community structure and N cycling rates in eastern hemlock forests. Can J For Res 29:630–645

Jones CG, Lawton JH (1995) Linking species and ecosystems. Chapman and Hall, New York

Karafiat H, Franz JM (1956) Studien zur Populationsdynamik der Tannenstammlaus *Adelges* (*Dreyfusia*) *piceae* (Ratz.). Zool Jahrb Syst 84:467–504

Kidd NAC (1990) The population dynamics of the large pine aphid, *Cinara pinea* (Mordv.) II. Simulation of field populations. Res Pop Ecol 32:209–226

Kindlmann P, Dixon AFG (1996) Population dynamics of a tree dwelling aphid: individuals to populations. Ecol Mod 89:23–30

Kindlmann P, Stadler B (2004) Linking aphid abundance and throughfall chemistry in a spruce forest. Ecol Mod (in press)

Klemola T, Hanhimäki S, Ruohomäki K, Senn J, Tanhuanpää M, Kaitaniemi P, Ranta H, Haukioja E (2003) Performance of the cyclic autumnal moth, *Epirrita autumnata*, in relation to birch mast seeding. Oecologia 135:34–361

Koricheva J, Larsson S, Haukioja E (1988) Insect performance on experimentally stressed woody plants: a meta-analysis. Annu Rev Entomol 43:195–216

Kunert G, Weisser WW (2003) The interplay between density- and trait mediated effects in predator–prey interactions: a case study in aphid wing polymorphism. Oecologia 135:304–312

Lawton JH (2000) Community ecology in a changing world. International Ecology Institute, Oldendorf/Luhe

Liebhold AM, Elkinton JS, Williams D, Muzika RM (2000) What causes outbreaks of gypsy moth in North America? Popul Ecol 42:257–266

Likens GE, Bormann FH (1995) Biogeochemistry of a forested ecosystem, 2nd edn. Springer, Berlin Heidelberg New York

Loehle C, Pechmann JK (1988) Evolution: the missing ingredient in system ecology. Am Nat 132:884–899

Loreau M (2001) Linking community, evolutionary and ecosystem ecology: another perspective on plant–herbivore interactions. Belg J Zool 131:3–9

Lovett GM, Ruesink AE (1995) Carbon and nitrogen mineralization from decomposing gypsy moth frass. Oecologia 104:133–138

Lovett GM, Reiners WA, Olson RK (1989) Factors controlling throughfall chemistry in a balsam fir canopy – a modeling approach. Biogeochemistry 8:239–264

Lovett GM, Nolan SS, Driscoll CT, Fahey TJ (1996) Factors regulating throughfall flux in a new New-Hampshire forested landscape. Can J For Res 26:2134–2144

Ludwig D, Jones DD, Holling CS (1978) Qualitative analysis of insect outbreak systems – spruce budworm and forest. J Anim Ecol 47:315–332

Matzner E (1988) Der Stoffumsatz zweier Waldökosysteme im Solling. Ber Forschungszentrums Waldökosysteme/Waldsterben Univ Göttingen A40:1–217
McClure MS (1983) Reproduction and adaptation of exotic hemlock scales (Homoptera, Diaspididae) on their new and native hosts. Environ Entomol 12:1811–1815
McClure MS (1989) Evidence of a polymorphic life-cycle in the hemlock woolly adelgid, *Adelges tsugae* (Homoptera, Adelgidae). Ann Entomol Soc Am 82:50–54
McClure MS (1991) Density-dependent feedback and population-cycles in *Adelges tsugae* (Homoptera, Adelgidae) on *Tsuga canadensis*. Environ Entomol 20:258–264
McClure MS, Salom SM, Shields KS (2001) Hemlock woolly adelgid. Forest Health Technology Enterprise Team, Morgantown, pp 1–14
McDowell WH, Likens G (1988) Origin, composition, and flux of dissolved organic carbon in the Hubbard Brook valley. Ecol Monogr 58:177–195
Michalzik B, Matzner E (1999) Dynamics of dissolved organic nitrogen and carbon in a central European Norway spruce ecosystem. Eur J Soil Sci 50:579–590
Michalzik B, Stadler B (2000) Effects of phytophagous insects on soil solution chemistry: herbivores as switches for the nutrient dynamics in the soil. Basic Appl Ecol 1:117–123
Michalzik B, Stadler B (2004) Canopy herbivores and organic matter fluxes: a key process in understanding soil organic matter dynamics. Geoderma (in press)
Michalzik B, Kalbitz K, Park JH, Solinger S, Matzner E (2001) Fluxes and concentrations of dissolved organic carbon and nitrogen – a synthesis for temperate forests. Biogeochemistry 52:173–205
Müller T, Müller M, Behrendt U, Stadler B (2004) Diversity of culturable phyllosphere bacteria on beech and oak: effects of lepidopterous larvae. Microbiol Res 158:291–297
Orwig DA, Foster DR (1998) Forest response to the introduced hemlock woolly adelgid in southern New England, USA. J Torr Bot Soc 125:60–73
Pedersen LB, Bille-Hansen J (1995) Effects of nitrogen load to the forest floor in sitka spruce stands (*Picea sitchensis*) as affected by difference in deposition and spruce aphid infestations. Water Air Soil Pollut 85:1173–1178
Piirainen S, Finer L, Starr M (1998) Canopy and soil retention of nitrogen deposition in a mixed boreal forest in eastern Finland. Water Air Soil Pollut 105:165–174
Prescott CE (2002) The influence of the forest canopy on nutrient cycling. Tree Physiol 22:1193–1200
Pschorn-Walcher H, Zwölfer H (1958) Preliminary investigations on the *Dreyfusia* (*Adelges*) populations, living on the trunk of the silver fir. Z Angew Entomol 42:241–277
Qualls RG, Haines BL (1991) Fluxes of dissolved organic nutrients and humic substances in a deciduous forest. Ecology 72:254–266
Reynolds BC, Hunter MD (2001) Responses of soil respiration, soil nutrients, and litter decomposition to inputs from canopy herbivores. Soil Biol Biochem 33:1641–1652
Royama T (1984) Population-dynamics of the spruce budworm *Choristoneura fumiferana*. Ecol Monogr 54:429–462
Royama T (1996) Analytical population dynamics. Chapman and Hall, London
Scheurer S (1964) Zur Biologie einiger Fichten bewohnender Lachnidenarten (Homoptera, Aphidina). Z Angew Entomol 53:153–178
Schowalter TD (2000) Insect ecology: an ecosystem approach. Academic Press, San Diego
Schulze ED (2000) The carbon and nitrogen cycle of forest ecosystems. Springer, Berlin Heidelberg New York
Seastedt TR, Crossley JDA (1984) The influence of arthropods on ecosystems. BioScience 34:157–161
Seastedt TR, Crossley DA, Hargrove WW (1983) The effects of low-level consumption by

canopy arthropods on the growth and nutrient dynamics of black locust and red maple trees in the southern Appalachians. Ecology 64:1040–1048
Selås V (1997) Cyclic population fluctuations of herbivores as an effect of cyclic seed cropping of plants: the mast depression hypothesis. Oikos 80:257–268
Stadler B (1995) Adaptive allocation of resources and life-history trade-offs in aphids relative to plant quality. Oecologia 102:246–254
Stadler B, Müller T (1996) Aphid honeydew and its effect on the phyllosphere microflora of *Picea abies* (L) Karst. Oecologia 108:771–776
Stadler B, Müller T (2000) Effects of aphids and moth caterpillars on epiphytic microorganisms in canopies of forest trees. Can J For Res 30:631–638
Stadler B, Müller T, Sheppard L, Crossley A (2001a) Effects of *Elatobium abietinum* on nutrient fluxes in sitka spruce canopies receiving elevated nitrogen and sulphur deposition. Agric For Entomol 3:253–261
Stadler B, Solinger S, Michalzik B (2001b) Insect herbivores and the nutrient flow from the canopy to the soil in coniferous and deciduous forests. Oecologia 126:104–113
Stadler B, Dixon AFG, Kindlmann P (2002) Relative fitness of aphids: effects of plant quality and ants. Ecol Lett 5:216–222
Stadler B, Kindlmann P, Smilauer P, Fiedler K (2003) A comparative analysis of morphological and ecological characters of European aphids and lycaenids in relation to ant attendance. Oecologia 135:422–430
Stadler B, Müller T, Orwig DA, Cobb RC (2004) Hemlock woolly adelgid in New England forests: canopy impacts transforming ecosystem processes and landscapes in New England forests. Ecosystems (in press)
Sudd JH, Sudd ME (1985) Seasonal changes in the response of wood ants (*Formica lugubris*) to sucrose baits. Ecol Entomol 10:89–97
Swank WT, Waide JB, Crossley JDA, Todd RL (1981) Insect defoliation enhances nitrate export from forest ecosystems. Oecologia 51:297–299
Tukey HB Jr (1970) The leaching of substances from plants. Annu Rev Plant Physiol 21:305–324
Varley GC, Gradwell GR (1970) Recent advances in insect population dynamics. Annu Rev Entomol 15:1–24
Visser ME, Holleman LJM (2001) Warmer springs disrupt the synchrony of oak and winter moth phenology. Proc R Soc Lond B 268:289–294
Watt AD, McFarlane AM (2002) Will climate change have a different impact on different trophic levels? Phenological development of winter moth *Operophtera brumata* and its host plants. Ecol Entomol 27:254–256
Wiens JA (1995) Landscape mosaics and ecological theory. Chapman and Hall, London
Wiens JA, Stenseth NC, van Horne B, Ims RA (1993) Ecological mechanisms and landscape ecology. Oikos 66:369–380
Williams DW, Liebhold AM (1995) Herbivorous insects and global change: potential changes in the spatial distribution of forest defoliator outbreaks. J Biogeogr 22:665–671
Williams DW, Liebhold AM (2000) Spatial scale and the detection of density dependence in spruce budworm outbreaks in eastern North America. Oecologia 124:544–552
Wu J, Loucks OL (1995) From balancing of nature to hierarchical patch dynamics: a paradigm shift in ecology. Q Rev Biol 70:439–466
Yao I, Akimoto S (2001) Ant attendance changes the sugar composition of the honeydew of the drepanosiphid aphid *Tuberculatus quercicola*. Oecologia 128:36–43
Yao I, Akimoto SI (2002) Flexibility in the composition and concentration of amino acids in honeydew of the drepanosiphid aphid *Tuberculatus quercicola*. Ecol Entomol 27:745–752

Zhang Q-B, Alfaro RI (2003) Spatial synchrony of the two-year cycle budworm outbreaks in central British Columbia, Canada. Oikos 102:146–154
Zoebelein G (1954) Versuche zur Feststellung des Honigtauertrages von Fichtenbeständen mit Hilfe von Waldameisen. Z Angew Entomol 36:358–362

Section IV
Methods: Reducing, Enhancing and Simulating Insect Herbivory

12 Simulating Herbivory: Problems and Possibilities

J. Hjältén

12.1 Summary

Much of our knowledge of plant–herbivore interactions are based on results from experiments applying mechanical damage to plants. However, in recent years, the use of simulated herbivory has been criticized and several problems identified. The aim of this chapter is therefore to identify the most obvious advantages and disadvantages of using simulated insect herbivory and suggest ways to avoid some of the problems or alternative ways to conduct the experiments, e.g. by using natural herbivores caged on specific plant parts (clip cages).

It is clear from the literature that simulated herbivory often fails to induce plant responses that are important for complex biotic interaction, e.g. interactions between plants, different insect herbivore species and their predators/parasites. This strongly suggests that if the aim of a study is to identify and understand the ecological role of these complex interactions and ecosystem processes we need to use natural herbivory. Simulated herbivory may, however, still be an appropriate method for investigating simple biotic interactions, e.g. the effects of herbivory on plant growth and survival or physiological processes in plants that deter or limit further herbivory by the same herbivore. Simulated herbivory might also be a useful tool for dissecting damage into its functional parts, i.e. mechanical wounding and elicitor application.

The study of more complex biotic interactions, e.g. plant-mediated competition or interactions involving a third tropic level, requires the activity of real herbivores. Nevertheless, we should be aware that even the use of natural herbivory is associated with potential problems, e.g. clip cages can reduce radiation, increase leaf temperature, reduce leaf expansion and restrict realistic interactions with predators and parasites, and this should be taken into account when designing experiments and interpreting results.

Ecological Studies, Vol. 173
W.W. Weisser and E. Siemann (Eds.)
Insects and Ecosystem Function

12.2 Introduction to the Problem

Numerous studies have used simulated herbivory to mimic natural damage caused by herbivores (for references see also Lehtilä and Boalt, Chap. 13, this Vol.). Mechanically induced damage has been used to determine the effects of both insect and mammalian herbivores on plant growth, biomass partitioning, reproduction, morphology and survival (e.g. Paige and Whitham 1987; Bach 1994; Hjältén et al. 1993; Escarré et al. 1996; Bergman 2002; Pilson and Decker 2002). Furthermore, simulated herbivory has been used to determine physiological responses of plants following damage, and the resulting effects on biotic interactions (Danell et al. 1985; Hjältén 1999; Honkanen et al. 1999; Krupnick et al. 2000; Tscharntke et al. 2001). Thus, studies involving simulated herbivory have had a profound influence on our understanding of plant–animal interactions. The use of simulated herbivory is popular because it has several advantages over natural herbivory in empirical studies, the principal benefits being that it is simple to perform, and both the damage itself and the confounding factors are much easier to control (Tiffin and Inouye 2000).

However, in recent years, the validity of using simulated herbivory has been questioned (Baldwin 1990; Tiffin and Inouye 2000). Plant responses to herbivory are very complex and there are several problems associated with simulating herbivory, including doubts about the ability of researchers to mimic natural damage well enough to induce 'natural' plant responses, e.g. responses affecting plant growth, survival and simple defensive reactions that may reduce further herbivory by the same herbivore. In this respect, there is a considerable difference between the relatively simple form of damage imposed by browsing and grazing mammals and the diverse methods by which phytophagous insects exploit their food plants (e.g. leaf-feeders, phloem and xylem suckers, gallers, leaf-miners, stem-borers and seed-eaters). Furthermore, in recent years, it has become apparent that simulated herbivory may fail not only to induce plant responses that are essential components of more complex natural biotic interactions between plants, different insect herbivore species and their predators/parasites (Baldwin and Preston 1999; Agrawal 2000b), but also to capture essential ecosystem processes related to, e.g., nutrient cycling. Thus, reliance on studies of simulated herbivory could lead to conclusions that are not valid under natural conditions.

In this chapter I will try to identify the most obvious advantages and disadvantages of using simulated insect herbivory as a tool to study biotic interactions, and suggest ways to avoid some of the problems associated with the use of mechanical damage.

12.3 Advantages of Simulated Herbivory

The reason why simulated herbivory has been (and still is!) used so extensively in studies of plant–herbivore interactions is, of course, that it has several advantages compared to natural herbivory (Table 12.1). These advantages are briefly described below.

One obvious advantage is that mechanical damage enables the experimenter to control the type, timing and degree of damage to the experimental plants. Different types of herbivore damage have different effects on plant growth, biomass partitioning and plant chemistry (Strauss 1991; Hjältén et al. 1993; Agrawal 2000a; Fig. 12.1). In natural environments, plants are usually damaged to different degrees by several different herbivore species simultaneously, which makes it very difficult to evaluate the relative effects of different herbivores on plant characters and biotic interactions. By simulating herbivory by different herbivores both separately and collectively this problem can be solved.

Even more important is that use of simulated herbivory makes it possible to obtain an exact measurement of the degree of damage (e.g. biomass removed) simply by weighing the removed biomass. Consequently, the plants' compensatory abilities or tolerance can be measured exactly (Oesterheld and McNaughton 1991; Hjältén et al. 1993; Bergström and Danell 1995; Sadras 1998), which is much more difficult with natural herbivore damage (Tiffin and Inouye 2000). Other studies have used simulated damage to evaluate the importance of the timing of damage on plant growth, reproduction and concentration of secondary metabolites (Obeso and Grubb 1994; Escarré et al. 1996; Krupnick et al. 2002). For example, Obeso and Grubb (1994) found that

Table 11.1. Summary of the advantages and disadvantages of simulated herbivory

Advantages	– Simple to perform – Control of the type of damage – Control of the magnitude and timing of the damage – Control of biotic and abiotic confounding effects
Disadvantages	– Difficult to mimic many types of natural herbivore damage – Difficult to mimic the distribution and timing of natural herbivore damage – Often impossible to induce natural plant responses without detailed knowl – edge of system-specific biotic and abiotic interactions – May not capture complex biotic interactions – May not capture herbivore effects on soil processes that are important for ecosystem function

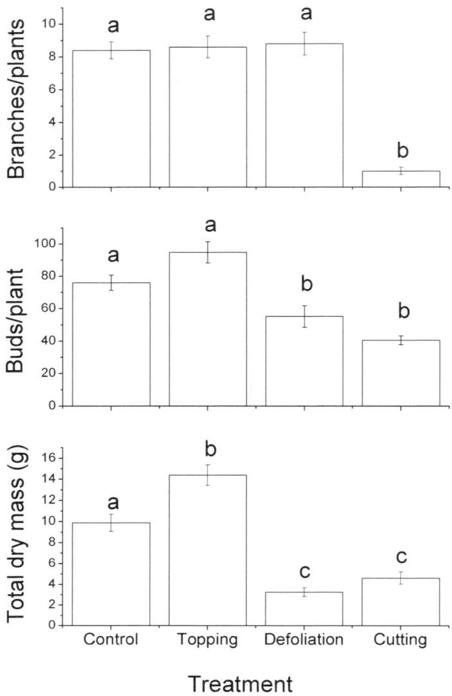

Fig. 12.1. Effects of different types of simulated herbivory on morphology and growth of juvenile birches. *Bars with different letters* denote statistically significant differences ($P < 0.05$)

late defoliation was much more detrimental for the annual *Senecio vulgaris* than early defoliation.

Nevertheless, one of the biggest advantages of using simulated insect herbivory is that it makes it possible to avoid bias resulting from unmeasured biotic or abiotic microenvironmental variables that can affect plant characters, herbivore density and/or preference, and higher trophic interactions (Rand 1999; Tiffin and Inouye 2000). For example, in situations where plant fitness and herbivore abundance vary in different types of microhabitat, estimates of plant responses to natural herbivory might be biased because plants growing in favourable microsites may receive more damage, but still grow better, than plants subjected to less herbivory that are growing in less favourable microsites (Tiffin and Inouye 2000).

Furthermore, resistance to insect herbivory varies among genetic plant families (genotypes) and different families will therefore receive different amounts of damage (Hjältén and Price 1996; Orians et al. 2000). If plant families also differ in characters other than resistance (e.g. growth and reproduction), or if the relative effect of herbivory on plant characters changes with the degree of damage, or differs between plant families, correlative measurements of herbivore effects on plant characters could lead to erroneous conclusions.

However, by randomly assigning mechanical damage to plants it is possible to avoid many of the problems discussed above (Baldwin 1990).

Finally, an important and obvious advantage of simulated herbivory is its convenience: it is simpler and less time-consuming to damage plants mechanically than to find naturally damaged plants or to introduce herbivores to selected plants. This might not be an advantage from the purely scientific perspective, but it may allow researchers to use bigger sample sizes in their studies.

12.4 Disadvantages of Simulated Herbivory

12.4.1 Simple Biotic Interactions

Although there are several advantages of simulating herbivory, the similarity of the observed effects of mechanical damage to natural responses and processes is highly debatable. Even simulating the effects of natural herbivory on simple biotic and abiotic interactions, e.g. the direct impact of herbivory on plant growth, survival, physiology and defensive responses against the herbivore causing the damage, is not straightforward since it must be properly mimicked with respect to both the nature and timing of the damage caused (Table 12.1).

Damage by mammalian herbivores on winter-dormant trees and shrubs is usually well defined and relatively easy to mimic compared to the diverse types of damage imposed by various functional groups of insects (e.g. leaf-feeders, phloem and xylem suckers, leaf-stem- and bud-gallers, leaf-miners, stem-borers, seed predators and frugivores) to parts of growing leaves or shoots. Among these functional groups of insects, the damage by externally chewing insects is, relatively speaking, the least difficult to simulate, but still a difficult task.

The response of winter-dormant trees and shrubs will probably to a lesser extent depend on the timing of the damage, or on the presence of herbivore saliva, which can be important in plants in a growing phase (Kahl et al. 2000; Agrawal 2002; Bergman 2002; Musser et al. 2002). This is mainly due to the physiological processes in plants slowing down during dormancy and basically stopping altogether in plants at high latitudes. This does not imply that timing is totally unimportant in dormant plants, but numerous studies have shown that timing of insect damage during the growing season influences the growth, reproduction and chemical expression of plants (e.g. Tolvanen et al. 1993; Honkanen and Haukioja 1994; Obeso and Grubb 1994; Sadras 1998; Thaler et al. 2002; Tiffin 2002) whereas similar studies on dormant plants are basically lacking. Although it is commonly assumed that the timing of damage is less important if trees are dormant, this still remains to be rigorously tested.

Even when the timing of the damage is not crucial, the problem of mimicking natural damage still remains, albeit for herbivores with sharp incisors, such as hares and voles, cutters may be appropriate (Hjältén et al. 1993) and for ungulates with less sharp teeth the herbivores' jaws can be used to inflict mechanical damage (Bergman 2002).

In contrast to vertebrate damage on winter-dormant plants, herbivore damage by insects to growing plant parts is often highly variable with respect to type, distribution, timing and severity. This makes herbivore damage to growing plant parts much harder to mimic. Several studies have also shown that plant responses to simulated and natural insect herbivory differ in terms of growth and chemical changes (Baldwin 1990; McCloud and Baldwin 1997; Agrawal 2002; see also Lehtilä and Boalt, Chap. 13, this Vol.). There are two major problems with mechanical damage: mechanical damage does not always accurately mimic natural damage, and some insect herbivores alter plant physiology by injecting growth substances during oviposition (e.g. gall formers) or saliva components while feeding (McCloud and Baldwin 1997; Agrawal 2000b). Even if damage by leaf chewers is relatively well defined, the use of special scissors or cutters has proven inadequate to properly mimic natural damage by such insects (Baldwin 1990; McCloud and Baldwin 1997; Agrawal 2000b). Agrawal (2000b) found that *Lepidium verginicum* individuals damaged by *Pieris rapae* were more resistant to damage by aphids and had higher rates of survival than mechanically damaged individuals. Bergman (2002) reported that willows subjected to mechanical damage grew significantly more branches if moose saliva was added to the wound. Furthermore, Agrawal (2002) found that natural herbivory by *Pieris rapae* on wild radish not only induced greater resistance to herbivory compared to mechanical damage, but also influenced the growth of progeny plants, probably because the naturally damaged plants produced larger seeds. Herbivory by gall-formers, miners, sucking and root-feeding insects is probably even harder to mimic, suggesting that in most cases it is very difficult or even impossible to accurately mimic natural damage, except (possibly) for very specific herbivores like seed-feeding or flower-head clipping insects (see, for instance, Pilson and Decker 2002).

Simulated herbivory makes it possible to control the distribution of damage, but we still need to consider the intra-plant distribution of natural damage when applying mechanical damage, because damage caused by herbivores is not randomly or regularly distributed (as is often the case with simulated herbivory) on plants. Instead, specific plant parts are preferred (Price et al. 1994; Waltz and Whitham 1997; Gomez and Gonzáles-Megías 2002). The distribution of damage also influences plant responses, probably due to plant modularity and source/sink regulations within the plants (Honkanen and Haukioja 1998). For example, Honkanen et al. (1999) found that variation in intra-plant distribution of defoliation in *Pinus sylvestris*, e.g. damage to apical or basal parts of the shoot, resulted in large differences in responses in both growth and secondary chemistry. Furthermore, depending on the plant parts consumed, one herbivore species might indirectly reduce the density of other herbivore species, either by unintentional predation or simply by removing valuable plant parts for other herbivores (Gomez and Gonzáles-Megías 2002). Unless the distribution of simulated damage is consistent with the distribution of natural herbivory, the obtained results could lead to incorrect or misleading conclusions.

The timing of damage and/or the length of time over which damage is done are also important factors to consider when applying mechanical damage. Simulated herbivory is often only applied once or on a few occasions, whereas natural insect herbivory, in most cases, is a continuous process that occurs throughout the growing season (and in many cases also during plant dormancy). Damage usually has the strongest effect on plant characters when it occurs early in the season (Krupnick et al. 2000), but in some situations late-season herbivory can also have strong effects (Tiffin 2002).

12.4.2 Complex Biotic Interactions

In recent years, it has become apparent that one of the main problems with mechanical damage is that it often fails to induce specific plant responses that play important roles in complex biotic interactions For example, McCloud and Baldwin (1997) convincingly showed that insect regurgitants play a profound role in the induction of defensive responses in native tobacco. Many of these responses are species-specific, i.e. different herbivores induce different plant responses (Voelckel and Baldwin, Chap. 17, this Vol.). For instance, Agrawal (2000a) found a high degree of specificity of induced resistance in wild radish to four caterpillar species. Some caterpillar species induced a general plant resistance to all herbivores, whereas at the other extreme, some caterpillar species did not induce any defensive plant responses at all. In addition, plants have been found to release volatiles that attract natural enemies (Turlings et al. 1998): a potentially very significant indirect response. Some plant volatiles may even induce defensive responses in neighbouring plants (see, for instance, Tscharntke et al. 2001; Baldwin et al. 2002). This clearly illustrates the immense complexity of interactions between plants, their herbivores and predators/parasites. Clearly, therefore, if mechanical damage fails to mimic natural damage sufficiently well, incorrect conclusions may be drawn regarding many complex biotic interactions, as well as the potential effect of herbivory on plant fitness in the system studied.

It is important to note that these complex interactions, involving plant-mediated competition between insect herbivores or indirect defences (herbivore-specific volatile signals released from the plant) involving a third tropic levels, would never have been detected if we had relied solely on simulated herbivory. Thus, it is absolutely essential that we use natural herbivory if we aspire to detect and understand these complex biotic interactions. Furthermore, we must realize that if we, e.g., fail to correctly estimate the relative importance of specific herbivores on plant population, due to simulated herbivory treatments failing to induce direct or indirect (mediated by a third trophic level) defensive responses in plants, this could lead to basic misinterpretation of ecosystem processes, such as whether or not plant populations mainly are top-down or bottom-up regulated.

Insect predators can have strong indirect effects not only on the populations of specific plant but also on plant diversity. Schmitz (2003) found that exclusion of spiders resulted in increased evenness in the plant community. However, it is also important to realize that different insect predators can generate contrasting trophic cascades. Snyder and Wise (2001) reported that during spring lycosid spiders reduced the densities of herbivorous beetles, thus increasing plant fruit production, whereas predatory beetles had much smaller effect. By contrast, during summer lycosid spiders strongly reduced plant yield possibly by intraguild predation on other predators, resulting in increased herbivore densities. These examples clearly show that if experiments relying on simulated herbivory fail to induce responses that affect complex biotic interaction, e.g. involving a third trophic level, we might end up with the wrong conclusion not only with regard to biotic interaction but also with respect to the basic questions on how the study system is regulated.

12.4.3 Basic Ecosystem Processes

One rarely addressed problem that could occur when applying mechanical damage to plants is that we may fail to consider herbivory effects that act on basic ecosystem processes. For example, selective browsing by mammalian and insect herbivores has been suggested to influence ecosystem functioning by changing species composition and soil processes, thus influencing nutrient availability (Pastor et al. 1993; Hobbs 1996; Kielland and Bryant 1998; Belowsky and Slade 2002). Furthermore, herbivory in itself can change the chemical properties of a plant, thus affecting litter properties and decomposition (Bardgett et al. 1998; Belowsky and Slade 2000; Hättenschwiler and Vitousek 2002; Hartley and Jones, Chap. 2, this Vol.; Wardle and Bardgett, Chap. 3, this Vol.). It has also been suggested that urine and faeces of mammalian as well as insect herbivores can have a strong effect on soil processes and by that also influence plant communities (Hobbs 1996; Reynolds and Hunter 2001). In some ecosystems, for example subarctic marshes, goose droppings might be very important since they can accelerate nutrient cycling and thus stimulate primary production (Cargill and Jefferies 1984). This effect is probably most important in systems with large losses of limiting nutrients during recycling, or where herbivores bring in nutrients from outside sources (see, for instance, de Mazancourt et al. 1998), but it is an important issue to consider in all ecosystems (Reynolds and Hunter 2001; Belowsky and Slade 2002). Part of these problems has been solved in some studies by simply adding natural faeces or urine, or synthetic alternatives (Reynolds and Hunter 2001; Persson 2003). Unfortunately, most studies involving simulated herbivory have ignored this potentially important feature of plant–animal interactions. This means that we may not capture effects that are essential for ecosystem processes and ecosystem function.

Finally, adult insect in taxa with phytophagous larvae may be important for ecosystem processes due to their role as pollinators (e.g. Després 2003), the most extreme case being *Ficus*–Agaonidae symbiosis and the *Yucca–Tegeticula* moth relationship (Dodd and Linhart 1994; Nefdt and Compton 1996). Thus, by simulating herbivory, we may overestimate the effect of herbivory on plant reproduction simply by missing the important role of herbivores as pollinators.

12.5 Conclusions and Suggestions for the Future

The appropriateness of using simulated herbivory largely depends on the question addressed or hypothesis tested. Simulated herbivory may be an appropriate method for investigating simple biotic interactions, e.g. the effects of herbivory on plant growth and survival or physiological processes in plants that deter or limit further herbivory by the same herbivore. Simulated herbivory might also be a useful tool for dissecting damage into its functional parts, i.e. mechanical wounding and elicitor application. However, it must be clearly acknowledged that results from experiments using simulated herbivory may not be generally applicable to herbivore effects on plant characteristics in all cases, or to complex biotic interactions under natural conditions. In the latter case, simulated herbivory is even more problematic (unless used in combination with natural herbivory) and alternative methods should be considered. The use of simulated herbivory may lead to misinterpretation regarding the importance of induced response in plants, affecting competitive interactions between both herbivores and higher trophic levels. This may lead to erroneous conclusion regarding the effect of herbivores on plant fitness, plant communities and ecosystem processes.

One of the principal reasons why researchers use simulated herbivory is, unfortunately, probably its convenience. It is simply much easier to control the degree, timing and distribution of damage using simulated herbivory compared to performing the same experiments with natural herbivory. Nevertheless, many studies, in both the field and experimental settings, have shown that herbivore damage can be controlled even if natural herbivores are used, by caging insect herbivores (e.g. in clip cages) on specific plant parts, or caging plant parts to keep out herbivores. The advantage of clip cages is, of course, that the damage is natural but also that the distribution, degree and timing of damage can be controlled. Assuming that the distribution and timing of damage in the natural environment are well documented and that the experiments are designed accordingly, the herbivory treatment should adequately mimic natural herbivory, inducing 'natural' responses and allowing conclusions to be drawn regarding both simple and complex biotic interactions. This method is, of course, more time-consuming, but allows more generalizations to be made regarding the effects of herbivory on plant characteristics as well as on other plants, herbivores and parasites/predators (cf. Schmitz, Chap. 14, this Vol.).

However, even the use of clip cages has disadvantages that we should be aware of. Clip cages reduce radiation and increase leaf temperature, thus increasing chlorophyll content and reducing leaf-soluble protein content, which could alter insect nutrition (Craft-Brandner and Chu 1999). In addition, clip cages can reduce leaf expansion which could lead to erroneous conclusions of herbivore effects on plant growth (Moore et al. 2003) and clip cages will probably also restrict realistic interactions with predators and parasites. These potential problems with clip cages should be taken into account when designing experiments and interpreting results.

In spite of this, the use of natural herbivory, by clip cages or otherwise, still has many advantages over simulated herbivory, since it potentially provides more detailed and accurate information on biotic interactions and ecosystem processes. It is therefore recommended that if the aim of a study is to explore and understand complex biotic interactions and ecosystem processes, researchers should at least always first consider using natural herbivory and only use simulated herbivory as a second option. Nevertheless, simulated herbivory should not be discarded from use in ecological research since it has several advantages over natural herbivory and as long as we are aware of the drawbacks/ limitations and restrict our conclusions accordingly it might still be a useful method.

However, it is recommended that mechanical damage should mainly be used to evaluate direct effects of herbivory on basic plant processes such as growth, reproduction and general physiological responses (e.g. changes in nutrient concentration and plant overall allocation to secondary compound) and not for processes involving complex biotic interactions (except as a complement to natural herbivory). Simulations should only be considered for insect herbivores inflicting well-defined damage to plants that is possible to mimic with accuracy (and great effort should be made to achieve this!), i.e. damage from externally leaf-feeding insects or seed-feeders. Finally, and most important, mechanical damage should only be used after natural herbivory has been considered and found to be more problematic, from a scientific perspective, than the use of mechanical herbivory.

Acknowledgements. I thank Per Hallgren and two anonymous referees for valuable comments on the manuscript and John Blackwell and Tuulikki Rooke for correcting the English. This study was financed by FORMAS.

References

Agrawal A (2000a) Specificity of induced responses in wild radish: causes and consequences for two specialist and two generalist caterpillars. Oikos 89:493–500

Agrawal A (2000b) Benefits and costs of induced plant defence for *Lepidium virginicum* (Brassicaceae). Ecology 81:1804–1813

Agrawal A (2002) Herbivory and maternal effects: mechanisms and consequences of transgenerational induced resistance. Ecology 83:3408–3415

Bach CE (1994) Effects of herbivory on growth and survivorship of sand-dune willow (*Salix cordata*). Ecol Entomol 19:303–309

Baldwin IT (1990) Herbivory simulations in ecological research. TREE 5:91–93

Baldwin IT, Preston CA (1999) The eco-physiological complexity of plant responses to insect herbivores. Planta 208:137–145

Baldwin IT, Kessler A, Halitschke R (2002) Volatile signalling in plant–plant-interactions: what is real? Curr Opin Plant Biol 5:1–4

Bardgett RD, Wardle DA, Yeates GW (1998) Linking above-ground and below-ground interactions: how plant responses to foliar herbivory influence soil organisms. Soil Biol Biochem 30:1867–1878

Belowsky G, Slade JB (2000) Insect herbivory accelerates nutrient cycling and increases plant production. Proc Natl Acad Sci USA 97:14412–14417

Belowsky G, Slade JB (2002) An ecosystem perspective on grasshopper control: possible advantages to no treatment. J Orthop Res 11:29–35

Bergman M (2002) Can saliva from moose, *Alces alces*, affect growth responses in the sallow, *Salix caprea*? Oikos 96:164–168

Bergström R, Danell K (1995) Effect of simulated summer browsing by moose on leaf and shoot biomass of birch, *Betula pendula*. Oikos 72:132–138

Cargill SM, Jefferies RL (1984) The effects of grazing by lesser snow geese on the vegetation of a sub-arctic salt marsh. J Appl Ecol 21:669–686

Craft-Brandner SJ, Chu C-C (1999) Insect clip cages rapidly alters photosynthetic traits of leaves. Crop Sci 39:1896–1899

Danell K, Huss-Danell K, Bergström R (1985) Interactions between browsing moose and two species of birch in Sweden. Ecology 66:1867–1878

de Mazancourt C, Loreau M, Abbadie L (1998) Grazing optimization and nutrient cycling: when do herbivores enhance plant production? Ecology 79:2242–2252

Després L (2003) Sex and pollen: the role of males in stabilising a plant–seed eater pollinating mutualism. Oecologia 135:60–66

Dodd RL, Linhart YB (1994) Reproductive consequences of interactions between *Yucca glauca* (Agavaceae) and *Tegeticula yuccasella* (Lepidoptera) in Colorado. Am J Bot 81:815–825

Escarré J, Lepart J, Sentuc JJ (1996) Effect on simulated herbivory in three old field Compositae with different inflorescence architecture. Oecologia 105:501–508

Gomez JM, Gonzáles-Megías A (2002) Asymmetrical interactions between ungulates and phytophagous insects: being different matters. Ecology 83:203–211

Hättenschwiler S, Vitousek PM (2002) The role of polyphenols in terrestrial ecosystem nutrient cycling. TREE 15:238–243

Hjältén J (1999) Interspecific variation in willow response to pruning: the effect on plant growth, survival and susceptibility to leaf-gallers. Ecoscience 6:62–67

Hjältén J, Price PW (1996) The effect of pruning on plant growth and sawfly population densities. Oikos 77:549–555

Hjältén J, Danell K, Ericson L (1993) Effect of simulated herbivory and intraspecific competition on the compensatory ability of juvenile birches. Ecology 74:1136–1142

Hobbs NT (1996) Modifications of ecosystems by ungulates. J Wildl Manage 60:695–713

Honkanen T, Haukioja E (1994) Why does a branch suffer more after branch-wide than after tree-wide defoliation? Oikos 71:441–450

Honkanen T, Haukioja E (1998) Intra-plant regulation of growth and plant–herbivore interactions. Ecoscience 5:470–479

Honkanen T, Haukioja E, Kitunen V (1999) Responses of *Pinus sylvestris* to simulated herbivory are modified by tree sink/source dynamics and by external resources. Funct Ecol 13:126–140

Kahl J, Siemens DH, Aerts RJ, Gaebler R, Kuehnemann F, Preston CA, Baldwin IT (2000) Herbivore-induced ethylene suppresses a direct defence but not a putative indirect defence against an adapted herbivore. Planta 210:336–342

Kielland K, Bryant JP (1998) Moose herbivory in taiga: effects on biochemistry and vegetation dynamics in primary succession. Oikos 82:377–383

Krupnick GA, Avila G, Brown KM, Stephenson AG (2000) Effects of herbivory on ethylene production and sex expression in *Curcurbita texana*. Funct Ecol 14:215–225

McCloud ES, Baldwin IT (1997) Herbivory and caterpillar regurgitants amplify the wound-induced increase in jasmonic acid but not nicotine in *Nicotiana sylvestris*. Planta 203:430–435

Moore JP, Taylor JE, Pual ND, Whittaker JB (2003) The use of clip cages to restrain insects reduces leaf expansion systematically in *Rumex obtusifolius*. Ecol Entomol 28:239–242

Musser RO, Hum-Musser SM, Eichenseer H, Peiffer M, Erwin G, Murphy JB, Felton GW (2002) Caterpillar saliva beats plant defenses. Nature 416:599–600

Nefdt RJC, Compton SG (1996) Regulation of seed and pollinator production in the fig wasp mutualism. J Anim Ecol 65:170–182

Obeso JR, Grubb PJ (1994) Interactive effects of the extent and timing of defoliation and nutrient supply on reproduction in a chemically defended annual *Senecio vulgaris*. Oikos 71:506–514

Oesterheld M, McNaughton SJ (1991) Effect of stress and time for recovery on the amount of compensatory growth after grazing. Oecologia 85:305–313

Orians CM, Griffiths M, Roche B, Fritz R (2000) Phenolic glucosides and condensed tannins in *Salix sericea*, *S. eriocephala* and their F1 hybrids: not all hybrids are created equal. Biochem Syst Ecol 28:619–632

Paige KN, Whitham TG (1987) Overcompensation in response to mammalian herbivory – the advantage of being eaten. Am Nat 129:407–416

Pastor J, Dewey B, Naiman RJ, McInnes PF, Cohen Y (1993) Moose browsing and soil fertility in the boreal forest of Isle Royale National Park. Ecology 74:467–480

Persson I-L (2003) Moose density and habitat productivity as drivers of ecosystem processes in northern boreal forests. PhD Thesis, Swedish University of Agricultural Sciences, Umeå

Pilson D, Decker KL (2002) Compensation for herbivory in wild sunflower: response to simulated damage by the head-clipping weevil. Ecology 83:3097–3107

Price PW, Clancy KM, Roininen H (1994) Comparative population dynamics of the galling sawflies. In: Price PW, Mattson WJ, Baranchikov YN (eds) The ecology and evolution of gall-forming insects. North Central Forest Experiment Station General

Tech Rep NC-174. US Department of Agriculture, Forest Service, St Paul, Minnesota, pp 1–11

Rand TA (1999) Effect of environmental context on the susceptibility of *Atriplex patula* to attack by herbivorous beetles. Oecologia 121:39–46

Reynolds BC, Hunter MD (2001) Responses of soil respirations, soil nutrients and litter decomposition to inputs from canopy herbivores. Soil Biol Biochem 33:1641–1652

Sadras VO (1998) Herbivory tolerance of cottons expressing insecticidal proteins from *Bacillus thuringiensis*: response to damage caused by *Helicoverpa* spp. and to manual bud removal. Field Crop Res 56:287–299

Schmitz OJ (2003) Top predator control of plant biodiversity and productivity in an old-field ecosystem. Ecol Lett 6:156–163

Snyder WE, Wise DH (2001) Contrasting trophic cascades generated by a community of generalist predators. Ecology 82:1571–1583

Strauss SY (1991) Direct, indirect, and cumulative effects of three native herbivores on a shared host plant. Ecology 72:543–558

Thaler JS, Fidantsef AL, Bostock RM (2002) Antagonism between jasmonate- and salicylate-mediated induced plant resistance: effects of concentration and timing of elicitation on defense-related proteins, herbivores, and pathogen performance in tomato. J Chem Ecol 28:1131–1159

Tiffin P (2002) Competition and time of damage affect the pattern of selection acting on plant defence against herbivores. Ecology 83:1981–1990

Tiffin P, Inouye BD (2000) Measuring tolerance to herbivory: accuracy and precision of estimates made using natural versus imposed damage. Evolution 54:1024–1029

Tolvanen A, Laine K, Pakonen T, Saari E, Havas P (1993) Effect of habitat and time of clipping on the recovery of the bilberry (*Vaccinium myrtillus*). Ann Bot Fenn 30:15–20

Tscharntke T, Thiessen S, Dolch R, Boland W (2001) Herbivory induced resistance and interplant signal transfer in *Alnus glutinosa*. Biochem Syst Ecol 29:1025–1047

Turlings TCJ, Lengwiler UB, Bernasconi ML, Wechsler D (1998) Timing of volatile emission in maize seedlings. Planta 207:146–152

Waltz AM, Whitham TG (1997) Plant development affects arthropod communities: opposing impact on species removal. Ecology 78:2133–2144

13 The Use and Usefulness of Artificial Herbivory in Plant–Herbivore Studies

K. Lehtilä and E. Boalt

13.1 Summary

Artificial damage is a popular method in plant–herbivore studies, because the use of real herbivores is often laborious and because it may be virtually impossible to use herbivores in many experimental setups. We made a literature search of studies that tested whether natural and artificial damage have similar effects on plants. Of 46 studies found, 33 (72 %) reported a significant difference between responses to artificial and natural herbivore damage in at least one of the statistical tests included. The studies contained 280 statistical tests, of which 99 (35 %) showed a significant difference between artificial and natural damage. Phytochemical responses to artificial and natural damage were different in 41 % of the statistical tests and 75 % of the studies found at least one significant difference. Plant resistance, measured as secondary damage, herbivore performance, fungal growth in damaged tissue or plant attractivity to parasitoids of herbivores, differed in 60 % of the statistical tests and 85 % of the studies had significant differences. Growth, reproduction and physiological responses to artificial and natural damage differed in 20–30 % of statistical tests and 50–83 % of studies had significant differences. Thus, studies on plant tolerance (growth and reproduction after damage) more often showed similar effects for artificial and natural damage than studies on plant resistance to herbivory, but even in tolerance studies artificial and natural damage often have different effects. Some studies indicated that application of herbivore saliva and careful imitation of timing and spatial pattern of damage helped in reaching the same effect with simulations and natural damage.

Ecological Studies, Vol. 173
W.W. Weisser and E. Siemann (Eds.)
Insects and Ecosystem Function

13.2 Introduction

In many experimental settings, artificial damage has several practical benefits over the use of real herbivores (Hjältén, Chap. 12, this Vol.). The extent of damage and the location of damaged parts can easily be controlled, and collateral damage to other than target tissues can be minimized. The removed biomass can be collected and measured. Furthermore, there is no need to collect and rear herbivores. The use of artificial damage enables efficient experimental designs, with balanced sample sizes of experimental groups and a low variation of treatment intensity within each experimental group. For these reasons, artificial damage is used more often in herbivory research than real herbivores.

Artificial damage does not, however, always adequately mimic natural damage (Baldwin 1990; Hjältén, Chap. 12, this Vol.). Many types of herbivory are not applicable for simulations. Damage by stem borers, miners, galling insects, root feeders or sucking insects is seldom tried to simulate. However, even when the apparent damage pattern is easy to reproduce, several characteristics of natural herbivory may be difficult to simulate, such as the timing of damage, herbivore host choice, location of damage within a plant, and subtle details of damage by herbivore mouth parts. Herbivore saliva can also play a role in plant responses (Walling 2000). The effects of trampling, defecating and urinating, dispersal of plant pathogens by herbivores, herbivore enemies and competitive release due to damage of neighbouring plants are seldom taken into account in herbivore simulations. The study of these and other effects of herbivory on plant community and ecosystem processes requires specific experimental designs. Indirect effects are often outside the scope of herbivore studies that use artificial damage, which usually concentrate on the hierarchical level of plant individuals.

In spite of many potential pitfalls, artificial damage continues to be a popular method in much of herbivory research. To control whether the effects of natural and artificial damage differ, several studies have included both types of treatments in experiments. Our aim is to review the tests that compare the plant response to natural and artificial herbivory. We examine how frequently natural and artificial damage give different responses. We also discuss what traits show differential responses to artificial and natural damage, and how the difference in response is affected by the type of herbivore and plant.

13.3 Material and Methods

We carried out a search from the Institute for Scientific Information (ISI) Web of Science database (www.isiknowledge.com, 1975–2002) with search words 'artificial damage', 'artificial herbiv*', 'simulated herbiv*', 'simulated damage', 'natural herbiv*' and 'natural damage'. We extended the search further by checking the reference lists of relevant articles that we found from the ISI database.

From articles, we recorded the method of artificial damage and the herbivore species of natural damage, type of response variables used, target tissue and sample size. We classified response variables as physiological and phytochemical responses, growth and reproduction responses, and resistance to secondary herbivory (other than chemical defence). From each statistical test, we checked whether there were pairwise significant differences among control, natural damage and artificial damage groups. In cases where there were several types of artificial damage, we chose the one that was closest to natural damage in pattern and timing. If there were no a posteriori tests, the results could usually not be used, but in some cases, a multiway ANOVA design enabled us to deduce the pairwise comparison among experimental groups. When applicable, we also used information from the studies that did not have any undamaged control group. We used treatment means and standard deviations to calculate the effect size as a standardized mean difference statistic $d = J(\overline{X}_N - \overline{X}_A)/s$, where $\overline{X}_N$ and $\overline{X}_A$ are the means for natural and artificial damage, respectively, s is the pooled standard deviation and J is a correction term that removes the bias due to small sample size (Gurevitch and Hedges 1993). The equation $d = 2J\sqrt{\chi^2/(N - \chi^2)}$, where N is the total sample size, was used for frequency data. The sign of d was assigned as negative if artificial damage had a stronger effect than natural damage to the expected direction from the control group, and positive otherwise.

We recorded whether the statistical tests of the same study belonged to the same experiment or to several independent experiments. Independent experiments were defined as distinct experiments with their own plant individuals, herbivores and separate statistical tests. Multiple years of data were considered as separate experiments if each year was tested separately, and one experiment if there was a common test of the whole data set. One experiment usually consisted of tests of many traits that were measured from the same plants. We made a general evaluation for each study and for each experiment of whether a significant difference in response to artificial and natural damage was found in any of the statistical tests. Thus, if one of the tests was significant, the whole study was assigned as showing a significant difference. When original studies discussed results of artificial and natural damage, differences were sometimes considered as irrelevant or minor, although they were statistically significant. We do not necessarily disagree with the interpretation of original tests, but to be consistent we followed the results of the statistical analyses.

The results are reported with studies, experiments and statistical tests as sampling units. It should be noted that only at the level of study, and possibly at the level of experiment, do the sample units represent observations that are independent in a statistical sense. Statistical tests belonging to the same experiment are associated, which has to be kept in mind when interpreting the testwise results. Testwise reports, however, give important insights into many of the questions discussed in the text. Furthermore, different experiments in the article are not independent in a strict sense, and the same caveat applies even to different studies with the same plant species. Similar to many ecological reviews, there is only a limited number of species in the data set and some popular model species are used in many studies. The only real solution to this problem of representativeness is to carry out more studies to cover a broad scale of taxa.

In some cases, the authors have known before or expected that the effects of artificial and natural damage differ. In the case of *Nicotiana–Manduca sexta*, Baldwin (1988) showed that artificial damage induces a stronger nicotine production than natural damage. Two later studies that continued to elucidate the system are included in this review (McCloud and Baldwin 1997; Kahl et al. 2000). Agrawal (1998), Agrawal et al. (1999) and Agrawal and Sherriffs (2001) compared real herbivory and rapid leaf clipping because earlier data suggested that clipping does not induce resistance. Turlings et al. (1990) used natural herbivory and artificial damage, with or without caterpillar regurgitants, to study volatile release and parasitoid attraction. These studies present a further test of natural and artificial damage, but there is a tendency for them to show a significant difference between the two damage types. A less clear case is the series of articles on birch herbivory (Haukioja and Neuvonen 1985; Hartley and Lawton 1987, 1991; Neuvonen et al. 1987; Hartley 1988; Hartley and Firn 1989; Hanhimäki and Senn 1992). Early studies showed a difference in response to artificial and natural damage. This was often confirmed in later studies, although it is not clear whether a specific result was a priori predicted. For some of the comparisons of the effect of natural and artificial herbivory, we present results with all studies included and with a data set without the studies where the authors apparently had a clear prior expectation of the result. In general, the results were approximately similar with both data sets.

13.4 Commonness of Differences Between Natural and Artificial Herbivory

We found 46 papers that tested the difference between responses to artificial and natural herbivore damage (Table 13.1). Thirty-three of the studies (72 %) reported a significant difference between responses to artificial and natural herbivore damage in at least one of the statistical tests included. Thirteen of the studies found no difference between responses to different damage methods. Many of the papers presented results from several independent experiments. When each experiment is evaluated separately, 50 of the total 73 experiments (68 %) showed a difference between artificial and natural herbivory and there was no difference in 23 experiments (32 %) in any of the tests of the experiment. Finally, we checked how many statistical tests showed a significant difference in responses. Of 280 statistical tests, 99 (35 %) showed a difference and 181 (65 %) showed no difference between responses to natural and artificial herbivory. The pattern was similar even if studies where a significant difference between natural and artificial damage was a priori expected (see Sect. 13.3) were not taken into account. Of the tests with no specific expectation of the result, 23 of 34 studies (68 %) and 77 of 238 statistical tests (32 %) showed a difference between natural and artificial damage. It is thus common, although not ubiquitous, that responses to natural and artificial herbivory differ.

Obviously, with a higher number of statistical tests within a study or an experiment, the probability that at least one of the tests is significant is increased. The studies including tests with a significant difference in response to artificial and natural herbivory contained more statistical tests than studies without any difference [Wilcoxon rank sum test, W=304.5, P=0.027; median and range of number of tests: studies showing a difference 4 (1...37) tests, studies with no difference 2 (1...5) tests].

Another factor potentially affecting the likelihood of finding a significant difference is the sample size. Ultimately, all ecological experiments would show a significant (although possibly biologically irrelevant) difference among experimental groups if the sample size is high enough, since there are always differences among ecological samples. Some of the studies had astonishing sample sizes, e.g. 2200 plants in Juenger and Bergelson (2000), showing a significant difference. There was no significant difference in sample sizes of experiments with negative and positive results [Wilcoxon rank sum test, W=587.5, P=0.886; median and range of sample size: significant difference 44 (12...2200), no difference 44 (12...105)].

The analysis of effect sizes (Table 13.1), however, suggested that there is a publication bias for too many studies with a low sample size showing a difference between artificial and natural damage. The relationship of sample size vs. effect size should reveal in a scatterplot a funnel-type pattern, with small sample sizes displaying more variation around the mean effect than large studies, but without any decreasing trends or without any large holes on the left side of the plot near the x-axis (Palmer 1999). There was a decreasing trend between sample size and effect size (Spearman correlation, R_s=–0.268, P=0.001, N=149). Further examination revealed that the publication bias was due to chemical response variables (R_s=–0.541, P=0.001, N=35). The other types of response variables did not show any association between sample size and effect size (resistance R_s=–0.136, P=0.465, N=31; growth R_s=0.005, P=0.977, N=45; reproduction R_s=0.046, P=0.791, N=35; physiological traits:

Table 13.1. References included in the review. General evaluation of difference in response to artificial and natural damage, response variables, plant and animal taxa and effect sizes

Reference	Difference in response to artificial and natural damage	Type of response variables	Response traits measured	Plant species	Plant family	Plant func-l tional group	Animal species	Animal taxon	Effect size (mean and range)[a] (%)
Insects; grasses									
Cabrera et al. (1995)	Yes	Chemical and physiological	Water potential; carbohydrate, protein, proline and chlorophyll content	*Hordeum vulgare*	Poaceae	Grass	*Schizaphis graminum*	Homoptera	340.4 (201.1...432.1)
Woodhead (1981)	Yes	Chemical	Phenolic content	*Sorghum bicolor*	Poaceae	Grass	*Chilo partellus*	Lepidoptera	
Capinera and Roltsch (1980)	Yes	Growth	Seedling size	*Triticum aestivum*	Poaceae	Grass	*Melanoplus sanguinipes*	Orthoptera	16.2 (−84.1...102.4)
Turlings et al. (1990)	No	Resistance	Parasitoid attractivity	*Zea mays*	Poaceae	Grass	*Spodoptera exigua*	Lepidoptera	88.2
Insects; herbs									
Bodnaryk (1992)	No	Chemical	Glucosinolate content	*Brassica napus*	Brassicaceae	Herb	*Phyllotreta cruciferae*	Coleoptera	43.9
Griffiths et al. (1994)	Yes	Chemical	Glucosinolate content	*Brassica napus*	Brassicaceae	Herb	*Delia floralis*	Diptera	
Mattiacci et al. (1994)	Yes	Resistance	Parasitoid attraction	*Brassica oleracea* var. *gemmifera*	Brassicaceae	Herb	*Pieris brassicae*	Lepidoptera	158.7 (127.8...189.6)
Doak (1991)	Yes	Growth	Shoot growth	*Epilobium latifolium*	Onagraceae	Herb	*Mompha albapalpella*	Lepidoptera	−141.9
Hammond and Pedigo (1981)	No	Physiological	Leaf water loss	*Glycine max*	Fabaceae	Herb	*Plathypena scabra*	Lepidoptera	
Ostlie and Pedigo (1984)	Yes	Physiological	Leaf water loss	*Glycine max*	Fabaceae	Herb	*Plathypena scabra, Trichoplusia ni*	Lepidoptera	

Poston et al. (1976)	No	Physiological	Net carbon exchange	*Glycine max*	Fabaceae	Herb	*Plathypena scabra*, *Cynthia cardui*	Lepidoptera	
Welter (1991)	No	Growth and physiological	Photosynthesis, plant growth	*Lycopersicon esculentum*	Solanaceae	Herb	*Manduca sexta*	Lepidoptera	−4.2 (−17.1...15.8)
Quinn and Hall (1996)	No	Growth	Leaf and nodule production	*Medicago sativa*	Fabaceae	Herb	*Hypera postica*	Coleoptera	−6.9 (−54.2...23.8)
Kahl et al. (2000)	No	Chemical	Volatile release	*Nicotiana attenuata*	Solanaceae	Herb	*Manduca sexta*	Lepidoptera	84.9
Baldwin (1988)	Yes	Chemical	Alkaloid content	*Nicotiana sylvestris*	Solanaceae	Herb	*Manduca sexta*	Lepidoptera	−155.0
McCloud and Baldwin (1997)	Yes	Chemical	Jasmonate acid and nicotine content	*Nicotiana sylvestris*	Solanaceae	Herb	*Manduca sexta*	Lepidoptera	11.7
Agrawal and Sherriffs (2001)	Yes	Resistance	Leaf damage, oviposition	*Raphanus raphanistrum*	Brassicaceae	Herb	*Pieris rapae*, *Phlytetreta* spp., *Oryctolagus cuniculus*	Coleoptera, Lepidoptera, Lagomorpha	18.9 (−35...55.5)
Agrawal et al. (1999)	Yes	Chemical and reproduction	Glucosinolates, flowering, seed set, pollen production	*Raphanus raphanistrum*	Brassicaceae	Herb	*Pieris rapae*	Lepidoptera	20.9 (0.2...42.7)
Agrawal (1998)	Yes	Resistance and reproduction	Subsequent insect damage, aphid abundance, seed set	*Raphanus sativus*	Brassicaceae	Herb	*Pieris rapae*	Lepidoptera	29.5 (10.4...50.1)
Hatcher et al. (1994)	Yes	Resistance	Fungal infection	*Rumex crispus*	Polygonaceae	Herb	*Gastrophysa viridula*	Coleoptera	
Hatcher et al. (1995)	Yes	Resistance	Fungal infection	*Rumex crispus*, *R. obtusifolius*	Polygonaceae	Herb	*Gastrophysa viridula*	Coleoptera	68.6 (−5.2...109.5)
Gavloski and Lamb (2000)	Yes	Growth and reproduction	Leaf length, seed production	*Sinapis alba*, *Brassica napus*	Brassicaceae	Herb	*Mamestra configurata*, *Phyllotreta cruciferae*, *Plutella xylostella*	Lepidoptera, Coleoptera	
Davis and Boyd (2000)	No	Chemical	Content of C, N, Ni and gluclosinolates	*Streptanthus polygaloides*, *S. insignus*	Brassicaceae	Herb	*Pieris rapae*	Lepidoptera	22.5 (−18.8...99)

Table 13.1. (*Continued*)

Reference	Difference in response to artificial and natural damage	Type of response variables	Response traits measured	Plant species	Plant family	Plant functional group	Animal species	Animal taxon	Effect size (mean and range)[a] (%)
Insects; shrubs and trees									
Hartley (1988)	Yes	Resistance and chemical	Feeding behaviour, phenolics	*Betula pendula*	Betula	Tree	*Apocheima pilosaria*	Lepidoptera	91.6
Hartley and Firn (1989)	Yes	Chemical	Phenolic content	*Betula pendula*	Betulaceae	Tree	*Apocheima pilosaria*	Lepidoptera	268.1 (193.1...361.4)
Hartley and Lawton (1987)	Yes	Resistance	Herbivore feeding preference	*Betula pendula*	Betulaceae	Tree	*Eriocrania* spp, several other insects	Lepidoptera	
Hartley and Lawton (1991)	Yes	Chemical	Phenolic content, protein precipitation ability	*Betula pendula*	Betulaceae	Tree	*Apocheima pilosaria*	Lepidoptera	0.0
Haukioja and Neuvonen (1985)	Yes	Resistance	Herbivore growth	*Betula pubescens*	Betulaceae	Tree	*Epirrita autumnata*	Lepidoptera	66.7 (66.3...67.1)
Hanhimäki and Senn (1992)	Yes	Resistance	Herbivore growth	*Betula pubescens* ssp. *tortuosa*	Betulaceae	Tree	*Epirrita autumnata*	Lepidoptera	14.3 (−0.2...28.8)
Neuvonen et al. (1987)	Yes	Resistance	Herbivore growth	*Betula pubescens*, *Prunus padus*, *Sorbus aucuparia*	Betulaceae, Rosaceae	Tree	*Epirrita autumnata*	Lepidoptera	74.0 (1.5...106.2)
Anderson et al. (2001)	Yes	Resistance	Feeding behaviour	*Gossypium hirsutum*	Malvaceae	Shrub	*Spodoptera littoralis*	Lepidoptera	23.5 (−6.9...54.5)
Anderson and Alborn (1999)	Yes	Resistance	Oviposition behaviour	*Gossypium hirsutum*	Malvaceae	Shrub	*Spodoptera littoralis*	Lepidoptera	
Heil et al. (2001)	Yes	Chemical	Extrafloral nectar production	*Macaranga tanarius*	Euphorbiaceae	Tree	Scarabeidae spp., *Xenocatantops humilis*	Coleoptera, Orthoptera	−170.8
Hodge et al. (2000)	No	Growth	Leaf retention	*Macropiper exelsum*	Piperaceae	Tree	*Cleora scriptaria*	Lepidoptera	5.4
Britton (1988)	No	Growth	Leaf area	*Pinus contorta*	Pinaceae	Tree	*Neodiprion*	Hymenoptera	2.0

Litvak and Monson (1998)	Yes	Chemical	Monoterpene production	*Pinus ponderosa*	Pinaceae	Tree	*Halisidota ingens*	Lepidoptera	95.6 (–33.7...505.6)
Faeth (1986)	Yes	Chemical	Protein and tannin content	*Querqus emoryi*	Fagaceae	Tree	Various insects	Lepidoptera, Coleoptera, Orthoptera, Hymenoptera	
Algae and their herbivores									
Pavia and Toth (2000)	Yes	Chemical	Phlorotannin content	*Ascophyllum nodosum*	Fucaceae	Brown alga	*Littorina obtusata*, *Idotea granulosa*	Gastropoda, Isopoda	144.3 (–17.6...318)
Toth and Pavia (2002)	No	Chemical and growth	Growth, phloro-tannin content	*Laminaria hyperborea*	Laminaria-ceae	Brown alga	*Lacuna vincta*, *Ansates pellucida*	Gastropoda	–3.6 (–48.3...40.8)
Mammals									
Teague (1988)	No	Growth	Plant size	*Acacia karroo*	Fabaceae	Herb	*Capra hircus*	Artiodactyla	
Reichman and Smith (1991)	No	Growth	Plant biomass	*Castor* sp.	Fabaceae	Herb	*Geomys bursarius*	Rodentia	–21.6
Howe et al. (1982)	Yes	Growth	Regrowth	*Lolium perenne*	Poaceae	Grass	*Sigmodon hispidus*	Rodentia	–230.5
Juenger and Bergelson (2000)	Yes	Growth and reproduction	Branch production, flowering, fruit production	*Ipomopsis aggregata*	Polemonia-ceae	Herb	*Thomomys* sp.	Rodentia	–6.9 (–45.2...43.2)
Paige (1992)	Yes	Growth and reproduction	Plant size, flowering, seed set	*Ipomopsis aggregata*	Polemonia-ceae	Herb	*Odocoileus hemionus*, *Cervus elaphus*	Artiodactyla	28.9 (–59.7...149.8)
Paige (1999)	Yes	Growth and reproduction	Root size, flower and fruit production	*Ipomopsis aggregata*	Polemonia-ceae	Herb	*Odocoileus hemionus*, *Cervus elaphus*	Artiodactyla	75.9 (–43.1...193.2)
Maschinski and Whitham (1989)	Yes	Reproduction	Seed set	*Ipomopsis arizonica*	Polemonia-ceae	Herb	*Spermophilus variegatus*, *Odocoileus hemiotus*	Rodentia, Artiodactyla	–36.6 (–85.8...15.5)

[a] Sign of effect size was assigned as negative if artificial damage had a stronger effect than natural damage to expected direction from control group, and positive otherwise

too low sample size for the test). It was not possible to further isolate a specific type of chemical studies behind the publication bias. Chemical studies with a high effect size and low sample size included several types of plants such as birch (Hartley and Firn 1989), pine (Litvak and Monson 1998), barley *Hordeum vulgare* (Cabrera et al. 1995) and the brown alga *Ascophyllum nodosum* (Pavia and Toth 2000). The response traits ranged from volatile compounds and phenolics to primary metabolites such as protein and carbohydrate content. The publication bias of chemical studies remained significant even if we removed the studies with prior expectation of a difference between natural and artificial damage (see Sect. 13.3). There may thus be missing chemical studies with a low sample size and negative result that have not been published. A negative result in this case means that artificial damage has a stronger effect on the expected direction than natural damage. Due to publication bias of the chemical studies, other results, where the whole data set is included, could also be biased. The results remained, however, very similar even when the chemical studies were removed from the analysis (difference of at maximum three percentage units; results not shown).

13.5 Strength of the Effect of Natural and Artificial Damage

Of 280 statistical tests, 155 showed a difference between the control and either natural damage, artificial damage, or both. In 56 tests, both damage types resulted in a similar and a significant difference to the control group. Assuming that statistical power of the experiments was sufficient to reveal biologically important differences, these 56 cases with a difference to the control group but no difference between damage methods are ideal to support the use of artificial damage. There were 106 tests with no differences between artificial damage, natural damage and the control group, and an additional 19 tests with no difference between two damage types but where the control group was missing. These results can be used to support the use of artificial damage.

Other types of test results are more problematic. In several tests, both damage methods resulted in a significant difference from the control group, but one of the treatments was significantly stronger than the other one. Whether this is acceptable depends on the goals of the study and on the actual difference between the treatments. In cases where both treatments differed from the control, natural damage had a significantly stronger effect in 30 tests and artificial damage in 10 tests.

Also, in other respects, natural damage often had a stronger effect than artificial damage. Of the total of 99 statistical tests showing a difference between natural and artificial damage, natural damage had a significantly stronger effect in 77 tests. In 43 of these 77 tests, there was no significant difference between the control group and the artificial damage treatment. There were only six tests where artificial damage produced a significant effect but natural damage did not. Thus, there was a considerable proportion of tests where the effect of herbivory would not have been detected if only artificial damage had been used.

Perhaps the most problematic cases are those where natural and artificial damage have an effect in opposite directions from the control group. Six tests (2 %) produced this result. Two of the tests used a chemical response variable, one test dealt with growth and three tests with reproduction.

13.6 Responses of Different Types of Response Traits to Artificial and Natural Damage

The effect of herbivory is mediated to other plant characteristics through physiological and chemical processes (Voelckel and Baldwin, Chap. 17, this Vol.). It is thus possible that physiological and phytochemical traits are more sensitive in showing the difference between responses to artificial and natural damage than other plant traits such as growth and reproduction. The review results give some support to the hypothesis that phytochemical response often differs between artificial and natural damage. Phytochemical response differed between artificial and natural damage in 75 % of the studies, 75 % of the experiments and 41 % of the tests. It must be noted, however, that chemical studies are difficult to interpret due to publication bias (see Sect. 13.4). It is possible that when sample size has been low, significant results have been published more often than non-significant results.

Most chemical traits used in the tests were secondary compounds with putative defensive function or enzymes involved in their biosynthesis. We found only three studies that tested the response of other types of chemical traits. Davis and Boyd (2000) did not find any difference between the effects of artificial and natural damage when studying nickel, carbon and nitrogen content and the carbon:nitrogen ratio after damage. Ni, C and N were studied for their possible role in resistance to herbivory. Faeth (1986) tested how leaf protein content is affected by damage. He found no difference in responses to artificial and natural damage. Cabrera et al. (1995), on the other hand, found a difference between the effects of different damage types on soluble protein content in leaves, and also on the levels of soluble sugar, free proline and chlorophyll. However, the method of artificial damage of Cabrera et al. (1995) was not an especially close imitation of natural herbivory. They used aphids as natural herbivores and artificial damage consisted of wounding leaves with microcapillary tips twice a day.

Physiological responses were more often similar for artificial and natural damage than phytochemical responses. Response variables included extrafloral nectar production (Heil et al. 2001), photosynthetic rate (Poston et al. 1976;

Welter 1991) and water status (Ostlie and Pedigo 1984; Cabrera et al. 1995). Production of extrafloral nectar was higher after artificial damage, or after jasmonic acid spraying, than after natural damage of similar extent (Heil et al. 2001). Otherwise physiological responses were closely similar after natural and artificial damage. Ostlie and Pedigo (1984) found a difference in water loss in the comparison of one of two herbivore species, but the difference disappeared 16 h after damage. Cabrera et al. (1995) also found differences in water status, but, again, wounding with microcapillary tips can hardly be expected to have the same effect on water potential as aphid damage.

Resistance to secondary herbivory was often different after artificial and natural damage. In 85 % of the studies, 84 % of the experiments and 60 % of the statistical tests, induced resistance of the damage methods was significantly different. Resistance studies include several cases where the test was replicated in several papers using the same study system, with a high probability of observing a difference between natural and artificial damage (see Sect. 13.3). However, significant differences between artificial and natural damage were even more common in the studies without a prior expectation of the result (results not shown). Resistance studies included herbivore bioassays of plant quality after damage (Haukioja and Neuvonen 1985; Hartley and Lawton 1987; Neuvonen et al. 1987; Hanhimäki and Senn 1992; Anderson and Alborn 1999; Anderson et al. 2001), fungal growth in damaged tissue (Hatcher et al. 1994, 1995), secondary damage on plants exposed to herbivores (Agrawal 1998; Agrawal and Sherriffs 2001) and plant attractivity to parasitoids of herbivores (Turlings et al. 1990; Mattiacci et al. 1994).

There may be an adaptive basis in the ability of plants, herbivores and herbivore enemies to discriminate between artificial and natural damage. From a plant's point of view, herbivore repellents are important after herbivore damage but not after other types of damage. Herbivores may benefit from either avoiding or favouring previously damaged leaves. For enemies of herbivores, it is even important to discriminate between damage types of different herbivore species (Takabayashi and Dicke 1996).

Plants often have a good tolerance to herbivory in terms of growth and reproduction (Strauss and Agrawal 1999; Stowe et al. 2000). It is expected that tolerance buffers growth and reproduction responses so that artificial and natural damage have more similar effect than other types of response variables. As predicted, growth and reproduction were less sensitive to different damage methods, but even for these response variables together, 59 % of studies and 26 % of tests showed a difference. Growth (23 % of the tests significantly different) seemed to be somewhat less sensitive to the choice of the damage method than reproduction (30 % of the tests significantly different).

13.7 Simulations of Mammalian and Invertebrate Herbivory

Because human tools are more of the size of mammalian jaws than insect mouthparts, one would expect that mammalian herbivory is easier to simulate than insect herbivory. The data set gives a possibility to test this hypothesis for tolerance traits, i.e. growth and reproduction after damage. For the other types of response traits, only invertebrates were used as natural damage agents in the papers of the data set. There were 7 studies, containing 57 statistical tests, with mammalian herbivory. In 5 studies and 22 statistical tests (39 %), there was a significant difference between the damage methods. Invertebrate herbivores were used in 10 tolerance studies; half of the studies showed a significant difference in a response to artificial and natural damage. Of 72 statistical tests of tolerance traits in invertebrate studies, 11 (15 %) showed a significant difference in a response.

According to these results, it is not easier to simulate mammalian damage than invertebrate damage. It must be noted, however, that only a few plant–vertebrate systems have been used in the studies of the data set. Four mammalian studies dealt with compensation and overcompensation of two *Ipomopsis* species, *I. aggregata* and *I. arizonica*. Natural and artificial damage often had a differential effect on growth and reproduction of *Ipomopsis* species, but it is not clear whether there were problems in sampling or in the experimental treatments of mechanically damaged plants (Bergelson et al. 1996; Paige 1999). The effect of natural damage may also be biased if the possibility that herbivores choose large plants could not sufficiently be taken into account in the statistical analysis (Bergelson et al. 1996). Mammalian saliva may, similar to insect saliva, affect plant growth and metabolism (Bergman 2002), but the effect of mammalian saliva was not tested in the current papers.

Invertebrate studies of tolerance consisted of a broader scope of plant–herbivore systems. Half of the studies did not observe any difference between artificial and natural damage. Simulated damage and natural herbivory of tobacco hornworm *Manduca sexta* had similar effects on growth and photosynthesis of tomato (Welter 1991). The effects of European pine sawfly *Neodiprion sertifer* on the growth of lodgepole pine (Britton 1988) and the effects of gastropod damage on the growth of the kelp *Laminaria hyperborea* (Toth and Pavia 2002) were similar to the effect of artificial damage. Quinn and Hall (1996) could artificially simulate the defoliation and denodulation of the legume *Medicago sativa* by alfalfa weevil *Hypera postica* and Hodge et al. (2000) showed that artificial damage and herbivory of *Cleora scriptaria*, a geometrid moth, had the same effect on leaf loss of *Macropiper excelsum*. In some of the studies, the lack of significant differences may be due to low sample size and few and insensitive response variables compared to mammalian studies. In two of the studies, a significant difference between two damage types was anticipated (Agrawal 1998;

Agrawal et al. 1999). Agrawal and Sherriffs (2001) studied the combination of vertebrate and insect herbivory. The damaging agents were lepidopteran larvae, whereas the response trait was secondary damage by *Pieris rapae*, flea beetles and rabbits. Although induction had a somewhat different effect on rabbit and insect herbivory, the general result was that artificial damage was a poor proxy of natural herbivory.

13.8 Attempts of Exact Simulation

Some studies have tested several methods of artificial damage to find the most adequate simulation. Gavloski and Lamb (2000) studied how damage by three herbivores of cruciferous plants with different damage patterns can be simulated. They used concentrated or dispersed defoliation of cotyledons of *Sinapis alba* and *Brassica napus*, with or without meristem removal, to simulate damage by the flea beetle *Phyllotreta cruciferae*, diamondback moth *Plutella xylostella* and bertha armyworm *Mamestra configurata*. There were some significant differences in leaf growth between artificial and natural damage when the appropriate artificial damage pattern was compared with natural damage, but only for the low damage level of *Mamestra configurata* and only for the first of the four observation dates, 8–9 days after damage. For the other herbivores and the later observation dates, responses to artificial and natural damage were similar. Both damage methods resulted in similar compensation in leaf growth, less compensation for the flea beetle damage than the other two herbivores, and better compensation by *S. alba* than by *B. napus*. Gavloski and Lamb (2000) thus showed that, in their system, when the amount of biomass removal and the pattern of removal (dispersed or concentrated, with or without meristem removal) are reproduced by punch or scissors, artificial damage produces acceptable results.

Gavloski and Lamb (2000) did not try to mimic the temporal pattern of herbivory. In contrast, Baldwin (1988) simulated the temporal, in addition to spatial, pattern of herbivory when he studied the effect of *Manduca sexta* damage on *Nicotiana sylvestris*. The response variable was the alkaloid content, which appeared to be more sensitive to the damage method than the growth and reproduction traits used by Gavloski and Lamb (2000). Damage increased the alkaloid content, and the increase was larger after simulated damage than after real herbivory. The response of artificial damage was closest to natural herbivory with a careful imitation of temporal and spatial pattern of *M. sexta* damage, where leaf removal was performed with microscissors at the same pace as damage by larvae. Later work has shown that both artificial and natural damage increase jasmonic acid concentrations in the leaves, but chemical compounds in herbivore saliva block the induction of alkaloids (McCloud and Baldwin 1997; Kahl et al. 2000; Musser et al. 2002).

After applying saliva to wounded tissue, artificial and natural damage resulted in similar induction of alkaloids.

In addition to alkaloid induction, oral secretions are important for the release of volatile compounds that attract enemies of herbivores to damaged plants (Turlings et al. 1990; Dyer et al. 1995; Alborn et al. 1997; Kahl et al. 2000). For instance, Turlings et al. (1990) studied whether plant odours from damaged plants attract *Cotesia marginiventris* parasitoids. They observed that plants damaged by *Spodoptera exigua* caterpillars released volatiles that were better attractants than volatiles from artificially damaged plants. When larval regurgitant was added to artificially damaged leaves, they were as attractive to parasitoids as *Spodoptera*-damaged leaves. The use of herbivore saliva in combination with artificial damage is thus a potential method to improve the quality of herbivore simulations. It is very difficult to collect enough insect saliva for large experiments, but treatments may become easier if the bioactive chemicals of saliva are identified (Moon et al. 1994; Alborn et al. 1997; Musser et al. 2002).

13.9 Conclusions

The literature review shows undeniably that artificial damage often has a different effect on plants than natural damage. Should ecologists for this reason avoid using artificial damage? Our review indicates that it is worthwhile to make an effort to use real herbivores whenever possible (see also Hjältén, Chap. 12, this Vol.). Our opinion, however, is that artificial damage still has its place in the toolbox of the ecologists. There were only a few cases, about 3 %, of the worst possible outcome where artificial and natural damage had opposite effects on plants. Furthermore, the difference in the effects of artificial and natural damage is not always undesirable but may give interesting insight into plant responses to herbivory. For instance, Agrawal (1998) and Agrawal et al. (1999) noted that clipping with scissors did not induce the production of secondary compounds, but natural damage did. Artificial damage can then represent the effect of biomass removal and natural damage the effect of both biomass removal and induction. Jasmonic acid spraying can be used to induce resistance without any biomass removal (Agrawal et al. 1999). Studies on the effects of oral secretions of herbivores (Dyer et al. 1995; Alborn et al. 1997; McCloud and Baldwin 1997; Musser et al. 2002; Voelckel and Baldwin, Chap. 17, this Vol.) are other examples of the importance of artificial damage as an experimental method.

At the community and ecosystem level, the most important effects that artificial damage should succeed in simulating correctly are often the ones affecting biomass and population dynamics of interacting species. Variation among response variables is important in this respect. The differences between artificial and natural damage were somewhat smaller for growth and reproduction than for the other types of response variables. There is thus hope that artificial damage has fairly similar effects to natural damage on the relative biomasses of different plant species in ecosystem studies. On the other hand, bioassays showed that herbivores often make a distinction between artificially and naturally damaged plants. The difference between natural and mechanical damage in induced responses may be important if secondary damage is extensive. A test of the adequacy of artificial damage for dominant plant species and their main herbivores would increase the reliability of the simulation method, but is very laborious in species-rich ecosystems. If artificial damage is used in ecosystem studies, it is of course important to take into account such problems as the difficulty of simulating many types of herbivore damage or other effects of herbivores than biomass removal, for instance impact on nutrient cycling (Hjältén, Chap. 12, this Vol.).

Ecosystem studies are typically interested in such system variables as measures of nutrient cycling and energy flow, productivity and biomasses of different trophic levels and functional types. We did not find controlled experiments that measure such characteristics and compare artificial and natural damage at the ecosystem level. Plant response traits nearest to the ecosystem variables are growth and reproduction responses, which are related to primary production and producer biomass. Because there was a tendency for the difference between responses to artificial and natural damage to decrease when the scope extended from cell and tissue level (chemical responses) to whole-plant level (growth and reproduction), it could be speculated that the difference between responses perhaps could be even smaller when the focus is at the level of ecosystems, while benefits of artificial damage – better control of damage level and experimental design – remain the same. Many types of feedback mechanisms, however, make it difficult to extrapolate from responses of individual plants to community and ecosystem level. On the other hand, if there are differences between responses at the ecosystem level, it opens up interesting possibilities. For instance, if natural damage induces herbivore defence and artificial damage does not, it could be possible to build up experimental designs to study whether induced secondary chemicals really play a role in the ecosystem context by decreasing herbivore pressure, affecting decomposition or by acting in some other way.

More generally, there are plenty of methods in biology that interfere with the natural function of biological systems. Interference often has an effect that can only partially be predicted and explained. For instance, plants are grown in growth chambers, greenhouses and common gardens, although it is known that biological processes work in a different way in artificial environments than in natural ones. Studies commonly discuss experimental artefacts as potential sources of error. Because plants can be exposed to both natural and artificial damage, it is possible to test whether the error is large or not.

We classified plant responses to growth, reproduction, resistance, chemical responses and physiological responses. These are emergent properties, so there is no general answer to the question of whether it is legitimate or not to use artificial damage. The review shows that it is good practice to carry out an experiment with both damage methods irrespective of the type of response variable, because when ecologists have tested the adequacy of herbivory simulations, in less than a third of the studies artificial and natural damage produced the same response.

Acknowledgements. We thank Evan Siemann, Wolfgang Weisser and an anonymous referee for comments on the manuscript. The study was financed by the Swedish Research Council and Östersjöstiftelsen.

References

Agrawal AA (1998) Induced responses to herbivory and increased plant performance. Science 279:1201–1202

Agrawal AA, Sherriffs MF (2001) Induced plant resistance and susceptibility to late-season herbivores of wild radish. Ann Entomol Soc Am 94:71–75

Agrawal AA, Strauss SY, Stout MJ (1999) Costs of induced responses and tolerance to herbivory in male and female fitness components of wild radish. Evolution 53:1093–1104

Alborn HT, Turlings TCJ, Jones TH, Stenhagen G, Loughrin JH, Tumlinson JH (1997) An elicitor of plant volatiles from beet armyworm oral secretion. Science 276:945–949

Anderson P, Alborn H (1999) Effects on oviposition behaviour and larval development of *Spodoptera littoralis* by herbivore-induced changes in cotton plants. Entomol Exp Appl 92:45–51

Anderson P, Jönsson M, Mörte U (2001) Variation in damage to cotton affecting larval feeding preference of *Spodoptera littoralis*. Entomol Exp Appl 101:191–198

Baldwin IT (1988) The alkaloidal responses of wild tobacco to real and simulated herbivory. Oecologia 77:378–381

Baldwin IT (1990) Herbivory simulations in ecological research. Trends Ecol Evol 5:91–93

Bergelson J, Juenger T, Crawley MJ (1996) Regrowth following herbivory in *Ipomopsis aggregata*: compensation but not overcompensation. Am Nat 148:744–755

Bergman M (2002) Can saliva from moose, *Alces alces*, affect growth responses in the sallow, *Salix caprea*? Oikos 96:164–168

Bodnaryk RP (1992) Effects of wounding on glucosinolates in the cotyledons of oilseed rape and mustard. Phytochemistry 31:2671–2677

Britton RJ (1988) Physiological effects of natural and artificial defoliation on the growth of young crops of lodgepole pine. Forestry 61:165–175

Cabrera HM, Argandoña VH, Zúñiga GE, Corcuera LJ (1995) Effect of infestation by aphids on the water status of barley and insect development. Phytochemistry 40:1083–1088

Capinera JL, Roltsch WJ (1980) Response of wheat seedlings to actual and simulated migratory grasshopper defoliation. J Econ Entomol 73:258–261

Davis MA, Boyd RS (2000) Dynamics of Ni-based defence and organic defences in the Ni hyperaccumulator, *Streptanthus polygaloides* (Brassicaceae). New Phytol 146:211–217
Doak DF (1991) The consequences of herbivory for dwarf fireweed: different time scales, different morphological scales. Ecology 72:1397–1407
Dyer MI, Moon AM, Brown MR, Crossley DA Jr (1995) Grasshopper crop and midgut extract effects on plants: an example of reward feedback. Proc Natl Acad Sci USA 92:5475–5478
Faeth SH (1986) Indirect interactions between temporally separated herbivores mediated by the host plant. Ecology 67:479–494
Gavloski JE, Lamb RJ (2000) Compensation by cruciferous plants is specific to the type of simulated herbivory. Environ Entomol 29:1273–1282
Griffiths DW, Birch ANE, Macfarlane-Smith WH (1994) Induced changes in the indole glucosinolate content of oilseed and forage rape (*Brassica napus*) plants in response to either turnip root fly (*Delia floralis*) larval feeding or artificial root damage. J Sci Food Agric 65:171–178
Gurevitch J, Hedges LV (1993) Meta-analysis: combining the results of independent experiments. In: Scheiner SM, Gurevitch J (eds) Design and analysis of ecological experiments. Chapman and Hall, New York, pp 378–425
Hammond RB, Pedigo LP (1981) Effects of artificial and insect defoliation on water loss from excised soybean leaves. J Kansas Entomol Soc 54:331–336
Hanhimäki S, Senn J (1992) Sources of variation in rapidly inducible responses to leaf damage in the mountain birch–insect herbivore system. Oecologia 91:318–331
Hartley SE (1988) The inhibition of phenolic biosynthesis in damaged and undamaged birch foliage and its effect on insect herbivores. Oecologia 76:65–70
Hartley SE, Firn RD (1989) Phenolic biosynthesis, leaf damage, and insect herbivory in birch (*Betula pendula*). J Chem Ecol 15:275–283
Hartley SE, Lawton JH (1987) Effects of different types of damage on the chemistry of birch foliage, and the responses of birch feeding insects. Oecologia 74:432–437
Hartley SE, Lawton JH (1991) Biochemical aspects and significance of the rapidly induced accumulation of phenolics in birch foliage. In: Tallamy DW, Raupp MJ (eds) Phytochemical induction by herbivores. Wiley, New York, pp 105–132
Hatcher PE, Paul ND, Ayres PG, Whittaker JB (1994) Interactions between *Rumex* spp., herbivores and a rust fungus: *Gastrophysa viridula* grazing reduces subsequent infection by *Uromyces rumicis*. Funct Ecol 8:265–272
Hatcher PE, Ayres PG, Paul ND (1995) The effect of natural and simulated insect herbivory, and leaf age, on the process of infection of *Rumex crispus* L. and *R. obtusifolius* L. by *Uromyces rumicis* (Schum.) Wint. New Phytol 130:239–249
Haukioja E, Neuvonen S (1985) Induced long-term resistance of birch foliage against defoliators: defensive or incidental? Ecology 66:1303–1308
Heil M, Koch T, Hilpert A, Fiala B, Boland W, Linsenmair KE (2001) Extrafloral nectar production of the ant-associated plant, *Macaranga tanarius*, is an induced, indirect, defensive response elicited by jasmonic acid. Proc Natl Acad Sci USA 98:1083–1088
Hodge S, Keesing VF, Wratten SD (2000) Leaf damage does not affect leaf loss or chlorophyll content in the New Zealand pepper tree, kawakawa (*Macropiper excelsum*). N Z J Ecol 24:87–89
Howe JG, Grant WE, Folse LJ (1982) Effects of grazing by *Sigmodon hispidus* on the regrowth of annual rye-grass (*Lolium perenne*). J Mammal 63:176–179
Juenger T, Bergelson J (2000) The evolution of compensation to herbivory in scarlet gilia, *Ipomopsis aggregata*: herbivore-imposed natural selection and the quantitative genetics of tolerance. Evolution 54:764–777
Kahl J, Siemens DH, Aerts RJ, Gäbler R, Kühnemann F, Preston CA, Baldwin IT (2000) Herbivore-induced ethylene suppresses a direct defense but not a putative indirect defense against an adapted herbivore. Planta 210:336–342

Litvak ME, Monson RK (1998) Patterns of induced and constitutive monoterpene production in conifer needles in relation to insect herbivory. Oecologia 114:531–540
Maschinski J, Whitham TG (1989) The continuum of plant responses to herbivory: the influence of plant association, nutrient availability, and timing. Am Nat 134:1–19
Mattiacci L, Dicke M, Posthumus MA (1994) Induction of parasitoid attracting synomone in Brussels sprouts plants by feeding of *Pieris brassicae* larvae: role of mechanical damage and herbivore elicitor. J Chem Ecol 20:2229–2247
McCloud ES, Baldwin IT (1997) Herbivory and caterpillar regurgitants amplify the wound-induced increases in jasmonic acid but not nicotine in *Nicotiana sylvestris*. Planta 203:430–435
Moon AM, Dyer MI, Brown MR, Crossley DA (1994) Epidermal growth factor interacts with indole-3-acetic acid and promotes coleoptile growth. Plant Cell Physiol 35:1173–1177
Musser RO, Hum-Musser SM, Eichenseer H, Peiffer M, Ervin G, Murphy JB, Felton GW (2002) Caterpillar saliva beats plant defences. Nature 416:599–600
Neuvonen S, Haukioja E, Molarius A (1987) Delayed inducible resistance against a leaf-chewing insect in four deciduous tree species. Oecologia 74:363–369
Ostlie KR, Pedigo LP (1984) Water loss from soybeans after simulated and actual insect defoliation. Environ Entomol 13:1675–1680
Paige KN (1992) Overcompensation in response to mammalian herbivory: from mutualistic to antagonistic interactions. Ecology 73:2076–2085
Paige KN (1999) Regrowth following ungulate herbivory in *Ipomopsis aggregata*: geographic evidence for overcompensation. Oecologia 118:316–323
Palmer AR (1999) Detecting publication bias in meta-analysis: a case study of fluctuating asymmetry and sexual selection. Am Nat 154:220–233
Pavia H, Toth GB (2000) Inducible chemical resistance to herbivory in the brown seaweed *Ascophyllum nodosum*. Ecology 81:3212–3225
Poston FL, Pedigo LP, Pearce RB, Hammond RB (1976) Effects of artificial and insect defoliation on soybean net photosynthesis. J Econ Entomol 69:109–112
Quinn MA, Hall MH (1996) Compensatory growth response of the legume, *Medicago sativa*, to defoliation and denodulation. Entomol Exp Appl 78:243–252
Reichman OJ, Smith SC (1991) Responses to simulated leaf and root herbivory by a biennial, *Tragopogon dubius*. Ecology 72:116–124
Stowe KA, Marquis RJ, Hochwender CG, Simms EL (2000) The evolutionary ecology of tolerance to consumer damage. Annu Rev Ecol Syst 31:565–595
Strauss SY, Agrawal AA (1999) The ecology and evolution of plant tolerance to herbivory. Trends Ecol Evol 14:179–185
Takabayashi J, Dicke M (1996) Plant–carnivore mutualism through herbivore-induced carnivore attractants. Trends Plant Sci 1:109–113
Teague WR (1988) The response of *Acacia karroo* plants to defoliation by hand compared to defoliation by goats. J Grassl Soc S Afr 5:122–124
Toth GB, Pavia H (2002) Lack of phlorotannin in the kelp *Laminaria hyperborea* in response to grazing by two gastropod herbivores. Mar Biol 140:403–409
Turlings TCJ, Tumlinson JH, Lewis WJ (1990) Exploitation of herbivore-induced plant odors by host-seeking parasitic wasps. Science 250:1251–1253
Walling LL (2000) The myriad plant responses to herbivores. J Plant Growth Regul 19:195–216
Welter SC (1991) Responses of tomato to simulated and real herbivory by tobacco hornworm (Lepidoptera: Sphingidae). Environ Entomol 20:1537–1541
Woodhead S (1981) Environmental and biotic factors affecting the phenolic content of different cultivars of *Sorghum bicolor*. J Chem Ecol 7:1035–1047

14 From Mesocosms to the Field: The Role and Value of Cage Experiments in Understanding Top-Down Effects in Ecosystems

O.J. Schmitz

14.1 Summary

Ecologists routinely use field experiments to gain predictive insights about ecosystem structure and function. However, experiments conducted at a whole ecosystem scale are often too crude to provide the detailed causal understanding needed for prediction. Ecologists instead try to gain such causal understanding by conducting experiments in small-scale enclosures or cages. Enclosure cages offer the fine-scale resolution and control needed to isolate certain combinations of species and derive a detailed understanding of interaction mechanisms among the species. However, the applicability of such insights to predicting whole ecosystem function depends critically on satisfying some key design criteria. These criteria include ensuring that enclosure experiments are conducted in natural field environments, as opposed to artificial laboratory settings; that behaviour of mobile species is not seriously hampered; and that experiments are conducted over time scales that represent species' natural life cycles. I detail here how enclosure cage experiments can be an effective tool in an endeavour to predict effects of perturbations on whole ecosystem function. I first provide a detailed explanation of the design of enclosure cages used in studies of herbaceous plants and arthropod herbivores and predators. I also provide guidelines for conducting experiments in ways that do not seriously distort the experimental conditions from those of natural environments. I then illustrate how cage experiments lead to predictive insights using an example from my own research on trophic interactions among spider predators, leaf-chewing and sap-feeding insect herbivores and grass and herb plants in a New England meadow ecosystem. I show that fulfilment of critical design criteria can allow one to isolate the dominant predator and herbivore species in this ecosystem and determine the nature and strength of top-down control of plant species composition and biomass production. I then show how predictions about top-down control of plant species diversity and biomass production are tested and confirmed using large-scale,

Ecological Studies, Vol. 173
W.W. Weisser and E. Siemann (Eds.)
Insects and Ecosystem Function

unenclosed field plots that experimentally manipulate both predator and herbivore trophic levels of the meadow ecosystem.

14.2 Introduction

Ecology has a long tradition of scientific discovery in which individuals have made in-depth field observations and measurements and then devised explanations for patterns among the measured variables (Hairston 1990). However, this approach by itself is insufficient for answering the pressing contemporary problem of *predicting* the consequences of natural and anthropogenic perturbations (e.g. biodiversity loss, invasive species, atmospheric N deposition, enrichment of atmospheric CO_2) on ecosystem function. Such predictive insight requires detailed understanding of the *causal* linkages between the myriad components of an ecosystem and ecosystem function. Impelled by this need to become more predictive, ecology has thus increasingly taken a more experimental tack to obtain the needed cause-effect insights (Hairston 1990; Werner 1998; Raffaelli and Moller 2000).

Experiments are powerful research tools because they afford control over extraneous variables and through replication they can lead to reasonably precise insights (Raffaelli and Moller 2000). Despite such reasons to embrace an experimental approach, it must be recognized that experiments are effectively contrivances (Skelly 2002). There is a challenge, therefore, to design and execute experiments in ways that provide meaningful insights about ecosystem function. After all, those aspects of experimentation (control, precision and replication) that offer the potential for powerful insights can also lead to outcomes that are simply artifacts of the experimental design itself rather than revelations about nature (Carpenter 1996; Werner 1998; Skelly 2002).

To illustrate this point, consider a hypothetical example (based on a composite of several cases reported in the literature) in which a researcher is interested in understanding the impact of insect herbivores on long-term plant standing crop biomass in a series of grasslands of similar age and size. Suppose that the experiment is conducted by spraying a set of randomly selected plots within each of the fields with an arthropod pesticide at regular intervals for several years to remove or reduce the abundance of insect herbivores. At the same time, another set of randomly selected plots in each field remain as untreated natural controls. Let us assume, for the sake of argument, that the researcher replicated the experiment sufficiently that it affords a high degree of statistical power (i.e. low likelihood of type II errors). This design then has all the hallmarks of good ecological experimentation (Hairston 1990): it has randomly assigned and well-replicated treatments and controls; it is repeated in different locations; and it is a long-term study. Suppose that the study revealed that there was no significant difference in plant biomass between pesticide and control plots. The consequent interpretation of this result is that herbivorous insects did not exert any dominant effect on plants and so the set of study fields are largely controlled by bottom-up (emphasizing nutrient cycling and plant competition) rather than to top-down (emphasizing trophic interactions, especially herbivory) processes.

Closer consideration of this case, however, suggests that the experiment provides an equivocal result that is directly related to the researcher's implicit, initial conception of the structure of the grassland ecosystem. That is, it was assumed that the grassland ecosystem contained only two trophic levels – plants and herbivores – or that only those two trophic levels were potentially important in determining function. There is, however, an alternative conception, namely that the ecosystem is comprised of three functionally important trophic levels: plants, arthropod herbivores and arthropod predators. If we take a classic conceptualization of a trophic chain (Hairston et al. 1960; Rosenzweig 1973; Oksanen et al. 1981; Carpenter et al. 1986) then top predators in three-level systems containing predators, herbivores and plants should have an indirect positive effect on plants by virtue of limiting the abundance of herbivores consuming plants. Experimental removal of top predators to create two-level systems containing only herbivores and plants will lead to increased herbivore abundance and an attendant decline in plant biomass. Removal of herbivores to create one-level systems should again lead to high levels of plant biomass. It is entirely conceivable then that if predators are very effective at limiting herbivore abundance and herbivore consumption of plants, plant biomass in three-level and one-level systems may be quite similar, i.e. there may be little or no detectable differences in plant biomass.

This conception thus leads to a decidedly different conclusion about the outcome of the experiment. Specifically, because the pesticide is not selective to the herbivore trophic level, its application may have effectively removed two trophic levels of the field system (arthropod predators and herbivores) simultaneously. Consequently, the system could be top-down rather than bottom-up controlled, but one cannot tell because a critical treatment is missing. To obtain an unequivocal result, the researcher should have created an additional treatment that isolated the direct effects of herbivory on plants from the indirect effects of predators on plants mediated by herbivory – something that cannot be achieved through a broad-spectrum pesticide application.

Clearly, there is an undeniable need in the above case for circumscribing or isolating certain components of the study system in order to elucidate the direct and indirect chains of causality and identify the correct mechanism governing function. Circumscribing focal sets of ecosystem components usually involves some means of enclosing or excluding those components. This is the domain of enclosure or cage experiments which leads to the purpose of this chapter. My aim here is to show why and how enclosure experiments are often critical to providing solid working insight into ecosystem functioning.

This is not to suggest that enclosure experiments are the only means by which one can come to understand ecosystem function. Enclosure experiments are a method. Like all methods, their utility in providing insight rests squarely on the nature of the questions that motivate their use (Werner 1998). Thus, this chapter is not about a methodology per se but rather how to use the method in a productive endeavour to understand the role of herbivores in natural ecosystems.

The idea of enclosing components of ecosystems to understand function is admittedly not universally accepted. Indeed, there are many arguments, some vehement, against the use of enclosures or cage experiments in ecology (see also Hjältén, Chap. 12, this Vol.; Lehtilä and Boalt, Chap. 13, this Vol.; Voelckel and Baldwin, Chap. 17, this Vol.):

- Systems of any kind that contain artificial complements of populations or communities do not have any semblance of reality (Wise 1993).
- Choice of setting for an enclosure experiment can cause the experimental outcome to be an artifact of the venue itself (Skelly 2002).
- Enclosures unrealistically constrain movement of species (MacNally 2000) and their small scale may distort natural dynamics (Carpenter 1996).
- The time scales of enclosure experiments exclude or distort important features of communities and ecosystems (Carpenter 1996; Raffaelli and Moller 2000).
- Enclosure experiments create greater amounts of semantic and deductive problems than enlightenment (MacNally 2000).

Faced with such strong arguments, one might be inclined to abandon the use of enclosure methods entirely. I would propose, however, that such forceful arguments stem from indiscriminate use of enclosures in ecological experimentation. Thus we can turn the motivation for these criticisms on its head and use the criticisms themselves as guidelines for careful experimental design in the future. More importantly, however, it is crucial that one has a larger conceptual framework in mind that sets the context for the use of any kind of experimental approach (Werner 1998; Carpenter 1999). It is without the bigger-picture context that experiments usually end up providing unrealistic, irrelevant and even diversionary results (Carpenter 1996; Skelly 2002).

Fisher (1997) provides a useful distinction that allows one to formulate the context for cage experiments: investigations of ecology *in* ecosystems vs. ecology *of* ecosystems. *In*-ecosystem investigations focus on elucidating mechanisms of species behaviour, population processes, species interactions or measurements of ecosystem process rates (Fisher 1997; Carpenter 1999). In this case, the ecosystem provides the setting within which one completes a comparatively reductionist investigation on subcomponents of the whole ecosystem. In essence, one examines 'pieces of the whole'. *In*-ecosystem studies allow for the use of well-replicated experiments that allow one to draw conclusions based on a high degree of statistical power. *Of*-ecosystem investigations examine how the entire ecosystem responds to human or natural perturbations (Fisher 1997; Carpenter 1999). These studies are concerned with large-scale effects such as loss of trophic levels, land use history, global change or nutrient loadings on net biogeochemical cycling, species composition and productivity of the ecosystem as a whole (Carpenter 1999). *Of*-ecosystem investigations are usually conducted in unenclosed settings and at a scale that is less amenable to experimental replication than *in*-ecosystem studies (Carpenter 1999). Thus, *of*-ecosystem studies aim to understand the functioning of the whole. Both kinds of investigation have their advantages and drawbacks, but they are usually viewed as wholly separate enterprises with different research goals [see *Special Feature*, Ecology 77:663–715 (1996)]. I demonstrate here that enlisting both approaches in a single research programme can lead to powerful, causal insights into the response of whole ecosystems to perturbations. In essence, I argue that *in*-ecosystem studies should be used as a prelude to obtaining *of*-ecosystem insight.

As stated above, it is difficult to obtain a predictive understanding of ecosystem function without first understanding how different ecosystem components fit together to determine function. Strategically, cage experiments, which are a crucial method in this endeavour, become part of the *in*-ecosystem phase of inquiry. I will illustrate here, by way of example from my own and related research, how cage experiments can be used productively to complement a single research programme aimed at obtaining an *of*-ecosystem understanding of function.

14.3 Research Approach

The long-term goal of my research is to identify the important determinant(s) of the structure and function of New England meadow ecosystems. These systems, like many in ecology, are comprised of many species linked together in highly interconnected networks (Paine 1988; Polis and Strong 1996). When I began, there was very little information about the degree to which each species interacted with others in this ecosystem and their collective effects on ecosystem function. To begin developing a tractable understanding of structure and function I began with the simple conceptualization that this complex system could be abstracted in terms of important drivers of function and that I only needed to build in greater complexity when the conceptualization became deficient. I therefore began with the reigning view in ecology at that time that ecosystem dynamics can be explained by the interplay between top-down (emphasizing the role of top predators) and bottom-up (emphasizing the role of competition for nutrients and abiotic conditions) factors (Leibold 1989, 1996; Hunter and Price 1992; Power 1992; Schmitz 1994; Polis and Strong

1996; Joshi et al., Chap. 6, this Vol.; Janssen and Sabelis, Chap. 9, this Vol.). One could proceed directly with large-scale experiments that manipulate the presence or structure of higher trophic levels of a whole ecosystem and determine if those manipulations have any bearing on plant biomass, productivity and nutrient cycling. While this simple experiment would lead to some understanding of effect, it does not lend itself well to obtaining predictive insight into which ecosystem component(s) cause the effect. As I show below, the potential for interactive effects among ecosystem components dictates the need to begin with a more reductionist approach in which one endeavours to understand how the pieces of the ecosystem fit together to influence the functioning of the whole.

My students and I thus began with a systematic, reductionist evaluation of the effects of top-down and bottom-up factors on species interactions and abundances – an *in*-ecosystem investigation. The insight from this research was then used to motivate a long-term perturbation experiment that could lead to a predictive *of*-ecosystem understanding of structure and function. Below, I detail the various steps taken in this endeavour and illustrate how cage experiments, used strategically, provided the necessary insights to allow prediction.

14.4 *In*-Ecosystem Investigation Using Enclosure Experiments

14.4.1 Natural History: Knowing the Players in the System

One of the most important points in the list of criticisms above is that cage experiments often contain artificial complements of populations or communities and thus do not have any semblance of reality (Wise 1993). One way to guard against this is to consider what reality is in a particular study system before doing any experimentation. Specifically, this means that one must obtain a solid working understanding of the natural history of the field system being studied.

The study system contains 18 species of herbaceous plants, although fewer coexist locally. Sampling, by clipping 50 1-m^2 plots, sorting plants to species, drying at 60 °C and weighing, revealed that the dominant (>90.6 % biomass) species are the herbs *Solidago rugosa*, *Solidago graminifolia*, *Potentilla simplex*, *Daucus carota* and *Trifolium repens* and the grass *Poa pratensis* (Fig. 14.1). Other herb species include *Asclepias syriaca*, *Solidago altissima*, *Solidago canadensis*, *Trifolium agrarium*, *Chrysanthemum leucanthemum*, *Aster novaeangliae*, *Fragaria canadensis*, *Erigeron canadensis*, *Hieracium aurantiacum* and *Rhudbeckia hirta* and the grasses *Phleum pratense* and *Hol-*

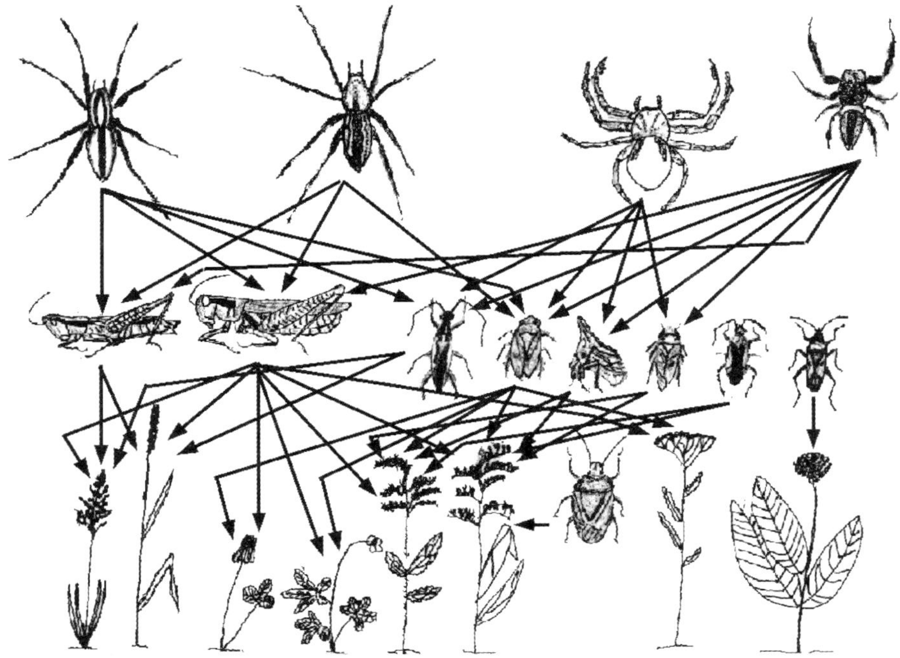

Fig. 14.1. Depiction of dominant plant, herbivore and predator species in the old-field ecosystem examined during 2 years of field research. *Arrows* indicate feeding links that were observed during the course of natural history observations in the field. Plant species, from *left* to *right*, are the grasses *Poa pratensis* and *Phleum pratense* and the herbs *Trifolium repens*, *Potentilla simplex*, *Solidago rugosa*, *Solidago altissima*, *Daucus carota* and *Asclepias syriaca*. The herbivores, from *left* to *right*, are the specialist grasshopper *Chorthippus curtipennis*, the generalist grasshopper *Melanoplus femurrubrum*, the grass-specialist plant bug *Leptopterna dolobrata*, the planthoppers *Campylenchia latipes* and *Stichtocephala festina*, the generalist pentatomid *Acrosternum hilare* (*below Stichtocephala*), the generalist spittle bug *Philaenus spumaris* (to the *right* of *Stichtocephala*), the *Solidago*-specialist plant bug *Lopidea media* and finally the milkweed bug *Lygaeus kalmii*. There are several species of hunting spider predators: from *left* to *right*, the wolf spider *Rabidosa rabida*, the nursery web spider *Pisaurina mira*, the crab spider *Misumena vatia* and the jumping spider *Phidippus rimator*

cus sp. Faunal sampling within 25 1-m^2 areas using a combination of direct observation, standard sweep net surveys and an electric-powered backpack aspirator indicates that there are two herbivore insect feeding guilds (sap-feeders and leaf-chewers) present each year (Fig. 14.1). The sap-feeding guild, comprised mainly of the generalist spittle bug *Philaenus spumaris*, the *Solidago*-specialist plant bug *Lopidea media*, the planthoppers *Campylenchia latipes* and *Stichtocephala festina*, the generalist pentatomid *Acrosternum hilare*, the milkweed bug *Lygaeus kalmii* and a grass-specialist plant bug *Lep-*

topterna dolobrata, is present during early June to mid/late July. The leaf-chewing guild, comprised of the specialist grasshopper *Chorthippus curtipennis* and the generalist grasshopper *Melanoplus femurrubrum*, begins in early/mid-July (this overlapping with the sap-feeding guild) and persists to late September. The field contains several species of hunting spider predators including the nursery web spider *Pisaurina mira*, the jumping spider *Phidippus rimator*, the wolf spider *Rabidosa rabida* and the crab spider *Misumena vatia*, which are dominant in the vegetation layer of the field during the entire summer period. These spiders are observed to consume both sap-feeding and leaf-chewing insects. Mammalian herbivores such as woodchucks *Marmota monax* do not reside in the field. Deer (*Odocoileus virginianus*) visit the field infrequently and selectively eat flower heads, leaving erect stems in the midst of flower patches. Such characteristic damage was not observed in our plots. Live trapping revealed that meadow voles *Microtus pennsylvanicus* exist at exceeding low densities (~7 individuals/ha, equivalent to 0.0028 animals per experimental plot) and thus have little or no impact on the vegetation.

The key observation here is that there are two fairly distinct insect herbivore feeding guilds (sap-feeders and leaf-chewers) present each year. These guilds may have widely different effects on plant species because of their feeding ecology and the degree to which predators influence their abundance. Consequently, there are several contingent ways that top-down effects could be transmitted through each guild to influence the abundance of plants and subsequent ecosystem function (Table 14.1). First, neither guild could have any net effect on ecosystem structure and function, in the presence or absence of predators, in which case one concludes that the field system is entirely bottom-up controlled (Table 14.1). Second, strong top-down effects might be transmitted through only one feeding guild but not the other (Table 14.1). In this case, one of the feeding guilds will play a dominant role in shaping ecosystem structure and function. Finally, top-down effects could be transmitted through both feeding guilds (Table 14.1). However, those effects could be antagonistic, i.e. one guild influences a particular group of plant species and the other guild influences another group. If groups of plants compete for resources then the effects due to one feeding guild could be reversed by the other feeding guild. In this case, there may be no net measurable effect of top-down manipulations on the ecosystem in the aggregate. However, it would be incorrect to conclude that the system was bottom-up controlled in this case. Finally, top-down effects might be transmitted through both feeding guilds in the same way (Table 14.1). Hence, the top-down effects involving one feeding guild could synergistically enhance those of the other feeding guild.

The potential for these contingencies means that the two feeding guilds must be examined in isolation of each other before we can draw reliable conclusions about their integrated effects on natural ecosystem function. Because there is some temporal overlap in their presence in the field, there is, to my reckoning, no other way to isolate their individual effects than through the

Table 14.1. Contingent possibilities for top-down control of the field system involving two insect herbivore feeding guilds

Case	Sap-feeding	Leaf-chewing	Conclusion
1	No	No	Ecosystem not top-down controlled
2	Yes	No	Top-down control due to sap-feeders
3	No	Yes	Top-down control due to leaf-chewers
4	Yes	Yes	Both guilds exhibit top-down control, but effects are antagonistic, giving appearance of no net top-down effects
5	Yes	Yes	Both guilds exhibit top-down control and effects are synergistic, giving rise to strong top-down effects

use of enclosure experiments. Thus, this was the approach that was taken in my research programme.

14.4.2 Enclosure Cages: Design and Biophysical Properties

Enclosures cages used in my research are standard for research on insects in grassland ecosystems (Ritchie and Tilman 1992; Belovsky and Slade 1993, 1995; Schmitz 1993, 1994). The cages are constructed of 1-m-high aluminium insect screening fastened at the base to a 10-cm-wide strip of aluminium sheet metal (Fig. 14.2). This cage design remains standard regardless of initial choice of basal area of the cages (e.g. 0.1, 0.25, 0.5 or 1 m^2). The choice of basal area of the cage depends very much on the size of the focal herbivore species used in experiments (see also Sect. 14.4.3.3 below). For example, enclosure cage experiments for small mammalian herbivores are possible, but they require such a large basal area that it is logistically impossible to conduct and monitor a highly replicated experiment. Moreover, such experiments cannot yield a high level of insight into interaction mechanisms. Thus, for all practical purposes, enclosure cage experiments are best suited to experiments involving insect herbivores and their arthropod predators. However, cages larger than 1 m^2 are difficult to census without seriously disturbing the vegetation and animals within the enclosure. Thus enclosure cages are best suited for experiments with insect herbivores that range in size from grasshoppers (2.5 cm length) or smaller. This allows animals to be stocked at realistic densities within a small basal area (1 m^2). To create self-sustaining mesocosms in the field, enclosure cages are secured by sinking the sheet metal beneath the soil surface and by fastening the sides of cages to wooden stakes. Cages are sealed at the top by folding the sheet aluminium screen over on itself (Fig. 14.2).

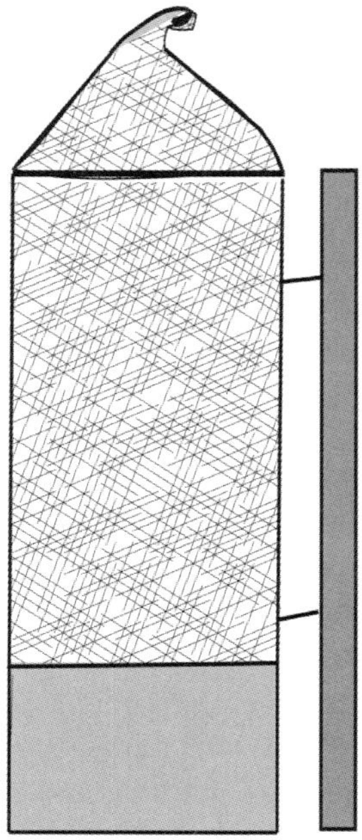

Fig. 14.2. Depiction of experimental enclosure cages and the biophysical attributes inside the cages and in matching open-field plots. Cages are constructed of aluminium screening attached to an aluminium sheet metal base by folding the sheet metal and screen over on itself and flattening the fold by hammering. The sheet metal is then sunk beneath the soil surface to enclose natural vegetation within the field. To increase upright stability, the cages are attached with wiring to a wooden stake that is pounded into the ground adjacent to the cage. Animal species are stocked to the cages from the open top and the top is then folded over on itself and sealed with binder clips, creating experimental mesocosm ecosystems

To evaluate the effects of aluminium screen enclosures on field conditions we compared incoming solar radiation, ambient air temperature and soil moisture in random 0.1-m^2 field plots and in matching experimental cages with a basal area of 0.1 m^2 (O.J. Schmitz and A.P. Beckerman, unpubl.). Solar radiation and temperature were measured using electronic sensors coupled to a data logger (LICOR, USA). The sensors were attached to a small platform (0.25 cm^2) elevated 0.5 m above the ground. Platforms were centred in the open field plot or in the cage. The platform height placed the solar radiation meter in the direct path of the sun above the vegetation. The temperature sensor was housed in a shade cup and was suspended from the platform into the upper third of the vegetation canopy. Solar radiation and temperature readings were taken every 15 min for 24 h over three representative days (full sun, partially overcast, completely overcast). Soil moisture was measured using a soil moisture probe coupled to the LICOR data logger. We augured 25-cm-deep holes into the centres of an additional set of 0.1-m^2 field plots and cages. We then placed PVC tubing into the holes. Relative soil moisture was measured as electrical conductivity at 6-, 12- and 18-cm depths within the PVC tubing. We also measured an index of biomass production of the vegetation by clipping all aboveground vegetation in another set of 0.1-m^2 field plots and cages. After 40 days, the plots were again clipped and all live biomass was dried at 60 °C and weighed.

We observed a 50 % reduction in solar radiation between open plots and enclosure cages (Fig. 14.2). There were, however, no differences in ambient air temperature or soil moisture between open plots and field cages (Fig. 14.2). We also observed a 17 % reduction in plant biomass production in enclosure cages relative to field plots (Fig. 14.2). The implication of this observation is that plants in cages could recover more slowly from herbivore damage, potentially causing cage experiments to produce slightly stronger measures of species interaction strengths than experiments in open field plots. However, I show (see Sect. 14.4.5 below) that cages do not inflate estimates of interaction strengths in this study system.

Arthropod species interactions in grassland or meadow ecosystems also can be sensitive to changes in biophysical conditions, especially moisture and temperature. For example, a 20–40 % water deficit during the growing season can lead to systematic changes in plant responses to trophic manipulations relative to higher levels of water or rainfall (Ovadia and Schmitz 2004). Thus, enclosure cages that alter moisture regimes could give outcomes that are idiosyncratic to the experimental setup. If drying is a concern, rainfall and soil moisture should be monitored periodically (say every 3–4 weeks). Water can be restored by watering cages at amounts matching ambient rainfall levels. One can guard against water deficits by using cages made of aluminium screening, as opposed to say polyethylene bags, which allows water to freely enter the enclosed microsites (see data in Fig. 14.2).

In addition, arthropod species are ectothermic and are thus sensitive to small changes in ambient air temperature. Enclosure cages that create shading conditions and reduce temperatures by as little as 5 °C can lead to dramatic reductions in feeding activity (Chase 1996) and life-cycle development (O.J. Schmitz and A.P. Beckerman, unpubl. data), both of which alter the strength of top-down effects on the ecosystem relative to the natural state (Chase 1996, O.J. Schmitz and A.P. Beckerman, unpubl. data). Again, care must be taken to monitor and maintain ambient air temperatures in the enclosures within 0–3 °C of natural air temperature. Such conditions are easily maintained by building enclosures with aluminium screening and by placing the enclosures in field locations that represent typical daily insolation (e.g. place enclosures in the middle of the field as opposed to shadier locations near field edges).

14.4.3 Considerations for the Design of Cage Experiments

14.4.3.1 Artificial Complements of Populations or Communities in Enclosure Cages Are Not Realistic

As mentioned above, we have used natural history insights to identify the temporal complement of species in the meadow ecosystem to avoid using artificial complements of species. Cages are placed directly over natural vegetation in the field. However, the small basal area of cages could lead to considerable heterogeneity in plant composition among cage microsites. To guard against such biases, one should sample the relative abundance (e.g. percentage cover) of dominant species (e.g. dominant herbs and grasses) at potential cage locations to ensure that plant composition is fairly homogeneous among cage sites before sinking cages into the soil. I recommend that dominant plants should not deviate by more than 10–15 % of the mean at any single cage site. However, it has been our experience that replicating treatments 10–15 times also helps to protect against type II errors in the event that the abundance of a dominant plant species deviates by more than 20 % among cage locations.

In preparing cages for experiments, we remove all animals occurring naturally in the vegetation layer of a cage location. Then the focal animal species are stocked to the cages at densities that accord with their natural field densities. Field densities are sampled at the time of stocking using sweep nets and electric-powered backpack aspirators. This ensures that the cage environments can be considered as field (as opposed to artificial) mesocosms.

14.4.3.2 Experimental Outcome Could Be an Artifact of the Venue

It has been demonstrated that the venue for an experiment can distort the outcome of an experiment and the subsequent interpretation about important drivers of ecosystem function (Skelly 2002). To guard against the effects of artificial venues, we have routinely conducted our experiments in the same fields in which the animals and plants reside or using vegetation taken directly from those fields (e.g. Schmitz 1993, 1994, 1998; Beckerman et al. 1997; Rothley et al. 1997; Schmitz et al. 1997; Uriarte and Schmitz 1998). When the experimental venue is matched with the natural field system, plant biomass in treatments containing a similar complement of species as in natural field conditions (i.e. three-trophic-level treatments containing plants, herbivores and predators) calibrates very closely to natural plant abundance in similar-sized, randomly sampled plots in the field (e.g. Schmitz 1993, 1994; Chase 1996).

14.4.3.3 Enclosures Unrealistically Constrain Movement of Species

Studies aimed at understanding the mechanism of interaction among species are challenging to design because examining the full range of species interactions with mobile individuals often requires a scale of resolution that is too large to quantify precisely and reliably the mechanisms of interactions among individuals and among species (Schmitz 1998). The best one can usually hope for using a completely unenclosed natural field setting is to quantify overall net trophic effects (e.g. overall net effect of predators on the plant trophic level) at the expense of a detailed mechanistic understanding of the pathways through which the effects pass (Schmitz 1998). Obtaining detailed understanding, as pointed out above, requires enclosing species to facilitate detailed observations.

The sizes of enclosure cages used in our studies are the largest possible to account for all individuals of the study species and still make precise measurements of direct and indirect interactions. The cage sizes may preclude the possibility of localized emigration. Therefore, the strengths of measured effects could be inflated relative to natural field conditions, even though the qualitative mechanism driving the interactions may be the same (e.g. see Finke and Denno 2002: note, however, that there was also a difference in experimental venue between the laboratory and field experiment in this case). One can guard against inflation of interaction strengths in cages by ensuring that species densities in enclosure cages approximate natural field densities (see below). Although this may not capture the full range of variability in effects that one may observe in a natural system, especially those arising from spatially explicit species interactions, it at least approximates the average response one might expect within a natural field (Schmitz 1998).

There are, however, two kinds of disturbances that could alter the behaviour of animal species in enclosure cages relative to natural field conditions because the species are confined. The first is related to taking routine censuses of cage populations throughout the field season to estimate herbivore or predator species survivorship. For example, frequently sampling grasshoppers in cages can cause them to respond as though they were subject to high predation risk, even when predators are not included in the enclosure cages. Thus, observer effects during censuses can confound the treatment effects. It has been our experience that sampling cages every 4–5 days is sufficient to remove any observer effects on grasshopper behaviour. Moreover, care should be taken during periods between the censuses to avoid walking through the experimental array of cages. The second disturbance comes from placing cages with different treatments too close together. In this case, cues from treatments containing predators could influence adjacent treatments that exclude predators, again causing prey to behave in ways not intended by the predator exclusion treatment. Our experience based on visual observations of animals under field conditions (e.g. Schmitz and Suttle 2001) and behavioural observations in terraria (Rothley et al. 1997; Schmitz et al. 1997; Sokol-Hessner and Schmitz 2002) suggests that a cage spacing of at least 1–1.5 m is sufficient to prevent confounding of one treatment by cues from an adjacent treatment.

14.4.3.4 Time Scale of Enclosure Experiments Exclude or Distort Important Features of Communities and Ecosystems

Time scale is obviously an issue for any experiment and careful design should tune the duration of the experiment to the life cycle of the species being examined. The norm for most experiments on trophic interactions in terrestrial systems is in the order of 4 months or less, regardless of whether they involve enclosures or not (Schmitz et al. 2000). For experiments involving herbivorous insect species, and annual and perennial herbaceous plant species, this time scale is sufficient to cover their full generation time. Yodzis (1995) argues that experiments ought to cover two full generations in order to derive reliable results about natural system function. I will show below, however, that this stringent criterion may not be necessary to obtain reliable insights from cage experiments involving arthropods in seasonal environments that are intended to inform the design of *of*-ecosystem experiments.

14.4.4 Mechanistic Insights from Enclosure Cage Experiments

Below I provide a synopsis of research conducted over the last 10 years that was aimed at unravelling whether or not the old-field system was top-down controlled, and if so, to identify which of the herbivore guilds were involved (Table 14.1) and the nature of the top-down effect. The experimental signature for a top-down effect will vary depending on that nature of predator–herbivore interactions. Removal of top predators from three-level systems containing predators, herbivores and plants could lead to a decline in plant biomass if the predator has a positive indirect effect on plants, mediated by say a reduction in herbivore density or a reduction in foraging activity to avoid predation risk (Schmitz 1998). Removal of top predators could have a negative indirect effect on certain plants (e.g. those providing safety) if predators cause herbivores to switch from feeding on highly nutritious plants to seeking refuge in and foraging on less nutritious but structurally complex plants that allow them to minimize exposure to predation risk (Schmitz 1998). The complete design needed to unravel the nature and sign of the direct and indirect effects is as follows: one-level plants only; two-level plants and herbivores; and three-level plants, herbivores and carnivores.

14.4.4.1 Identifying the Potential for Top-Down Control

We examined the role of predation on the damage sap-feeding herbivores inflict on old-field plants across a natural productivity and successional gradient (Uriarte and Schmitz 1998). We found that the ability of predators to control herbivore damage on plants varied with productivity. Top-down control was strong in sites that were highly productive through fertilization, while plant competitive interactions may have obscured any pattern of top-down control in intermediate and low productivity sites (Uriarte and Schmitz 1998). Most of the fields in the study area are in an early–mid successional state. We thus concluded that top-down effects were unlikely to be transmitted through the sap-feeding guild under natural conditions in our study fields.

Experiments with the leaf-chewing herbivores revealed the emergence of top-down effects (Schmitz 1998). Further experimentation, in which the grass specialist and generalist grasshoppers were systematically isolated in different cages, revealed that the top-down effects only transmitted through the generalist grasshopper. In other words, the effects of the generalist species swamped out the effects of the specialist when the two species coexisted (Schmitz 1998).

This series of experiments indicated that contingency 3 (Table 14.1) is the likely explanation for trophic control in this system. However, instead of stopping with this insight, we continued experimenting to identify the nature of the top-down effect. As will be shown in the description of *of*-ecosystem experiments below, knowing how top-predator effects are transmitted lower down in the food chain is important because the exact transmission mechanism can affect the nature and sign of the indirect effect of predators on resources (Schmitz et al. 2000) and thus can affect the reliability of predictions about the outcome of perturbation experiments on ecosystems (Schmitz 2003). The case in point here regards the tension between the classic perspective that top-down effects arise from changes in herbivore density (density-mediated effects), and the more contemporary perspective that top-down effects might be transmitted by consumer behaviour (trait-mediated effects) in addition to, or in place of, density-mediated effects. The classical tri-trophic theory outlined above, and related constructs (Hairston et al. 1960; Rosenzweig 1973; Oksanen et al. 1981; Carpenter et al. 1986) are all based on the assumption that the indirect effect of predators on resources is mediated by changes in consumer density due to direct predation. It is well known in ecology that consumers also respond to predators by behaving in ways that reduce contact with them in order to reduce the risk that they will be captured (Sih 1980; Mangel and Clark 1986; Lima and Dill 1990).

It is conceivable that changes in consumer behaviour, arising from the need to avoid predators, can cause food web effects that are similar in sign and strength to those propagated by reductions in consumer density. For instance, each direct predation event involves a single consumer and a single resource item. During that same time period, the risk introduced by the mere presence of that predator could have more widespread effects by causing many individuals to become increasingly vigilant at the expense of foraging (Schmitz et al. 1997). This persistent level of vigilance can have lasting effects, via changes in prey energy budgets and hence survival, whose impact at the population and community level equals that of direct predation. Theory that explicitly embodies such anti-predator behaviour of consumers (Abrams 1984, 1992, 1995) predicts a wider range of indirect effects in food chains than does classical theory.

Teasing apart the importance of behaviour and density effects in communities is, necessarily, a multiscale problem because the investigation must transcend three organizational hierarchies: individual, population and community (Levin 1998; Ovadia and Schmitz 2002). A complete picture of the interplay among these organizational hierarchies in the old-field study system required a systematic evaluation of the spider impacts at the morphological and behavioural level of grasshoppers, at the population level of grasshoppers and at the entire ecosystem level. This research was completed using the dominant players, the hunting spider *P. mira*, the generalist grasshopper herbivore *M. femurrubrum*, and the grass and herb resources of the grasshopper.

At the morphological level, examinations were conducted to verify that the species of spider predator used in the experiments was indeed capable of capturing and subduing grasshopper prey. Laboratory feeding trials and field observations of spider hunting were conducted to ascertain which size classes of grasshopper prey were vulnerable to the spider predators.

Behavioural-level examinations were conducted to evaluate the changes in generalist grasshopper vigilance and diet selection in response to the presence of spider predators. This is the level at which predators have a direct effect on the trade-off between costs and benefits of foraging through their effects on grasshopper foraging time budgets (vigilance) and plant resource selection.

Predators are capable of inducing both direct and indirect effects on prey at the population level. They can induce direct effects by killing grasshoppers and reducing their density. They can have indirect effects on density because their very presence can cause behavioural changes of grasshoppers (time budget and diet shifts) that could increase grasshopper mortality through starvation. Thus, the population-level examinations independently quantified the effects of predation and predation risk on the population size of grasshopper prey. This was accomplished using two predation treatments that changed the modality of predator effects. In one treatment (predator spiders), spiders were capable of capturing and subduing prey. In another treatment (risk spiders), direct predation was 'switched-off' by rendering spiders incapable of capturing prey while still allowing them to wander throughout the canopy in search of prey.

Finally, at the ecosystem level, the extent to which spider effects on plants were transmitted by changes in grasshopper behaviour relative to changes in grasshopper density was quantified using the same predation treatments used to examine population-level effects.

To measure killing success, individual grasshoppers with known body size (mass and length) were presented to *P. mira* spiders of different known body size in small glass terraria. These feeding trials revealed that the maximum size of grasshopper prey killed is linearly related to spider body size. *P. mira* spiders could capture and subdue grasshoppers up to 1.3 times their own body size. In the laboratory, all of the spiders used in the feeding trials killed small, early developmental stage juvenile grasshoppers (early instars), 50 % of them killed larger, older developmental stage juvenile grasshoppers (late instars) and 10 % of them killed an adult grasshopper within 10-day feeding periods (Schmitz and Suttle 2001). In the natural field setting, *P. mira* successfully captured a late instar and an adult in the course of 17 observed predation attempts on a variety of prey (Schmitz and Suttle 2001).

Behavioural trials were conducted in screen terraria in order to facilitate focused observations on individual grasshoppers when in the presence and absence of spiders (Schmitz et al. 1997). The terraria were placed in a field setting in a line with uniform exposure to the sun. Each terrarium contained a 40-cm^2 piece of fresh sod cut from the old-field in which the grasshoppers occurred naturally. Grasshoppers were randomly assigned to a control (no predator) or predator treatment and allowed to acclimatize to this experimental environment for 1 day prior to taking measurements. These trials revealed that grasshoppers exhibited some periodicity in their foraging behaviour throughout the day, with high peaks of feeding separated by periods of resting. However, there were shifts in this feeding periodicity between no predator and predator treatments (Schmitz et al. 1997). In the presence of predators, grasshopper foraging activity became protracted to the middle of the day relative to the no predator control. This activity shift makes biological sense. Grasshoppers forage during the hotter parts of the day. This is a period when spiders have difficulty coping with heat stress because they do not have a thick cuticle to prevent desiccation. Indeed, field observations of spiders revealed that they become dormant during the middle part of the day and seek shade underneath leaves of plants (O.J. Schmitz and K.B. Suttle, unpublished data).

This activity shift by grasshoppers is costly, however, because grasshoppers can incur a greater risk of mortality from heat stress itself or heat stress can cause uncoordinated movement and torpor that could increase their vulnerability to predation at later time periods (Schmitz et al. 1997}. Also, the change in feeding periodicity results in a 65-min (18 %) reduction in daily feeding time.

Grasshoppers shifted their resource selection in experimental field cages in which grasshoppers, plant resources and spider predators occurred at natural field densities. In the absence of predators, grasshoppers preferred grasses to herbs, with a diet comprised of 70 % grass (Rothley et al. 1997). With the experimental addition of *P. mira* spiders, grasshoppers reduced the proportion of grasses in the diet to 44 %. They also reduced their total daily dry mass intake by 10 % owing to decreased total feeding time due to vigilance (Rothley et al. 1997}. The increased preference for herbs appears to be a solution to a predation-risk foraging trade-off. Grasshoppers may be seeking refuge from predation in herbs because of the leafy structural complexity afforded by this vegetation and this may decrease spider capture success relative to grasses (Beckerman et al. 1997) The reduction in feeding time and diet shift is costly: it translates into a 25 % reduction in estimated daily energy intake (Rothley et al. 1997), thus increasing the chance of starvation.

Population-level studies were conducted in a field setting using experimental cages. Two sets of experiments were run in two consecutive summers with different size classes for prey. The first experiment began with early instar (2nd and 3rd) grasshopper nymphs and it was terminated each year when all the nymphs were fully molted into 4th instar nymphs (a 24-day duration). The second experiment was conducted in a completely new set of mesocosms and began with 4th instar nymphs. The experiment continued through 5th instars and adults until early autumn when an insect-killing frost naturally terminated the experiment (60 days). Grasshopper populations in each cage were censused biweekly for the duration of the experiments.

To isolate the relative importance of behavioural effects from density effects, two types of experimental spider treatments were used in addition to a no spider control. In one treatment (predation spiders), the spiders were free to hunt and capture their prey within the experimental cages. In the second treatment (risk spiders), spiders were free to hunt prey, but they were prevented from capturing prey by gluing together their killing mouth parts (chelicerae) with surgical cement. Behavioural trials in the laboratory indicated that gluing the spider mouth parts had no adverse effects on movement behaviour of these treated spiders relative to untreated controls (Schmitz et al. 1997). In the wild, hunting spiders have the capacity to survive 50 or more days without eating (Foelix 1996). Indeed, live risk spiders were frequently recovered at the end of the experiments (Schmitz et al. 1997; Schmitz 1998).

In experiments involving early instar nymphs, both spider treatments resulted in a population decline of grasshoppers, relative to a no-predation control. The magnitude of reduction was 29 % for the predation spider treatment and 20 % for the risk spider treatment. Even though there are significant density reductions in this experiment, the mechanism largely driving this response is risk effects, i.e. increased starvation as a consequence of predator presence. The mortality arising from direct predation is largely compensatory to mortality, arising from starvation caused by the need for the grasshoppers to be vigilant and undergo a diet shift. In other words, spiders did not cause additional direct mortality to grasshopper cage populations. Rather, they only consumed the proportion of the population equivalent to that which was destined to die from starvation.

The experiment involving late instar and adult grasshoppers revealed that neither predator treatment had a significant effect on prey density relative to the control. Recall, however, that these spiders were capable of killing larger prey such as late instar and adult grasshoppers. Moreover, predation events were observed in the cages during the course of completing biweekly censuses (Beckerman et al. 1997; Schmitz 1998). This lack of a net density effect implies that direct and indirect predation effects were entirely compensatory to natural mortality of individuals in the absence of predation. This absence of a strong density effect also implies that top-down effects should be largely if not entirely driven by trait-mediated interactions. If true, then one would expect that grass and herb biomass in one-level systems should be comparatively high. The addition of the generalist grasshopper to create two-level systems should result in a strong reduction in grass biomass and slight reduction in herb biomass owing to grasshopper diet preferences. Adding *P. mira* spiders to create a three-level system should cause a relaxation of herbivore pressure on grass and intensification of pressure on herbs. Thus, grass abundance should be higher in a three-level than a two-level system (a positive indirect effect of top predators) and herb abundance should be lower in a three-level than a two-level system (a negative indirect effect of top predators).

The population-level experiment described above had three of four treatments necessary to test for these kinds of indirect effects: a two-trophic-level treatment (plants and grasshoppers) and three-trophic-level treatments (plants, grasshoppers and risk or predation spiders). The food web-level experiment was 'piggy-backed' on the population-level experiment by adding a plant-only (one-level web) treatment to the experimental design. Both grass and herb abundance in each of the treatments was monitored in addition to grasshopper abundance. This experiment was conducted twice each year. First, early instar grasshoppers were stocked to cages and the experiment was terminated when they became late instars. A new experiment in a new set of mesocosms was then begun with late instars and this experiment was terminated in early autumn when adults were dying from a killing frost. The community-level experiments revealed that the observed cascading effects were largely, if not completely, a consequence of diet shifts by grasshoppers in response to predation risk. For both developmental stages of grasshopper, top predators had an indirect positive effect on grasses. Experiments with earlier instar nymphs revealed little or no statistically detectable effect on herb abundance, possibly because of the interplay between direct diet shift and indirect density reduction due to starvation from predation risk (Schmitz et al. 1997). In other words, herbivores may have shifted their diet, but at the same time there were fewer herbivores available to damage the plants. Experiments with later developmental stages revealed that grasshoppers in two-level treatments caused a reduction in both grass and herb biomass relative to a one-level control. Spiders caused a net increase in grass biomass and a reduction in herb biomass relative to a two-level treatment. Moreover, the effect was identical for predation spider and risk spider treatments. These results provide support for the a priori expectation of positive and negative indirect effects of predators on grass and herb biomass.

It is noteworthy that if we examine top predator effects on individual plant species, it turns out that we can be even more precise in our understanding of top-down effects. Indeed, the effects can be found largely in two dominant plant species, the grass *Poa pratensis* and the highly productive and competitively dominant herb *Solidago rugosa* (Schmitz 2003). Moreover, the abundance of *S. rugosa* is inversely related to the diversity (specifically evenness) of other herb species in the field.

This insight then leads to some interesting predictions of the effects of top predators on plant diversity and productivity in this field system. Indeed, complexity in this system can be distilled down into some straightforward principles. Top-down effects transmitted through the leaf-chewing guild should predominate in the ecosystem as a whole. Specifically, the dominant interaction shaping the structure of the ecosystem is the trait-mediated interaction involving *P. mira* spiders, *M. femurrubrum* grasshoppers, *P. pratensis* grass and *S. rugosa* herb. *P. mira* should have a strong diversity-enhancing effect on plants by causing *M. femurrubrum* grasshoppers to switch from feeding on *P. pratensis* to feeding heavily on *S. rugosa* (Fig. 14.3). This suppression of *S. rugosa* abundance should, over the long term, cause other less

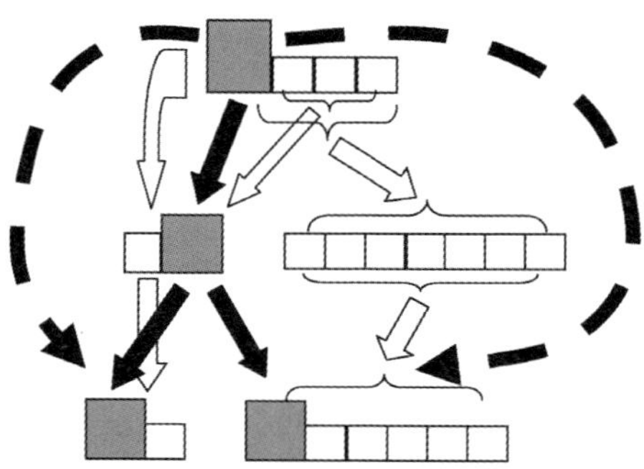

Fig. 14.3. Schematic depiction of predator (*top*), herbivore (*middle*) and plant (*bottom*) species presented in Fig. 14.1. Each *square* represents a unique species associated by guild or functional group, e.g. leaf-chewing (*left*) vs. sap-feeding (*right*) herbivores; grasses (*left*) vs. herbs (*right*). *Large dark squares* represent dominant players in the ecosystem as determined by trophic interaction strengths and ecosystem structure (Schmitz 2003). *Dark solid arrows* indicate strong direct interactions, *hollow arrows* indicate weak direct interactions and *dark dashed arrows* indicate indirect effects. Dominant species, from *top* to *bottom*, are *Pisaurina mira* (spider), *Melanoplus femurrubrum* (grasshopper), *Poa pratensis* (grass) and *Solidago rugosa* (herb)

productive herb species that are intolerant to shading by *S. rugosa* to proliferate (i.e. increase plant species diversity). This amounts to a multitrophic-level variant of the classic keystone predation hypothesis (Paine 1966; Leibold 1996) in which *P. mira* acts as an *indirect* keystone predator (Schmitz 2003). One consequence of the shift in species abundance is that an important ecosystem function, biomass production in the plant trophic level as a whole, should decline.

14.4.5 *Of*-Ecosystem Studies: Testing the Reliability of Mechanistic Insights from Cage Experiments

The ultimate arbiter of whether or not the cage experiments provide predictive insight into the effects of perturbations on long-term ecosystem structure and function is to test the prediction using an *of*-ecosystem phase of inquiry. To accomplish this, I return to the pesticide experiment described in Section 14.2. Recall that in order to test for top-down effects on whole ecosystem function, I argued that it was insufficient simply to apply a pesticide (to remove higher trophic levels) and compare this treatment with an unmanipulated control. Rather, one must systematically remove one trophic level at a time. I therefore tested the indirect keystone predation hypothesis (top-down control) with a multiyear (1999–2001) field experiment that systematically manipulated the number of trophic levels in 2×2-m open field plots. I measured the effects of each manipulation on plant diversity and productivity.

The experiment consisted of two treatments and a control replicated ten times using a randomized blocks design (Schmitz 2003). The control was the natural field system containing plants, herbivores and predators: hereafter the three-trophic-level treatment. I created a predator exclusion treatment (hereafter the two-trophic-level treatment containing herbivores and plants) by surrounding 2×2-m plots with a 45-cm-wide aluminium sheet-metal barrier, buried 5–10 cm in the ground. All arthropod predators within the exclusion plots were removed by hand at the beginning of the experiment. Semi-weekly inspections of plots were conducted to ensure the persistent absence of arthropod predators. Finally, a pesticide treatment was created to prevent damaging herbivory (hereafter, the one-trophic-level treatment containing only plants). I used the organic pyrethroid insecticide Fenvelerate (which is widely used by fruit growers). Organic pyrethroids have no demonstrable adverse effects on plant growth (see review in Root 1996). The pesticide was applied using a backpack sprayer at a concentration of 0.21 l m^{-2} every 15 days. This watering level is two orders of magnitude lower than the mean rainfall level received by all plots for the same period (Ovadia and Schmitz 2004) and should thus not lead to a systematic watering bias on plant production in the pesticide treatment.

Sampling prior to initiating the experiment revealed that initial plant community structure (number of plant species and percentage of area covered by a species) was statistically similar among treatment plots (Schmitz 2003). The sheet metal barrier used to exclude predators could also introduce a bias in plant production by altering microclimatic conditions relative to fully open plots. Before adding an additional sheet metal 'control' plot and the time required to sample that control, we can first evaluate if such a control is necessary by quantifying biophysical variables (photosynthetically active radiation, PAR, wind speed and ambient air temperature) important to plant growth (Drake et al. 1989) in a representative enclosed and a paired open plot. PAR was measured using a LI-COR LI-250 light meter coupled to a LI-190SA quantum sensor. Wind speed was measured using a Kestrel 1000 wind meter accurate to 0.3 m s^{-1}. Ambient air temperature was measured using a Cole-Parmer Digi-Sense 8523 thermistor thermometer coupled to an air probe accurate to 0.1 °C. All measurements were made at four elevations above ground (0, 20, 40 and 60 cm). Comparison of PAR, average daily wind speed and temperature revealed no systematic bias in microclimate between open and enclosed plots (Schmitz 2003) consistent with similar evaluations for other field enclosures (Drake et al. 1989), indicating that the sheet metal did not introduce biases on plant production.

Each year, individual plant species within plots were sampled nondestructively once per month throughout the growing season (May to September) using the 1.6×1.6-m sampling quadrat. I measured the number of plant species present in a plot and percentage of plot area that each species covered. I estimated plant species diversity using two indices: species richness and evenness. Species richness (S) is the standard measure in studies of diversity/productivity relationships (Tilman et al. 2001). Plant species richness does not account for changes in species abundances alluded to above. Top predators in my system should indirectly alter the proportional representation of those plant species whose abundance would otherwise be suppressed by the dominating effect of *S. rugosa*. Species evenness is one measure that combines the two mechanistic factors – dominance and plant species abundance – that are important in this and other old-field systems (Wilsey and Potvin 2000).

I estimated dry plant biomass production (g $plot^{-1}$ day^{-1}), an index of productivity, by calculating the differences in plant biomass in plots between consecutive sampling periods in 2001 (31 May and 19 June; and 19 June and 19 July). The differences in biomass for each of the two sampling periods were divided by the intervening number of days to estimate rate. These two periods represent independent growth conditions (wetter spring, drier summer) in the field and so enable an assessment of the consistency of the productivity response. I converted nondestructive sample estimates of percentage cover into estimates of dry biomass using regression equations. To do this, mean height of a species in a plot was measured concurrently with each sampling of plant species cover using a meter stick on five random plants of a species.

Plant species biomass was then estimated using multiple regression equations describing prior calibrations of plant species dry mass in relation to species percentage cover and species height (Schmitz 2003).

14.4.5.1 Direct and Indirect Effects of Top Predators

The experiment revealed that there were significant, cumulative treatment effects on percentage *P. pratensis* cover and percentage *S. rugosa* cover. Three years of predator exclusion resulted in significantly lower *P. pratensis* and *S. rugosa* cover relative to the one-trophic-level, plant-only treatment (Schmitz 2003). Sustained predator presence (three-trophic-level treatment) resulted in increased *P. pratensis* cover and decreased *S. rugosa* cover relative to the predator exclusion treatment (Schmitz 2003). This indirect positive effect of predators on *P. pratensis* and negative indirect effect on *S. rugosa* is consistent with results from the cage experiments described above but observed at a fourfold larger scale and over a threefold longer period. There are two important conclusions to draw from this result.

First, contrary to recommendations (Yodzis 1995), it was not necessary to run the short-term experiments for more than one generation of the focal species in order to obtain reliable predictive insight about top-down effects in *of*-ecosystem processes. I should add, however, that this conclusion may only apply to seasonal systems such as this in which each of the animal species in question completes its life cycle within the course of a single field season and thus has non-overlapping generations. In other words, the long-term dynamics of the unperturbed field system can be viewed as a concatenated string of short-term events in which major density-dependent processes operate within a season leading to little carryover of density-dependent effects from one season to the next (Schmitz 2000). Perturbations should lead to changes in ecosystem state, but the new state should be reached within 2–3 years, according to simulation models of the field system (Schmitz 2000). This rate of change appears to be borne out by the *of*-ecosystem field experiment (Schmitz 2003).

Second, I evaluated whether or not the interaction strengths in the cage experiments were indeed stronger than in the field by estimating the strength of the top-down effect of predators on plants as the log ratio $[\ln(Np_+/Np_-)]$of total estimated July 2001 dry plant biomass in plots in the presence (Np_+) and absence (Np_-) of predators (Schmitz et al. 2000). The strength of predator effects on plants was estimated to be 0.62±0.107 (95 % confidence interval, CI) (Schmitz 2003). This value is very similar to those estimated for the cage experiments in my system: 0.56±0.30 (95 % CI) for experiments involving early instar grasshoppers and 0.04±0.21 (95 % CI) for experiments involving late instars and adults (Schmitz et al. 2000). Thus, cage experiments do not de facto inflate estimates of interaction strengths.

14.4.5.2 Top Predator Effects on Plant Diversity and Productivity

Manipulation of trophic structure caused significant shifts in plant species richness and evenness combined (Schmitz 2003). A posteriori statistical analysis revealed that this significance was entirely due to treatment effects on plant species evenness: there was no significant treatment effect on plant species richness (Schmitz 2003). Species richness did not change in the experimental system because most of the plant species sampled after 3 years existed in all plots at the onset of experimentation, and there were no new, invading herb or grass species given the successional stage of the system. The experiment revealed, however, that plant species evenness was significantly higher in three-trophic-level plots than in either the two-trophic-level or one-trophic-level treatments (Schmitz 2003). In addition, there was a significant decrease in estimated plant productivity with increasing number of trophic levels in both sampling periods (Schmitz 2003). Thus, top predators had a demonstrated, important effect on the structure and functioning of the natural ecosystem. They caused a shift in plant species diversity by increasing the proportional representation of less productive herb species. This translated into a decline in plant-trophic-level biomass production.

Despite the high species richness in all trophic levels in this system, top-down effects appeared to transmit linearly down a chain comprised of a dominant predator and herbivore species to the plant species (Fig. 14.3). Furthermore, I was able to use mechanistic insight about the chain of interactions between the dominant species derived from *in*-ecosystem cage experiments to begin predicting reliably the effect of manipulating higher trophic levels on *of*-ecosystem plant species composition and plant biomass production.

References

Abrams PA (1984) Foraging time optimization and interactions in food webs. Am Nat 124:80–96

Abrams PA (1992) Predators that benefit prey and prey that harm predators: unusual effects of interacting foraging adaptations. Am Nat 140:573–600

Abrams PA (1995) Implications of dynamically-variable traits for identifying, classifying, and measuring direct and indirect effects in ecological communities. Am Nat 146:112–134

Beckerman AP, Uriarte M, Schmitz OJ (1997) Experimental evidence for a behavior-mediated trophic cascade in a terrestrial food chain. Proc Natl Acad Sci USA 94:10735–10738

Belovsky GE, Slade JB (1993) The role of vertebrate and invertebrate predators in a grasshopper community. Oikos 68:193–201

Belovsky GE, Slade JB (1995) Dynamics of some *Montana* grasshopper populations: relationships among weather, food abundance and interspecific competition. Oecologia 101:383–396

Carpenter SR (1996) Microcosm experiments have limited relevance for community and ecosystem ecology. Ecology 77:677–680
Carpenter SR (1999) Microcosm experiments have limited relevance for community and ecosystem ecology: reply. Ecology 80:1085–1088
Carpenter SR, Kitchell JF, Hodgson JR (1986) Cascading trophic interactions and lake productivity. Bioscience 35:634–639
Chase J (1996) Abiotic controls of trophic cascades in a simple grassland food chain. Oikos 77:495–506
Drake GG, Leadley PW, Arp WJ, Nassiry D, Curtiss PS (1989). An open-top chamber for field studies of elevated atmospheric CO_2 concentration on saltmarsh vegetation. Funct Ecol 3:363–371
Finke DL, Denno RF (2002) Intraguild predation diminished in complex-structured vegetation: implications for prey suppression. Ecology 83:643–652
Fisher SG (1997) Creativity, idea generation and the functional morphology of streams. J N Am Benth Soc 16:305–318
Foelix RF (1996) The biology of spiders. Harvard University Press, Cambridge, MA
Hairston NG (1990) Ecological experiments: purpose, design and execution. Cambridge University Press, Cambridge
Hairston NG, Smith FE, Slobodkin LB (1960) Community structure, population control, and competition. Am Nat 94:421–425
Hunter MD, Price PW (1992) Playing chutes and ladders – heterogeneity and the relative roles of bottom-up and top-down forces in natural communities. Ecology 73:724–732
Leibold MA (1989) Resource edibility and the effects of predators and productivity on the outcome of trophic interactions. Am Nat 134:922–949
Leibold MA (1996) A graphical model of keystone predators in food webs: trophic regulation, of abundance, incidence, and diversity patterns in communities. Am Nat 147:784–812
Levin SA (1998) Ecosystems and the biosphere as complex adaptive systems. Ecosystems 1:431–436
Lima SL, Dill LM (1990) Behavioral decisions made under the risk of predation. Can J Zool 68:619–640
MacNally R (2000) Modeling confinement experiments in community ecology: differential mobility among competitors. Ecol Model 129:65–85
Mangel M, Clark CW (1986) Towards a unified foraging theory. Ecology 67:1127–1138
Oksanen L, Fretwell SD, Arruda J, Niemela P(1981) Exploitation ecosystems in gradients of primary productivity. Am Nat 118:240–262
Ovadia O, Schmitz OJ (2002) Linking individuals with ecosystems: experimentally identifying the relevant organizational scale for predicting trophic abundances. Proc Natl Acad Sci USA 99:12927–12931
Ovadia O, Schmitz OJ (2004) Weather variation and trophic interaction strength: sorting the signal from the noise. Oecologia (in press)
Paine RT (1966) Food web complexity and species diversity. Am Nat 100:65–73
Paine RT (1988) Food webs: road maps of interaction or grist for theoretical development? Ecology 69:1648–1654
Polis GA, Strong DR (1996) Food web complexity and community dynamics. Am Nat 147:813–846
Power ME (1992) Top-down and bottom-up forces in food webs – do plants have primacy? Ecology 73:733–746
Raffaelli D, Moller H (2000) Manipulative field experiments in animal ecology: do they promise more than they can deliver? Adv Ecol Res 30:299–337
Ritchie ME, Tilman D (1992) Interspecific competition among grasshoppers and their effect on plant abundance in experimental field environments. Oecologia 89:524–532

Root RB (1996) Herbivore pressure on goldenrods (*Solidago altissima*): its variation and cumulative effects. Ecology 77:1074–1087
Rosenzweig ML (1973) Exploitation in three trophic levels. Am Nat 107:275–294
Rothley KD, Schmitz OJ, Cohon JL (1997) Foraging to balance conflicting demands: novel insights from grasshoppers under predation risk. Behav Ecol 8:551–559
Schmitz OJ (1993) Trophic exploitation in grassland food webs: simple models and a field experiment. Oecologia 93:327–335
Schmitz OJ (1994) Resource edibility and trophic exploitation in an old-field food web. Proc Natl Acad Sci USA 91:5364–5367
Schmitz OJ (1998) Direct and indirect effects of predation and predation risk in old-field interaction webs. Am Nat 151:327–342
Schmitz OJ (2000) Combining field experiments with individual-based modeling to identify the dynamically-relevant organizational scale in a field system. Oikos 89:471–484
Schmitz OJ (2003) Top predator control of plant diversity and productivity in an old-field ecosystem. Ecol Lett 6:156–163
Schmitz OJ, Suttle KB (2001) Effects of top predator species on direct and indirect interactions in a food web. Ecology 82:2072–2081
Schmitz OJ, Beckerman AP, O'Brien KM (1997) Behaviorally mediated trophic cascades: effects of predation risk on food web interactions. Ecology 78:1388–1399
Schmitz OJ, Hambäck PA, Beckerman AP (2000). Trophic cascades in terrestrial systems: a review of the effects of top carnivore removals on plants. Am Nat 155:141–153
Sih A (1980) Optimal behavior: can foragers balance two conflicting demands? Science 210:1041–1043
Skelly DK (2002) Experimental venue and estimation of interaction strength. Ecology 83:2097–2101
Sokol-Hessner L, Schmitz OJ (2002) Aggregate effects of multiple predator species on a shared prey. Ecology 83:2367–2372
Tilman D, Reich PB, Knops J, Wedin D, Mielke T, Lehman C (2001) Diversity and productivity in a long-term grassland experiment. Science 294:843–845
Uriarte M, Schmitz OJ (1998) Trophic control across a natural productivity gradient with sap-feeding herbivores. Oikos 82:552–560
Werner EE (1998) Ecological experiments and a research program in community ecology. In: Resetarits WJ, Bernardo J (eds) Experimental ecology: issues and perspectives. Oxford University Press, Oxford
Wilsey BJ, Potvin C (2000) Biodiversity and ecosystem functioning: importance of species evenness in an old field. Ecology 81:887–892
Wise DH (1993) Spiders in ecological webs. Cambridge University Press, Cambridge
Yodzis P (1995) Food webs and perturbation experiments: theory and practice. In: Polis GA, Winemiller KO (eds) Food webs: integration of patterns and dynamics. Chapman and Hall, New York

15 Reducing Herbivory Using Insecticides

E. Siemann, W.P. Carson, W.E. Rogers and W.W. Weisser

15.1 Summary

Insecticides are a vital tool for manipulating insect herbivory, but limitations of the method can result in erroneous conclusions about the relationships among herbivores, plant population dynamics, plant community composition and ecosystem processes. In particular, direct effects of insecticide applications on plants or ecosystem processes, effects on non-target organisms and indirect effects via plant competition may cause ecologists to misjudge the importance of insect herbivores in ecosystems. A survey of published studies showed that most investigators considered some of the more likely artifacts, but few were thorough in testing for such artifacts. Data on insect damage and insect abundance are particularly useful for establishing a causative role for insect herbivore suppression in insecticide effects.

15.2 Basic Concepts

Insecticides are an important tool for studying the role of insect herbivores in terrestrial ecosystems. Because insect herbivores are typically small and mobile, physically excluding them from replicated experimental areas in the field is difficult. The main methods of physical exclusion of aboveground insect herbivores (cages, netting and screens) also exclude larger herbivores, such as mammals, which can make it difficult to assess the role of insect herbivores independently (but see Schmitz, Chap. 14, this Vol.). In addition, any method of physical exclusion may change microclimate and light levels. Physical exclusion of belowground herbivores in field experiments is more problematic. So long as ecologists are careful about the inferences they draw from insecticide experiments and are mindful of the limitations of the method, chemical exclusion of insects remains one of the most valuable methods to investigate the role of insects in terrestrial ecosystems.

Ecological Studies, Vol. 173
W.W. Weisser and E. Siemann (Eds.)
Insects and Ecosystem Function

15.3 Using Insecticides to Infer the Role of Herbivores

Applications of insecticides can reveal the role of insect herbivores in an unmanipulated ecosystem by examination of the assembly, dynamics, structure and functioning of an ecosystem with insect herbivores absent or present at low densities. Still, there are some inherent difficulties in determining the relationships among components in an intact complex system by examining the same system with a single component removed. These include multiple potential paths of causation as well as emergent properties or higher-order interactions in complex systems that may be difficult or impossible to infer by analyzing the behaviour of the system with some components missing (Billick and Case 1994). There are, however, some general principles that apply to interpretation of insecticide experiments that have a tradition in the ecological literature and that have a solid conceptual basis.

Increases in peak standing crop with insecticide applications are usually equated with an effect of insect herbivores on net productivity in the unmanipulated system (Fig. 15.1). In the short term, such increases give insight into the magnitude of the effect of insect herbivory on net productivity within the existing suite of plant species present at the start of the study. The magnitude of the response of net plant productivity to insect exclusion with insecticides

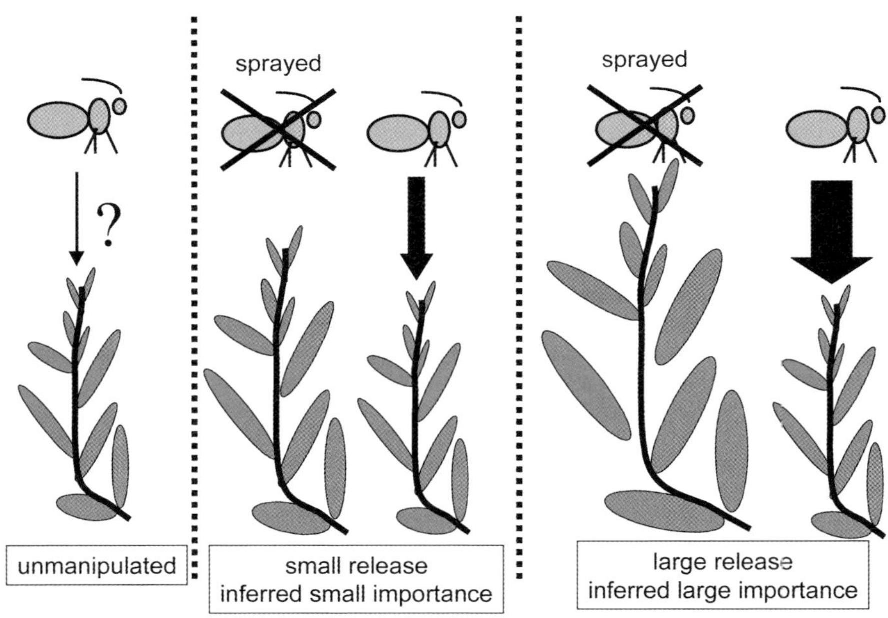

Fig. 15.1. Ecologists infer the importance of herbivores in an unmanipulated ecosystem by measuring the response to herbivore removal

should be positively related to the magnitude of net productivity reduction due to insect herbivores in the unmanipulated ecosystem.

If relative impacts of herbivores on plants differ (see below) and insect suppressions continue long enough for plant species composition to respond, changes in productivity may also reflect the indirect effects insect herbivores have on net productivity by favouring or excluding plant species that differ in attributes such as nutrient use efficiency (Crawley 1997; Wardle 2002). Changes in plant species composition with insecticide application are often used to infer relative impacts of insect herbivores on different groups or species of plants when insects are present (Fig. 15.2). Depending on the study, this may include investigations into the relative impact of insect herbivores on different functional groups of plants (i.e. grasses vs. forbs) or different species of plants. Stronger responses of plant mortality, growth and/or reproduction to suppression of insect herbivores (i.e. releases) are equated with relatively larger negative impacts of insect herbivores on plants in the absence of sprays. Decreases in plant performance with insect herbivore suppression are attrib-

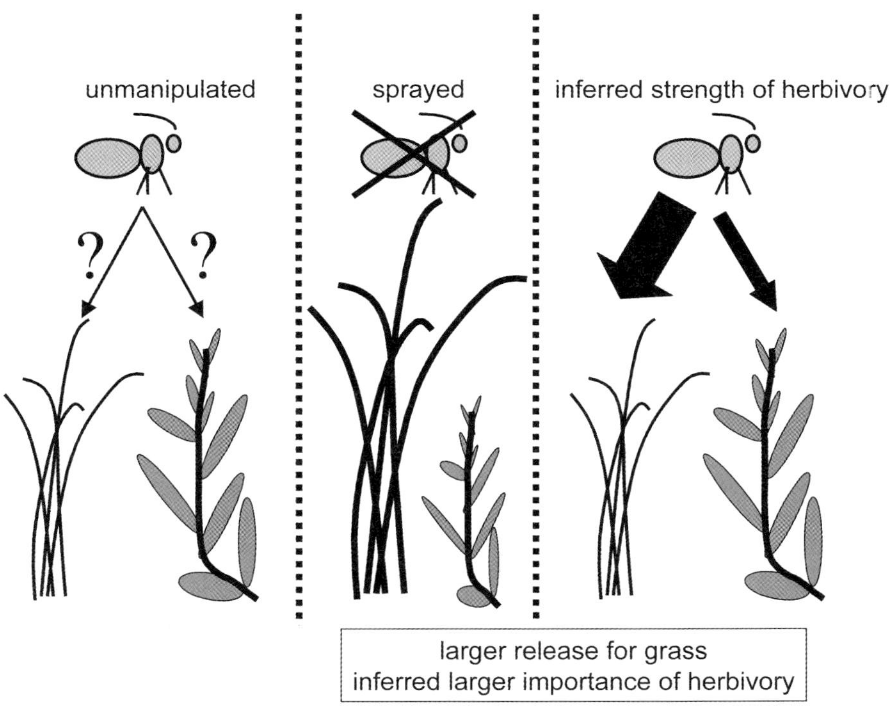

Fig. 15.2. Ecologists infer the relative impact of herbivores on different species of plants in an unmanipulated ecosystem by measuring their responses to herbivore removal. In the *middle section*, the large increase in the grass when herbivores are removed suggests that herbivores have a larger impact on the grass in the unmanipulated system

uted to negative indirect effects of herbivore removal via releases of competitors that are larger than the direct positive effects on the plant itself. Changes in plant growth and survival often are used to infer impacts of insect herbivores on plant population dynamics. If insect herbivores have similar influences on different groups or species of plants, whole community removals of insect herbivores will not lead to changes in plant community composition, which may result in the erroneous conclusion that they do not have strong interactions with plants in the unmanipulated community, even when the effects on individual groups or plants are large (Fig. 15.3). In such communities, however, there may be a strong response of peak standing crop or net plant productivity to insect exclusion. Indeed, there is an inverse relationship in grasslands between the strength of the responses of standing crop and species composition to exclusion of grazing mammals, with composition responses dominating at higher productivity (Chase et al. 2000).

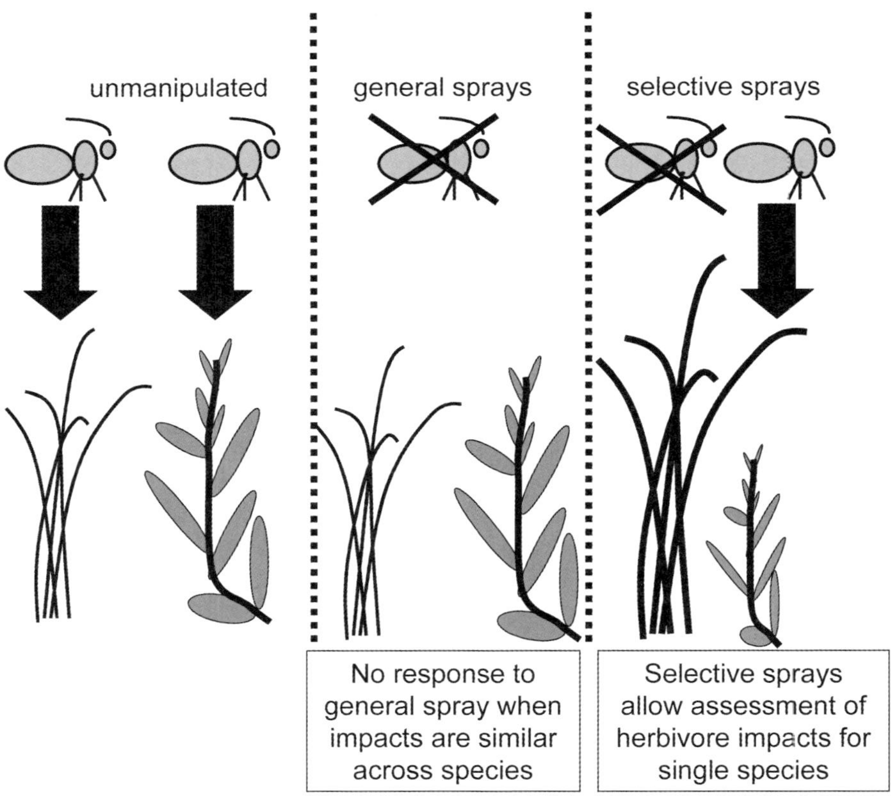

Fig. 15.3. Insecticide applications to whole communities versus individual species of plants may give different insights into the role of herbivores

More selective removals of insect herbivores from single species or groups of plants within an ecosystem may give a different picture from whole-system manipulations of insect herbivores (Siemann and Rogers 2003). The response of a single species of plant to whole community insect suppressions is proportional to the difference between its herbivore load and the herbivore loads of other plants in the community, while its response is proportional to the absolute herbivore loads of the other plants in the community if a single plant is grown in a herbivore-free state (Fig. 15.3). The use of methods other than general plot-wide applications has the potential to expand the inferences ecologists can draw from studies using insecticides to exclude insect herbivores.

15.4 Ghost of Herbivory Past

Herbivore suppression experiments may miss the effects of herbivory if a herbivore is able to reduce a host plant to low densities and/or restrict its distribution ('Ghost of Herbivory Past', Carson and Root 2000). For example, a beetle introduced to control St. John's Wort (*Hypericum perforatum*) drove the formerly abundant plant to shaded refugia where the plants suffered less attack (Harper 1977). A typical insect exclusion experiment in which plots are located independent of *Hypericum*'s distribution and randomly assigned to insecticide or control treatments could lead to the faulty conclusion that insect herbivores are not important in determining local plant community composition. Similar effects of herbivores on the distribution of their host plants have been shown for other species (e.g. Louda and Rodman 1996). In general, if experiments are short term or have plots containing few individuals of herbivore-susceptible plants at the start of the experiment, even large effects of herbivores on plant growth, survival and reproduction may be statistically undetectable.

To test how the number of replicates and the duration of a herbivore exclusion experiment might influence the likelihood of detecting a significant herbivore effect, we performed computer simulations of herbivore exclusion experiments where insecticide and control treatments were randomly allocated to field plots. We calibrated the model with demographic data on *Solidago canadensis* (formerly *S. altissima*; mortality and rhizome production) from long-term experiments with insect herbivore manipulations (Cain et al. 1991; Carson and Root 2000). We varied four starting conditions (*Solidago* density, strength of release from herbivory, number of experimental replicates, duration of experiment, Fig. 15.4A). At high densities of *Solidago* such as occur in the northeast United States (up to 20 plants m^{-2}, Carson and Root 2000), effects of herbivore suppression would be detected even in short-term experiments with modest decreases in mortality and increases in rhizome

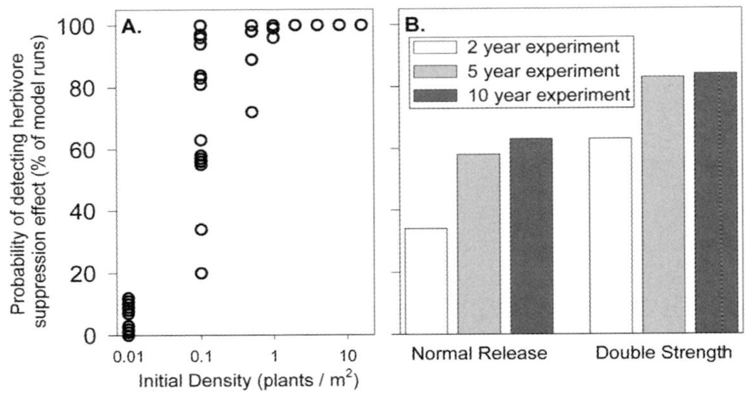

Fig. 15.4. Simulated experimental herbivore removal experiments. **A** Probability of detecting a significant effect of herbivores on plant population dynamics depended on the strength of release, number of experimental replicates and duration of study. **B** When host plants are at low densities (0.1 plants m^{-2}), a five-replicate study would often miss strong responses of individual plants

production with herbivore suppression (Fig. 15.4A). However, at the densities at which *S. canadensis* occurs in prairies in the midwestern United States (approximately 0.1 plants m^{-2}), even a 10-year, five-replicate field experiment with 5×5-m plots would miss the response of herbivore suppression 40 % of the time (Fig. 15.4B). The sensitivity of the model outcome to initial density (Fig. 15.4A) suggests that typical insect herbivore exclusion experiments may be appropriate only for detecting responses of plant species that occur at high densities (i.e. dominant taxa) unless experiments are run with many replicates for many years. If the effect of herbivores on a rare plant is the object of investigation, it may be advisable to add plants to plots (i.e. phytometers) or to locate plots in areas where the plant is present at higher densities.

15.5 Artifacts of Method May Masquerade as Release from Herbivory

15.5.1 What Types of Artifacts Are a Concern?

Because insecticides are deposited on plants and are toxic to insects, the primary way that insecticide applications are expected to alter ecosystems is through reductions in the population densities and feeding activities of herbivorous insects. However, insecticides may also have direct and indirect effects on other components of ecosystems and may therefore change plant

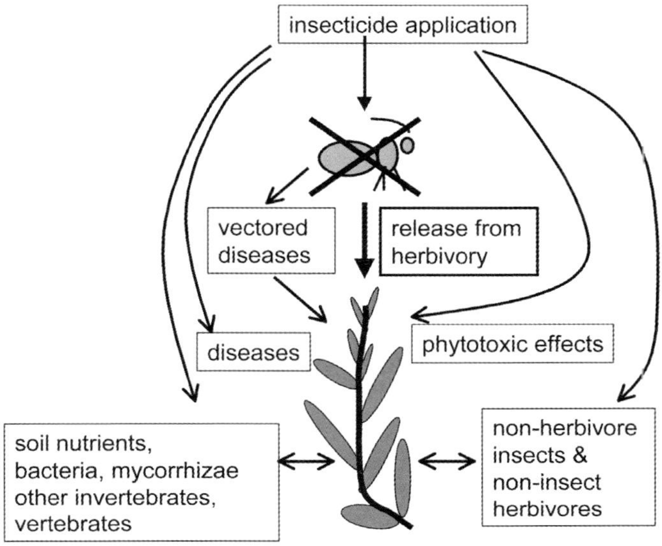

Fig. 15.5. Insecticide applications may affect ecosystems by a number of mechanisms

species composition and/or ecosystem functioning (Fig. 15.5). For instance, some insecticides may directly change the plant community because they are toxic to plants at some application rates and increase their mortality, decrease their growth or inhibit seed germination. Insecticides may also be toxic to or may change the activity of other components of the ecosystem such as non-herbivorous insects, non-insect herbivores, bacteria or fungi. Depending on the chemical composition of an insecticide and the inactive ingredients in the formulation, insecticide application may provide limiting nutrients. These side effects of insecticides are of concern, because they may make it difficult to attribute observed differences in ecosystem functioning between sprayed and unsprayed plots to the action of insect herbivores. Investigators can maximize their insights into the role of insect herbivores in terrestrial ecosystems by collecting appropriate data, performing additional studies if possible and considering alternative explanations.

15.5.2 Overview of Published Studies

We searched the literature for ecological studies in which insecticides were used in order to examine the relationship between insect herbivores and plant fitness, population dynamics, community structure or ecosystem processes. We generated an initial set of articles by searching for the words 'insecticide', 'insecticides', 'insecticidal' and 'exclusion' in the abstracts of articles in the following journals: *American Journal of Botany* (1914–Jan 2002), *American Nat-*

uralist (1867–Sept 2002), *American Midland Naturalist* (1909–July 2002), *Biotropica* (1969–June 2002), *Bulletin of the Torrey Botanical Club* (1870–1996), *Conservation Biology* (1987–Oct 2002), *Ecological Applications* (1991–Aug 2002), *Ecological Monographs* (1931–Aug 2002), *Ecological Entomology* (1991–Oct 2002), *Ecology* (1920–Sept 2002), *Environmental Entomology* (1991–Feb 2002), *Functional Ecology* (1987–Oct 2002), *Journal of Applied Ecology* (1964–Aug 2002), *Journal of Ecology* (1913–Oct 2002), *Journal of Tropical Ecology* (1985–Sept 2002), *Journal of Animal Ecology* (1932–Sept 2002), *Journal of the Torrey Botanical Society* (1997–Sept 2002), *Oecologia* (1991–Sept 2002), *Oikos* (1988–Aug 2002), *Plant Ecology* (1997–July 2002), *Vegetatio* (1991–1996). We read these articles and kept those that employed insecticide treatments in order to investigate the role of insects in determining ecosystem functioning, plant community composition or diversity or plant population dynamics. In total this represented 66 articles. This is not an exhaustive search of the literature for ecological studies that employed insecticides. Rather, it represents a body of articles that allows us to investigate the inferences ecologists draw in such studies and their consideration of potential methodological artifacts.

We classified each article with respect to the following criteria (see Appendix): (1) What response variable was the focus of the study? (2) What type of insecticide was used? (3) Were insect damage or insect abundance quantified? (4) Were toxic effects of insecticide on germination, plant survival or plant growth considered? (5) Was greater toxicity to higher trophic levels of insects considered? (6) Was toxicity to detritivorous insects, soil microbes or mycorrhizae considered? (7) Were fertilization effects of insecticide considered?

15.5.3 Quantification of Herbivore Damage

Additional data can clarify whether a reduction in insect herbivory is the mechanism underlying a plant or ecosystem response to insecticide application. In particular, it is critical to collect data on insect herbivore abundance and insect damage to plants in experimental units where insecticides have been applied and compare these units to those to which insecticides have not been applied. If a species or group of plants performs better in plots where insecticide has been applied, insect damage should be lower on insecticide-treated plants if reductions in insect herbivory are responsible. These types of data may be most appropriate for chewing or mining insects where damage can be carefully quantified. A large response of plants to insecticide application without documented decreases in insect damage makes it difficult to have confidence that reduced insect herbivory is the cause. On the other hand, collecting data on the abundance of insect herbivores can be useful for some groups, especially Hemipterans, which often can be easily counted on a plant but for which the damage cannot be as easily quantified independent of their abundance or plant vigour. As for insect damage data, insect herbivore abundance, due to changes in activity or in situ population densities, should be lower on insecticide-treated plants than on untreated plants if insect herbivores are driving plant responses to insecticide application. Of the studies we reviewed, 55 and 41 % of studies collected data on herbivore damage and abundance, respectively.

15.5.4 Phytotoxic Effects

Responses of plants, plant communities and ecosystems to insecticide applications may reflect toxic effects of insecticides on plants (i.e. phytotoxic effects). In particular, plants that decrease in abundance or vigour with insecticide applications may be more sensitive to toxic effects and plants that increase with applications may be less sensitive. In this case, an investigator could attribute differential responses of plants to insecticides to differences in the strengths of their releases from herbivory, when they may actually indicate differences in their susceptibility to phytotoxic effects of the insecticide sprays. In the extreme, phytotoxic effects could depress overall ecosystem productivity, but it is unlikely that an insecticide licenced for application in agricultural or horticultural settings would be that toxic to plants at recommended application rates. It is more likely that variations in the magnitude of toxic effects on different plant species may cause changes in species composition.

Phytotoxic effects could include suppression of germination or decreases in survival, growth and/or reproduction. For insecticides registered in the United States, the current requirement for labels is that they report phytotoxic effects if 'more than 25 percent of terrestrial plants show adverse effects on plant life cycle functions and growth such as germination, emergence, plant vigour, reproduction and yields when tested at the maximum label application rate or less' (Environmental Protection Agency 40 CFR Part 159, Reporting Requirements For Risk/Benefit Information). In the European Union, regulations under consideration call for reporting of phytotoxicity data when there is 'more than 50 % effect for one or more species at the maximum application rate' (Guidance Document on Terrestrial Ecotoxicology Under Council Directive 91/414/EEC, European Commission on Health and Consumer Protection). Toxic effects are usually for the product applied alone, with recommendations for applying products in combination based on chemical characteristics of active ingredients. The literature is one resource for ecologists to address the issue of phytotoxic effects.

Another way to disentangle phytotoxic effects and releases from herbivory is to conduct experimental toxicity trials in controlled conditions. For experiments with factorial treatments (such as a fungicide treatment), phytotoxicity trials should also be factorial in case sprays are more toxic in combination.

For instance, in one study pesticide sprays reduced seed germination in combination even though neither had a detectable effect when applied alone (Gange et al. 1992). A rigorous way to conduct these experiments is with a gradient of application rates that include concentrations less than and greater than the field application rates (i.e. perform a dose response experiment).

To test for phytotoxic effects on germination, seeds could be treated with the insecticides and germination rates measured with or without insecticide application. Differential suppression or stimulation of germination among plant species could potentially drive changes in species composition. This particular type of artifact would be a concern in studies at the community or ecosystem level that are of sufficient duration to have recruitment of new plants from seed. Twenty-eight percent of such studies we reviewed considered effects of their insecticide applications on seed germination rates. We did not find any examples of experiments that tested the effects of insecticide sprays on germination in field conditions.

To test for phytotoxic effects on survival, growth and reproduction, plants should be treated with insecticide in conditions of constant, ideally low, insect herbivory. This may be accomplished by growing plants in controlled laboratory or greenhouse conditions where herbivory is extremely low independent of insecticide applications. If plants are grown in pots where roots can be quantitatively recovered, measuring root mass in addition to aboveground mass is desirable in order to test for phytotoxic effects on root mass as well as possible changes in the relative allocation of growth above ground vs. below ground in response to sprays. This may be particularly valuable if field experiments only measure aboveground biomass or growth. In studies at the community or ecosystem level, phytotoxic effects on survival, growth or reproduction could obscure the relationship between a plant's response to insecticide applications and the relative importance of herbivory for that plant species. In studies of the responses of a single species to insecticide application, phytotoxic effects could lead to an underestimation of the release from herbivory, with the indirect positive effect via reduced insect herbivory being obscured by a direct negative effect. Of the studies we reviewed, 58 % of the studies of plant communities, 29 % of the studies of ecosystem processes and 44 % of the studies of individual plant responses considered toxic effects of insecticides on plant growth, survival or reproduction. Only 52 % of these studies conducted experimental tests of their insecticide applications on the species of plants in their studies or referenced literature for the species they studied. The rest cited toxicity data for other species of plants, typically agricultural crops. Many ecologists suppress insects throughout the growing season, resulting in more applications per year than the experiments they reference.

Overall, ecologists have given insufficient attention to the possible role of phytotoxic effects in their studies. This is not to say that they are likely to have compromised many studies, but rather that greater attention to this possibility would strengthen the causal link from insecticide applications to insect herbivore exclusion to plant and ecosystem responses. In particular, we would suggest caution for studies in which application rates or frequencies are greater than those recommended on the product label and for studies of plants for which no recommendation for closely related species is given on the product label.

15.5.5 Insecticides May Be Toxic to Several Groups of Insects

Insecticides may change plant composition or productivity by changing the abundance or activities of insects and non-insect arthropods that are not herbivores, such as pollinators, predators, parasitoids and detritivores. Some relevant data on toxicity to pollinators is readily available because insecticides list their toxicity to honeybees (*Apis mellifera*) on their label in Australia, the United States and in most European countries. Still, this may not predict lethal affects on other pollinators that differ in behaviour, physiology or size (Johansen 1972) and does not predict sublethal effects such as lower visitation rates (NRCC 1981). There is little evidence that plant populations or communities are pollinator limited, but reduced pollinator activity may impact the reproductive output of individual plants.

Differences in the susceptibility of insect herbivores and higher trophic levels (predators and parasitoids) to insecticide applications and variable rates of recolonization or population regrowth following applications can obscure the path from insecticide application to plant responses. In the extreme, insecticides may increase insect herbivore abundance if they are more toxic to predators or parasitoids than to insect herbivores (Spencer and Norman 1952). Of the studies we reviewed, 36 % discussed the relative toxicity of their insecticide treatments to herbivores vs. predators or parasitoids. Data on toxicity to higher trophic levels is rarely on a product label, but this information can be found in the literature for many insecticides (Coats et al. 1979 is a good example). Recently adopted guidance documents in the European Union recommend reporting of toxic effects on a standard parasitoid and predator arthropod species [*Aphidius rhopalosiphi* (Hymenoptera, Braconidae) and *Typhlodromus pyri* (Acarina: Phytoseiidae)], but unfortunately no insect predator is tested. Nevertheless, it is not just toxicity per se that is important for insecticide effects on higher trophic levels, but also the level of exposure. A substance that is potentially toxic might not reduce the population size of a predator and parasitoid when it is present only in minute quantities in living herbivores and is not taken up directly from the plant. Conversely, even a low level of toxicity might have great effects on higher trophic levels when exposure is high. Data on the relative toxicity of insecticides to herbivores vs. predators and parasitoids will assist ecologists in choosing an insecticide that is highly toxic to the herbivores they wish to control while simultaneously limiting experimental artifacts mediated through changes in other insect groups.

15.5.6 Effects of Insecticides on Non-Arthropods

Insecticides may alter the feeding choices of non-arthropod herbivores. Some pesticides, such as nicotine (insecticide), thiram (fungicide) and ziram (fungicide), are sold to discourage mammal browsing. DDT has been used as both an insecticide and a rodenticide. Some insecticides are marketed to repel (Proxpur) or kill (Fenthion) birds . However, the effect of most insecticides on the palatability of plants to non-arthropod herbivores is unknown. Thus, changes in non-arthropod activity are a possible artifact of insecticide applications in ecological studies. One way to limit such artifacts is to avoid insecticides that are known repellents or that are known to be extremely lethal to non-arthropods. Vertebrate feeding may influence insect herbivores by stimulating plant regrowth and increasing palatability (Du Toit et al. 1990; Bailey and Whitham 2002) or may decrease palatability by inducing defence production (Young and Okello 1998; Shimazaki and Miyashita 2002). In some cases, the defence compounds produced in response to vertebrate browsing may attract specialist insect herbivores (Martinsen et al. 1998). Changes in vertebrate feeding rates might reflect a direct effect of insecticides on vertebrate feeding activity or they may reflect changes in vertebrate feeding in response to reduced insect herbivory. For ecosystems where mammal herbivory is less important, these concerns may not be applicable. It is advisable to collect data on feeding rates of non-arthropod herbivores if possible to quantify their contribution to changes in plants or ecosystem processes.

Insecticides may also have effects on the survival and feeding activity of slugs and snails which may often be important herbivores (Crawley 1997). Some compounds have both molluscicidal and insecticidal activity (e.g. bendiocarb, methiocarb, Azinphos-methyl) and caution should be used in interpreting results when such compounds are used. However, since they belong to two widely used classes of insecticides (carbamates, organophosphates) changes in mollusc feeding are a potential issue for many studies.

15.5.7 Effects of Insecticides on Soil Organisms

Insecticides may be toxic to soil organisms that are critical for soil nutrient cycling or above-/belowground interactions. One group of bacteria that are a particular concern, with regard to producing spurious effects of insecticides, are those responsible for nitrogen transformations. In one study of the effects of 54 pesticides on denitrification and nitrification in the soil, 6 significantly stimulated denitrification, 8 significantly inhibited denitrification and 19 significantly reduced nitrification (Pell et al. 1998). There is a potential for such effects to masquerade as effects of insect herbivores on nutrient cycling via changes in their host plants. Since these same changes in soil nutrients may themselves change plant growth and survival, disentangling the chain of cause and effect may be difficult. Of the studies we reviewed, 22 % considered effects of insecticide application on the soil biota. Currently, there is no requirement for effects on soil organisms or soil nutrient cycling to appear on product labels so the main source of such information is the scientific literature. Recently adopted EU guidance documents recommend testing of pesticides for toxic effects on earthworms, standard soil arthropods (collembolans or mites) and rates of litter decomposition, soil nitrification and carbon mineralization (Guidance Document on Terrestrial Ecotoxicology Under Council Directive 91/414/EEC, European Commission on Health and Consumer Protection). Consideration of effects would strengthen all studies of insect herbivores, but it is especially important in studies on belowground herbivores in which insecticide is soaked into the soil or incorporated into the soil at the beginning of the study.

Some insecticides, especially those in the carbamate and organophosphate classes, are also nematicides. In studies that use insecticides that are also nematicides, there is the potential to overestimate the importance of insect herbivory because observed releases reflect both insect and nematode impacts. Three insecticides used in the studies we reviewed are also registered as nematicides: aldicarb, carbofuran and isazophos. In one study using aldicarb the effect of insecticide application is attributed entirely to insect suppression (Norris 1997). In all the other studies using these compounds, effects of insecticide application on plants are more properly ascribed to the combined effect of belowground insect herbivore and nematode suppression (all of the carbofuran and isazophos studies). More generally, because interactions in the soil food web are important for ecosystem processes but are largely unexplored (Wardle 2002), it is difficult to foresee how much the measured response variables such as plant biomass or plant species composition are influenced by alterations in the soil food web. For aboveground manipulations, it is possible to reduce the effects on the soil food web by minimizing the amount of insecticide that enters the soil.

15.5.8 Nutrient Inputs May Facilitate Plant Growth

Many insecticides may contain significant concentrations of limiting nutrients (Table 15.1). In particular, carbamates, organophosphates and pyrethroids often contain significant concentrations of nitrogen, and organophosphates always contain phosphorus. For some classes of insecticides, every compound that has been used in ecological studies is free of nitrogen and phosphorus (chlorinated hydrocarbons, flavonoids and organochlorines). The only other

Table 15.1. Properties of insecticides used in surveyed ecological studies. Number of times used represents the number of papers in our literature survey that used the chemical. Percentages of nitrogen and phosphorus are percent by weight of active ingredient. Amounts of nitrogen (*N amt*) and phosphorus (*P amt*) (in mg m^{-2} $year^{-1}$) are at the maximum recommended rate or the maximum rate used in a paper we reviewed

Chemical	Class	No. of times used	N (%)	P (%)	N amt	P amt
Aldicarb	Carbamate	1	14.73	0.00	29.60	0.00
Carbaryl	Carbamate	7	6.96	0.00	22.28	0.00
Carbofuran	Carbamate	4	6.33	0.00	12.66	0.00
Endosulfan	Chlorinated hydrocarbon	1	0.00	0.00	0.00	0.00
Rotenone	Flavonoid	1	0.00	0.00	0.00	0.00
Aldrin	Organochlorine	2	0.00	0.00	0.00	0.00
Chlordane	Organochlorine	5	0.00	0.00	0.00	0.00
DDT	Organochlorine	2	0.00	0.00	0.00	0.00
Lindane	Organochlorine	2	0.00	0.00	0.00	0.00
Acephate	Organophosphate	1	7.65	16.92	10.19	22.53
Azinphos-methyl	Organophosphate	2	13.24	9.76	27.41	20.21
Chlorpyrifos	Organophosphate	9	3.99	8.84	3.20	7.07
Diazinon	Organophosphate	3	9.21	10.18	14.73	16.29
Dimethoate	Organophosphate	7	6.11	13.52	40.32	89.20
Fenitrothion	Organophosphate	1	5.05	11.18	16.17	35.76
Fonofos	Organophosphate	1	0.00	12.58	0.00	10.06
Isazophos	Organophosphate	1	13.39	9.88	133.94	98.77
Malathion	Organophosphate	10	0.00	9.38	0.00	11.72
Omethoate	Organophosphate	1	6.57	14.53	16.43	36.34
Deltamethrin	Pyrethroid	1	2.77	0.00	0.05	0.00
Esfenvalerate	Pyrethroid	1	3.34	0.00	0.33	0.00
Fenvalerate	Pyrethroid	3	3.34	0.00	1.11	0.00
Fluvalinate	Pyrethroid	2	5.57	0.00	0.07	0.00
Permethrin	Pyrethroid	1	0.00	0.00	0.00	0.00
Resmethrin	Pyrethroid	1	0.00	0.00	0.00	0.00
Pyrethrum mix	Pyrethroid	1	NA	NA	NA	NA

elements present in the active ingredients of any product used in ecological studies are C, H, Cl, O, F, Br and S, none of which is typically a limiting nutrient in terrestrial ecosystems. Even at the maximum rates applied in ecological studies, the amounts of nitrogen and phosphorus added were small. The maximum amounts were 130 mg m^{-2} $year^{-1}$ for nitrogen and 100 mg m^{-2} $year^{-1}$ for phosphorus (isazophos; as used by Wardle and Barker 1997). Background rates of nitrogen and phosphorus cycling in terrestrial ecosystems are typically two and one order of magnitude higher, respectively, than even these amounts (Schlesinger 1997). Assessment of fertilization effects would be easier if investigators report their application rates in terms of active ingredient per plant or unit area per year. Based on maximum recommended application rates for the compounds used by ecologists in the studies we surveyed, we do not think that direct nutrient enrichment effects are an important source of artifacts in ecological studies with insecticides. Nevertheless, it is a good idea to calculate the amounts of nitrogen and phosphorus supplied to the system and to compare it to the natural pools of these nutrients.

Insecticides are usually applied in a mixture of inactive ingredients. For foliar applications, the main inactive ingredient is usually water often with a small amount of surfactant which aids dispersion of the water. In every study we examined, plants not treated with insecticide were sprayed with water to account for the water applied to insecticide-treated plants or plots. We do not think that small amounts of surfactant (i.e. soap) are likely to contribute to major methodological artifacts. For soil applications to control below-ground herbivores, insecticides are sometimes applied as a solution soaked into the soil. In all studies using this method, appropriate water-only controls were used. Some chemicals (especially chlorpyrifos) are applied as granules. These granules typically are composed of a low concentration of active ingredient (5 % or less) and a large proportion of sometimes specified inert ingredients. These materials are typically organic materials with low concentrations of nutrients, such as ground corn cobs or nut shells. The mass of inert ingredients is usually less than 5 g m^{-2} per application. Such a small amount of low nutrient organic material represents a small input.

15.5.9 Insect-Vectored Diseases

Removal of carbon and nutrients is not the only means whereby insect herbivores impact plant growth and survival. For instance, many important plant diseases are vectored by insects, especially viral diseases (Perring et al. 1999). Extremely large effects of insect suppression on plant survival with slight changes in herbivore damage and abundance are one indication that insect-vectored diseases may be magnifying the effect of insect herbivore suppression. Unfortunately, the most common vectors for viral diseases are sucking feeders (Perring et al. 1999), making it difficult to independently quantify herbivore consumption and host plant responses. Some diseases have characteristic symptoms which allow the investigator to associate mortality with disease (Mitchell 2003) and draw more accurate conclusions about the role of insect herbivores in the ecosystem. Insecticides that rapidly kill or repel herbivorous insects and those with residual action are those most likely to reduce the transmission of insect-vectored viral diseases. Synthetic pyrethroids are a class of insecticidal compounds that are particularly effective at reducing viral spread (Perring et al. 1999). Other diseases may not be vectored by insects but may opportunistically infect weakened plants and magnify the effect of insect herbivory on plant performance. Fungal diseases are more likely to fall into this category. Factorial insecticide and fungicide experiments have the potential to help tease apart the relationships among insect herbivory, fungal diseases and plant performance.

Insecticides may also influence the susceptibility of plants to fungal pathogens that are not vectored by herbivores. Mostly this is not due to the active ingredient itself, but to the surfactants used in many formulations. A proper control should therefore include spraying of the surfactant.

15.4.10 Community-Level Artifacts

One major drawback in the vast majority of toxicity studies is that the toxic effects are investigated in single-organism studies, e.g. by rearing sprayed and unsprayed plants in the absence of herbivores in the greenhouse, or by directly exposing a particular insect to the active substance. Because interactions in the aboveground and belowground food webs are complex, it is generally difficult to extrapolate from such single-organism studies to community-level effects. For example, if an annual plant shows a significant 5% reduction in seed set when sprayed, this may or may not have consequences for plant species composition in the longer term. To estimate community-level effects, community-level experiments should be performed. For example, entire plant communities could be grown in the greenhouse, or transplanted from the field, and the similar ecosystem variables could be measured in the greenhouse and in the field. For community transplants, a single insecticide spray may be sufficient to eliminate the resistant herbivore insect community. Thereafter, only one group of replicates would be subjected to further spraying while the control group would only receive the first spray. While such an approach is easier for some ecosystems than for others, it would possibly be more conclusive than single-organism studies. Another advantage might be the reduced effort needed to assess side effects. For example, even when the ecosystem under consideration consists of only 20 higher plants, it would be extremely time-consuming if not impossible to test each of these plants and the majority of soil organisms one-by-one in isolation, let alone in two-species combinations. We recommend that community-level tests for insecticide effects should become more common in ecological studies.

15.6 Are There Better Types of Insecticides?

There are several classes of insecticides that have not been used in ecological studies. These include antibiotic insecticides that are applied as a bait. They have the advantage of low toxicity to non-target organisms, but they must be consumed by the herbivore in order to be effective. They may be ineffective in small plots with high rates of recolonization and they are unlikely to be effective against a broad spectrum of insects with varied feeding habits. Insect growth regulators have many of the same advantages and disadvantages. Some insecticides are toxic to other groups of organisms such as fungi (dinitrophenol pesticides) or plants (arsenic pesticides), or are non-selective (methyl bromide), which limits their use in ecological studies of insect herbivory. Some classes of insecticides, such as formamidine insecticides (also known as amidine insecticides), have not been used although there do not appear to be any distinct reasons why they have been avoided by ecologists. Ecologists utilize a diversity of insecticides (Table 15.1). The list of products used will likely expand, but the new insecticides will demand the same caution in use and interpretation of results as those currently available. The ultimate aim, however, should be to use insecticides that are as selective as possible. Combining selective insecticides would allow us to draw inferences about the role of different groups of insect herbivores, and of the role of herbivores on particular plants in the community.

15.7 Conclusions

Ecologists should show caution in interpreting chains of causation from insecticide application to plant and ecosystem responses without careful examination of intermediate steps in the process. In particular, data on herbivores and herbivory are critical to any argument that insecticide treatments are impacting plants or ecosystem processes via changes in herbivory. Direct toxic effects of insecticides on plants and effects on soil nutrient cycling may also weaken the inferences ecologists can draw in their studies. Insecticides will continue to be a valuable tool for ecologists who study insect herbivores in terrestrial ecosystems.

Acknowledgements. E.S. and W.E.R. were supported by the US NSF (NSF DEB-9981654, DEB-0315796), EPA (R82-8903) and USDA (2003-35320-13498). W.P.C. was supported by the US NSF (DEB-9527729 and DEB-9515184). W.W.W. was supported by the Deutsche Forschungsgemeinschaft, through grants FOR 456, WE 2618/3-2618/5.

Appendix: Results of Surveyed Studies

Author	Year	Q1[a]: object	Q2[b]: insecticide	Q3a[c]: dam	Q3b[d]: abund	Q4a[e]: germ	Q4b[f]: surv	Q4 c[g]: growth	Q5[h]: preds	Q6a[i]: det	Q6b[j]: micro	Q7[k]: fert
Agnello et al.	1996	IG	Chlorpyrifos, Methomyl	Yes	Yes	No	No	No	No	No	No	No
Barrett	1968	ER	Carbaryl	No	No	Exp	Exp	Exp	Exp	Exp	No	No
Brown	1993	ER	Permethrin	No	Yes	No	No	No	Lit	No	No	No
Brown and Gange	1989a	SR, CC	Chlorpyrifos/Dimethoate	No	No	Lit	Lit	Lit	No	No	Lit	Lit
Brown and Gange	1989b	SR, CC	Chlorpyrifos	No	No	Exp	Exp	Exp	Lit	Lit	Lit	No
Brown et al.	1987	IG, IR	Malathion	No	No	No	Exp	Exp	No	No	No	Exp
Brown et al.	1988	IG, PG, SR, CC	Malathion	No	Yes	Lit	Lit	Lit	Exp	Lit	Lit	Lit
Cain et al.	1991	IG, IR	Fenvalerate	No	Yes	Lit	Lit	Lit	No	No	No	Lit
Calvo-Irabien and Islas-Luna	1999	IS	Monocrotophos	Yes	No	No	No	No	No	No	No	No
Cantlon	1969	PG	Malathion, DDT, Aldrin	No	No	No	No	No	No	No	No	No
Carson and Root	1999	CC, CP	Fenvalerate	Yes	Yes	Lit	Lit	Lit	No	No	No	No
Christensen and Whitham	1993	IR, IG, IS	Dimethoate	Yes	No	No	No	No	No	No	No	No
Cohen et al.	1994	HR	Deltamethrin	No	Yes	No	No	No	Exp	No	No	No
Dempster	1968	ER	DDT	No	No	No	No	No	Lit/exp	Lit/exp	Lit	No
Dominguez and Dirzo	1994	IR, IG, IS	Omethoate	Yes	No	Exp	Exp	Exp	No	No	No	Exp
Donaldson	1997	ER, PG	Malathion	No	No	No	No	No	No	No	No	No
Edwards et al.	1968	HR, ER	Dichloropropene, Formalin, Aldrin	No	Yes	No	No	No	Exp	Lit	Lit	No
Edwards et al.	1979	HR	Fonofos	No	Yes	No	No	No	No	No	No	No
Fox and Morrow	1992	IG	Carbaryl	Yes	No	No	Lit	Lit	No	No	Exp	No
Fraser and Grime	1997	CP, ER	(Chlorpyrifos/Diazinon), (Dimethoate/Permethrin)	Yes	No	Exp	Exp	Exp	No	No	No	Exp

Funderburk et al.	2000	IG	Lambda Cyalothrin, Spinosad, Aldicarb, Phorate	Yes	Yes	No	No	No	No	No	No	No
Ganade and Brown	1997	CC, IR, IG, SR	Chlorpyrifos	Yes	No	Lit	Lit	Lit	Lit	Lit	Lit	Lit
Gange et al.	1989	IG	Malathion	No	No	Lit	Lit	Lit	No	Lit	Lit	No
Gange et al.	1992	IR	Chlorpyrifos, Dimethoate, Iprodione	No	No	Exp	No	No	No	No	No	No
Gange et al.	1989	IR, IG, IS	Malathion	Yes	No	No	No	No	Lit	No	No	No
Hartley	1998	IG	Resmethrin	Yes	No	No	No	No	No	No	No	No
Ingham et al.	1986	ER	Carbofuran	No	No	No	No	No	Exp	Exp	Exp	No
Kaitaniemi et al.	1998	IG	Fenirothrion	Yes	No	No	No	No	No	No	No	No
Kelly and Dyer	2002	PG, IG, IR, IS	Orthene	Yes	No	No	No	No	Lit	No	No	No
Kleintjes	1997	IG	Phosphamidon	Yes	Yes	No	No	No	Exp	No	No	No
Louda	1982	IS, PG	Malathion, Lindane	No	No	Lit	Lit	Lit	No	No	No	No
Louda	1982	IR, PG, IS	Malathion, Lindane	No	No	No	No	No	No	No	No	No
Louda	1984	IG, IR	Rotenone	Yes	Yes	No	No	No	No	No	No	Lit
Louda and Potvin	1995	IG, IS, IR	Lindane	Yes	No	No	No	No	No	No	No	No
Louda and Rodman	1996	IG, IS, IR, PG	Pyrethrum	Yes	No	Lit	Lit	Lit	No	No	No	Lit
Malone	1969	ER, SR, PG, CC	Diazinon	No	No	No	No	No	Lit/exp	Exp	No	No
Maron	1997	IS	Malathion	Yes	No	No	No	No	No	No	No	No
Martens and Boyd	2002	IG	Aldicarb	Yes	No	No	No	No	No	No	No	No
Masters et al.	2001	HR, IG, IR	Chlorpyrifos	Yes	Yes	Lit	Lit	Lit	Lit	No	Lit	Lit
McBrien et al.	1983	PG, CC	Carbaryl, Malathion	No	Yes	No	No	No	No	No	No	No
McNaughton	1970	IG	Dimethoate	No	No	No	No	Exp	No	No	No	No
Messina et al.	2002		Malathion	Yes	Yes	No	Lit	Lit	No	No	No	Lit
Müller-Schärer and Brown	1995	PG, IG	Dimethoate, Chlorpyrifos	Yes	Yes	No	No	No	No	No	No	No
Norris	1997	CC, IG, PG	Aldicarb	Yes	No	No	No	No	No	No	No	No
Palmisano and Fox	1997	PG, IG, IR, IS	Lindane, Carbaryl	Yes	Yes	No	No	Exp	No	No	No	No
Parker et al.	1984	ER	Chlordane	No	Yes	Exp	Exp	Exp	Exp	Exp	Exp	Exp
Paynter et al.	1998	IS, IR, PG	Tau-Fluvalinate, Azinphos-methyl	Yes	No	Exp	Exp	No	No	No	No	Exp

Appendix (*Continued*)

Author	Year	Q1[a]: object	Q2[b]: insecticide	Q3a[c]: dam	Q3b[d]: abund	Q4a[e]: germ	Q4b[f]: surv	Q4 c[g]: growth	Q5[h]: preds	Q6a[i]: det	Q6b[j]: micro	Q7[k]: fert
Perfecto	1990	HR, ER	Carbofuran, Chlorpyrifos	Yes	Yes	No	Exp	Exp	Exp	No	No	No
Pysek	1992	IG, IR, IS	Metathion	Yes	No	No	No	No	No	No	No	No
Rees and Brown	1992	IG	Dimethoate, Chlorpyrifos	Yes	No	Lit	Lit	No	Lit	Lit	Lit	No
Risch et al.	1986	ER	Carbaryl	No	Yes	No	No	No	Exp	Exp	No	No
Root	1996	IG, IR	Fenvalerate	Yes	Yes	Lit	Lit	Lit/exp	No	No	No	Lit
Root and Gowan	1976	IG, HR	Carbaryl, Carbofuran	Yes	Yes	No	No	No	Exp	No	No	No
Santos and Whitford	1981	ER	Chlordane	No	Yes	No	No	No	No	Exp	Exp	No
Santos et al.	1984	ER	Chlordane	No	Yes	No	No	No	Lit	Exp	Exp	No
Santos et al.	1981	ER	Chlordane	No	Yes	No	No	No	No	No	No	No
Seastedt et al.	1983	IG, PG	Carbaryl	Yes	No	No	No	Exp	No	Exp	No	Exp
Sessions and Kelly	2002	CC, IG	Fluvalinate	Yes	No	No	No	No	No	No	No	No
Shure	1971	ER, SR, CP, PG, IR	Diazinon	No	No	Exp	Exp	Exp	Exp	No	No	No
Siemann and Rogers	2003		Esfenvalerate	Yes	Yes	No	Exp	Exp	Lit	Lit	Lit	Lit
Strong et al.	2000	HR, IG	Resmethrin	Yes	Yes	No	No	Lit	No	No	No	Lit
Traveset	1994	IR	Endosulfan	Yes	No	No	No	No	No	No	No	No
Wardle and Barker	1997	CP, SR, CC, IG	Isazophos	Yes	No	No	No	No	No	No	Exp	No
Weary and Merriam	1978	ER	Carbofuran	No	No	No	No	No	Lit	Exp	No	No
Whitford et al.	1981	ER	Chlordane	No	Yes	No	No	No	No	No	No	No
Whittaker and Warrington	1985	IG	Malathion	No	No	No	No	No	No	No	No	No

[a] *Q1*: What type of plant response variable was the focus of the study? *IG* Individual plant growth; *IS* individual plant survival; *IR* individual plant fitness/reproduction; *PG* plant population growth; *SR* plant community diversity (species richness); *CC* plant community composition; *CP* plant community productivity; *HR*, herbivore response; *ER*, ecosystem response

[b] *Q2*: What type of insecticide did they use? (active ingredients)

[c] *Q3a*: Did they quantify changes in insect damage with insecticide treatments?

[d] *Q3b*: Did they quantify changes in insect abundance with insecticide treatments?

[e] *Q4a*: Did they consider effects of insecticide on germination? *Lit* Yes, using a citation of another study; *Exp* yes, by performing their own tests; *No* no

[f] *Q4b*: Did they consider toxic effects of insecticide on plant survival?

[g] *Q4c*: Did they consider toxic effects of insecticide on plant growth?

[h] *Q5*: Did they consider greater toxicity to higher trophic levels?

[i] *Q6a*: Did they consider greater toxicity to detritivorous insects?

[j] *Q6b*: Did they consider effects on soil microbes (especially for nitrification) or mycorrhizae?

[k] *Q7*: Did they consider fertilization effects of insecticide?

References

Agnello AM, Reissig WH, Spangler SM, Charlton RE, Kain DP (1996) Trap response and fruit damage by obliquebanded leafroller (Lepidoptera: Tortricidae) in pheromone-treated apple orchards in New York. Environ Entomol 25:268–282

Bailey JK, Whitham TG (2002) Interactions among fire, aspen, and elk affect insect diversity: reversal of a community response. Ecology 83:1701–1712

Barkham JP (1980) Population-dynamics of the wild daffodil (*Narcissus pseudonarcissus*). 2. Changes in number of shoots and flowers, and the effect of bulb depth on growth and reproduction. J Ecol 68:635–664

Barrett GW (1968) The effects of an acute insecticide stress on a semi-enclosed grassland ecosystem. Ecology 49:1019–1035

Billick I, Case T (1994) Higher order interactions in ecological communities: what are they and how can they be detected? Ecology 75:1529–1543

Brown MW (1993) Resilience of the natural arthropod community on apple to external disturbance. Ecol Entomol 18:169–183

Brown VK, Gange AC (1989a) Differential effects of above- and below-ground insect herbivory during early plant succession. Oikos 54:67–76

Brown VK, Gange AC (1989b) Herbivory by soil-dwelling insects depresses plant-species richness. Funct Ecol 3:667–671

Brown VK, Leijn M, Stinson CSA (1987) The experimental manipulation of insect herbivore load by the use of an insecticide (Malathion) – the effect of application on plant-growth. Oecologia 72:377–381

Brown VK, Jepsen M, Gibson CWD (1988) Insect herbivory: effects on early old field succession demonstrated by chemical exclusion methods. Oikos 52:293–302

Cain ML, Carson WP, Root RB (1991) Long-term suppression of insect herbivores increases the production and growth of *Solidago altissima* rhizomes. Oecologia 88:251–257

Calvo-Irabien LM, Islas-Luna A (1999) Predispersal predation of an understory rainforest herb *Aphelandra aurantiaca* (Acanthaceae) in gaps and mature forest. Am J Bot 86:1108–1113

Cantlon JE (1969) The stability of natural populations and their sensitivity to technology. Brookhaven Symp Biol 22:197–205

Carson WP, Root RB (1999) Top-down effects of insect herbivores during early succession: influence on biomass and plant dominance. Oecologia 121:260–272

Carson WP, Root RB (2000) Herbivory and plant species coexistence: community regulation by an outbreaking phytophagous insect. Ecol Monogr 70:73–99

Chase JM, Leibold MA, Downing AL, Shurin JB (2000) The effects of productivity, herbivory, and plant species turnover in grassland food webs. Ecology 81:2485–2497

Christensen KM, Whitham TG (1993) Impact of insect herbivores on competition between birds and mammals for pinyon pine seeds. Ecology 74:2270–2278

Coats SA, Coats JR, Ellis CR (1979) Selective toxicity of three synthetic pyrethroids to eight coccinelids, a eulophid parasitoid and two pest chrysomelids. Environ Entomol 8:720–722

Cohen JE, Schoenly K, Heong KL, Justo H, Arida G, Barrion AT, Litsinger JA (1994) A food web approach to evaluating the effect of insecticide spraying on insect pest population dynamics in a Philippine irrigated rice ecosystem. J Appl Ecol 31:747–763

Crawley MJ (1997) Plant–herbivore dynamics. In: Crawley MJ (ed) Plant ecology, 2nd edn. Blackwell, Oxford, pp 401–474

Dempster JP (1968) The control of *Pieris rapae* with DDT. III. Some changes in the crop fauna. J Appl Ecol 5:463–475

Dominguez CA, Dirzo R (1994) Effects of defoliation on *Erythroxylum havanense*, a tropical proleptic species. Ecology 75:1896–1902
Donaldson JS (1997) Is there a floral parasite mutualism in cycad pollination? The pollination biology of *Encephalartos villosus* (Zamiaceae). Am J Bot 84:1398–1406
Du Toit JT, Bryant JP, Frisby K (1990) Regrowth and palatability of acacia shoots following pruning by African savanna browsers. Ecology 71:149–154
Edwards CA, Reichle DE, Crossley DAJ (1968) Experimental manipulation of soil invertebrate populations for trophic studies. Ecology 50:495–498
Edwards CA, Sunderland KD, George KS (1979) Studies on polyphagous predators of cereal aphids. J Appl Ecol 16:811–823
Fox LR, Morrow PA (1992) *Eucalypt* responses to fertilization and reduced herbivory. Oecologia 89:214–222
Fraser LH, Grime JP (1997) Primary productivity and trophic dynamics investigated in a north Derbyshire, UK, dale. Oikos 80:499–508
Funderburk J, Stavisky J, Olson S (2000) Predation of *Frankliniella occidentalis* (Thysanoptera: Thripidae) in field peppers by *Orius insidiosus* (Hemiptera: Anthocoridae). Environ Entomol 29:376–382
Ganade G, Brown VK (1997) Effects of below-ground insects, mycorrhizal fungi and soil fertility on the establishment of *Vicia* in grassland communities. Oecologia 109:374–381
Gange AC, Brown VK, Evans IM, Storr AL (1989) Variation in the impact of insect herbivory on *Trifolium pratense* through early plant succession. J Ecol 77:537–551
Gange AC, Brown VK, Farmer LM (1992) Effects of pesticides on the germination of weed seeds – implications for manipulative experiments. J Appl Ecol 29:303–310
Harper JL (1977) Population biology of plants. Academic Press, London
Hartley SE (1998) The chemical composition of plant galls: are levels of nutrients and secondary compounds controlled by the gall-former? Oecologia 113:492–501
Johansen CA (1972) Toxicity of field-weathered insecticide residues to four kinds of bees. Environ Entomol 1:393–394
Kaitaniemi P, Ruohomaki K, Ossipov V, Haukioja E, Pihlaja K (1998) Delayed induced changes in the biochemical composition of host plant leaves during an insect outbreak. Oecologia 116:182–190
Kelly CA, Dyer RJ (2002) Demographic consequences of inflorescence-feeding insects for *Liatris cylindracea*, an iteroparous perennial. Oecologia 132:350–360
Kleintjes PK (1997) Midseason insecticide treatment of balsam twig aphids (Homoptera: Aphididae) and their aphidophagous predators in a Wisconsin Christmas tree plantation. Environ Entomol 26:1393–1397
Louda SM (1982a) Distribution ecology – variation in plant recruitment over a gradient in relation to insect seed predation. Ecol Monogr 52:25–41
Louda SM (1982b) Limitation of the recruitment of the shrub *Haplopappus squarrosus* (Asteraceae) by flower and seed feeding insects. J Ecol 70:43–53
Louda SM (1984) Herbivore effect on stature, fruiting, and leaf dynamics of a native crucifer. Ecology 65:1379–1386
Louda SM, Potvin MA (1995) Effect of inflorescence-feeding insects on the demography and lifetime fitness of a native plant. Ecology 76:229–245
Louda SM, Rodman JE (1996) Insect herbivory as a major factor in the shade distribution of a native crucifer (*Cardamine cordifolia* A. Gray, bittercress). J Ecol 84:229–237
Malone CR (1969) Effects of Diazinon contamination on an old-field ecosystem. Am Midl Nat 82:1–27
Maron JL (1997) Interspecific competition and insect herbivory reduce bush lupine (*Lupinus arboreus*) seedling survival. Oecologia 110:284–290

Martens SN, Boyd RS (2002) The defensive role of Ni hyperaccumulation by plants: a field experiment. Am J Bot 89:998–1003
Martinsen GD, Driebe EM, Whitham TG (1998) Indirect interactions mediated by changing plant chemistry: beaver browsing benefits beetles. Ecology 79:192–200
Masters GJ, Jones TH, Rogers M (2001) Host-plant mediated effects of root herbivory on insect seed predators and their parasitoids. Oecologia 127:246–250
McBrien H, Harmsen R, Crowder A (1983) A case of insect grazing affecting plant succession. Ecology 64:1035–1039
McNaughton F (1970) Net primary production of sycamore (*Acer pseudoplatanus*) in western Scotland. J Appl Ecol 3:577–590
Messina FJ, Durham SL, Richards JH, McArthur ED (2002) Trade-off between plant growth and defense? A comparison of sagebrush populations. Oecologia 131:43–51
Mitchell CE (2003) Trophic control of grassland production and biomass by pathogens. Ecol Lett 6:147–155
Müller-Schärer H, Brown VK (1995) Direct and indirect effects of aboveground and belowground insect herbivory on plant-density and performance of *Tripleurospermum perforatum* during early plant succession. Oikos 72:36–41
National Research Council of Canada (1981) Pesticide–pollinator interactions. National Research Council of Canada, Ottawa, Ontario, 190 pp
Norris RF (1997) Impact of leaf mining on the growth of *Portulaca oleracea* (common purslane) and its competitive interaction with *Beta vulgaris* (sugarbeet). J Appl Ecol 34:349–362
Palmisano S, Fox LR (1997) Effects of mammal and insect herbivory on population dynamics of a native California thistle, *Cirisum occidentale*. Oecologia 111:413–421
Parker LW, Santos PF, Phillips J, Whitford WG (1984) Carbon and nitrogen dynamics during the decomposition of litter and roots of a Chihuahuan desert annual, *Lepidium lasiocarpum*. Ecol Monogr 54:339–360
Paynter Q, Fowler SV, Memmott J, Sheppard AW (1998) Factors affecting the establishment of *Cytisus scoparius* in southern France: implications for managing both native and exotic populations. J Appl Ecol 35:582–595
Pell M, Stenberg B, Torstensson L (1998) Potential denitrification and nitrification tests for evaluation of pesticide effects in soil. Ambio 27:24–28
Perfecto I (1990) Indirect and direct effects in a tropical agroecosystem: the maize–pest–ant system in Nicaragua. Ecology 71:2125–2134
Perring TM, Gruenhagen NM, Farrar CA (1999) Management of plant viral diseases through chemical control of insect vectors. Annu Rev Entomol 44:457–481
Pysek P (1992) Seasonal changes in response of *Senecio ovatus* to grazing by the chrysomelid beetle *Chrysomela speciosissima*. Oecologia 91:596–628
Rees M, Brown VK (1992) Interactions between invertebrate herbivores and plant competition. J Ecol 80:353–360
Risch SJ, Pimentel D, Grover H (1986) Corn monoculture versus old field: effects of low levels of insecticide. Ecology 67:505–515
Root RB (1996) Herbivore pressure on goldenrods (*Solidago altissima*): its variation and cumulative effects. Ecology 77:1074–1087
Root R, Gowan JA (1976) The influence of insecticides with differing specificity on the structure of the fauna associated with potatoes. Am Midl Nat 99:299–314
Santos PF, Whitford WG (1981) The effects of microarthropods on litter decomposition in a Chihuahuan desert ecosystem. Ecology 62:654–663
Santos PF, Phillips J, Whitford WG (1981) The role of mites and nematodes in early stages of buried litter decomposition in a desert. Ecology 62:664–669

Santos PF, Elkins NZ, Steinberger Y, Whitford WG (1984) A comparison of surface and buried *Larrea tridentata* leaf litter decomposition in North American hot deserts. Ecology 65:278–284

Schlesinger WH (1997) Biogeochemistry: an analysis of global change, 2nd edn. Academic Press, San Diego, 588 pp

Seastedt TR, Crossley DAJ, Hargrove WW (1983) The effects of low-level consumption by canopy arthropods on the growth and nutrient dynamics of black locust and red maple trees in the southern Appalachians. Ecology 64:1040–1048

Sessions L, Kelly D (2002) Predator-mediated apparent competition between an introduced grass, *Agrostis capillaris*, and a native fern, *Botrychium australe* (Ophioglossaceae), in New Zealand. Oikos 96:102–109

Shimazaki A, Miyashita T (2002) Deer browsing reduces leaf damage by herbivorous insects through an induced response of the host plant. Ecol Res 17:527–533

Shure DJ (1971) Insecticide effects on early succession in an old field ecosystem. Ecology 52:271–279

Siemann E, Rogers WE (2003) Herbivory, disease, recruitment limitation and the success of an alien tree species. Ecology 84:1489–1505

Spencer H, Norman PA (1952) Increases in citrus red mite infestations after the application of parathion sprays. Fla Entomol 35:87–90

Strong AM, Sherry TW, Holmes RT (2000) Bird predation on herbivorous insects: indirect effects on sugar maple saplings. Oecologia 125:370–379

Traveset A (1994) The effect of *Agonoscena targionii* (Licht) (Homoptera, Psylloidea) on seed production by *Pistacia terebinthus* L. Oecologia 98:72–75

Wardle DA (2002) Communities and ecosystems: linking the aboveground and belowground components. Princeton University Press, Princeton, NJ

Wardle DA, Barker GM (1997) Competition and herbivory in establishing grassland communities: implications for plant biomass, species diversity and soil microbial activity. Oikos 80:470–480

Weary GC, Merriam HG (1978) Litter decomposition in a red maple woodlot under natural conditions and under insecticide treatment. Ecology 59:180–184

Whitford WG, Repass R, Parker LW, Elkins NZ (1981) Effects of initial litter accumulation and climate on litter disappearance in a desert ecosystem. Am Midl Nat 108:105–108

Whittaker JB, Warrington S (1985) An experimental field-study of different levels of insect herbivory induced by *Formica rufa* predation on sycamore (*Acer pseudoplatanus*). 3. Effects on tree growth. J Appl Ecol 22:797–811

Young TP, Okello BD (1998) Relaxation of an induced defense after exclusion of herbivores: spines on *Acacia drepanolobium*. Oecologia 115:508–513

16 The Role of Insect Herbivores in Exotic Plant Invasions: Insights Using a Combination of Methods to Enhance or Reduce Herbivory

W.E. Rogers and E. Siemann

16.1 Summary

Exotic plant invasions are threatening biodiversity and altering fundamental ecosystem properties and processes worldwide. Escape from native insect herbivores is believed to be one of the primary causes contributing to the successful invasion of many introduced plants. With biotic homogenization increasing globally, studies that examine the effects of herbivores on both native and introduced species are essential for understanding the influence of exotic species invasions on community dynamics and ecosystem function. While collecting field observations and life history characteristics of an exotic plant can be useful, it is experimental manipulations that will most clearly reveal the mechanisms responsible for the dominance of an aggressive invasive species. Employing a variety of methodological approaches that both increase and decrease insect herbivory will best elucidate the population ecology and ecosystem impact of an exotic plant invader. There is a pressing need to develop effective management strategies to lessen the effects of exotic invaders on a variety of threatened species and imperiled ecosystems. Such experiments will not only increase basic ecological knowledge, but also provide useful insights to land managers pressed with addressing a large and growing problem with tremendous societal, economic and environmental costs.

16.2 Introduction

Numerous biases and inherent problems are associated with the different methods of experimentally manipulating herbivore damage (Hendrix 1988; Baldwin 1990; Osterheld and McNaughton 2000; Hjältén, Chap. 12, this Vol.; Lehtilä and Boalt, Chap. 13, this Vol.; Schmitz, Chap. 14, this Vol.; Siemann et

Ecological Studies, Vol. 173
W.W. Weisser and E. Siemann (Eds.)
Insects and Ecosystem Function

al., Chap. 15, this Vol.). Perhaps the best manner of dealing with these shortcomings is to concurrently perform a variety of experiments that approach herbivory questions using several of the methodologies described herein and compare and contrast the findings from these different studies. Together the results of several approaches should be richer and more reliable than any method used in isolation. In this chapter, we outline a variety of methodological techniques for assessing and comparing the effects of insect herbivory on exotic invaders and native plants. Specific reference will be made to studies we have conducted examining the role of herbivores in invasions of *Sapium sebiferum* in North America and Hawaii. In aggregate, these studies will provide other researchers with examples and a framework for pursuing questions related to the accumulation dynamics of herbivores on plants, mechanisms of community assembly and coevolutionary interactions of herbivores and their hosts.

16.3 The Role of Herbivores in Exotic Plant Invasions

Invasions by exotic plant species are considered to be one of the greatest contemporary and future threats to the integrity of ecosystems worldwide (Coblentz 1990; Soule 1990; Chapin et al. 2000; Pimentel et al. 2000). Despite the importance of the problem, ecologists are still in the early stages of understanding the mechanisms underlying exotic plant invasions. Nevertheless, because invasive plants typically experience low losses to herbivores in their introduced range (Elton 1958; Tucker and Richardson 1995; Yela and Lawton 1997; Maron and Vila 2001), the assertion that herbivores are important in mediating plant competition is nearly ubiquitous in the invasion literature (Groves 1989; Mooney and Drake 1989; Tucker and Richardson 1995; Williamson 1996; Keane and Crawley 2002). In general, insect herbivores need not consume a large amount of plant material to have a large effect on plant community composition, they need only reverse the outcome of competition (e.g. Louda et al. 1990; Grover 1994; Leibold 1996; Crawley 1997).

The Enemy Release Hypothesis predicts that when exotic plants are introduced with few or none of the specialist herbivores from their native habitat and are not a preferred choice of generalist herbivores in their introduced range they will suffer low rates of attack by enemies and thereby gain a competitive advantage over native plants (Schierenbeck et al. 1994; Williamson 1996; Keane and Crawley 2002; Wolfe 2002; DeWalt et al. 2004). With reduced damage, resources normally lost to enemies or used for the production of defences may be allocated to growth and/or reproduction by a plastic phenotypic response (Bazzaz et al. 1987; Tilman 1999; Alpert et al. 2000; Stowe et al. 2001; Schlichting and Smith 2002). Since relatively small amounts of leaf herbivory can have major detrimental effects on plant growth and survival (Marquis 1992), this release from enemies can greatly benefit exotic species competing with native plants.

16.4 Focal Plant Species

Chinese tallow tree [*Sapium sebiferum* (L.) Roxb, Euphorbiaceae] is a major invader throughout the southeastern United States (Bruce et al. 1997; Grace 1998; Siemann and Rogers 2003a). Originally introduced to North America in 1772, *Sapium* has become naturalized from the southern Atlantic coast to the Texas Gulf coast (Bruce et al. 1997). It aggressively displaces native plants and forms monospecific stands. First established in Texas in the early 1900s, in the past 50 years much of the coastal prairie, abandoned agricultural fields and floodplain forests along the Texas Gulf coast have been converted to *Sapium*-dominated woodlands (Bruce et al. 1997; Grace 1998). It is monecious, has insect-pollinated flowers from April to June and fruits ripen from August to November (Bruce et al. 1997). Seeds are dispersed by many bird species. It is a deciduous tree that loses its leaves in autumn and has range limits largely determined by winter temperatures and aridity (Bruce et al. 1997). Rapid growth, colourful autumn foliage, abundant flowers and seeds rich in oils have encouraged widespread plantings that readily escape from cultivation.

16.5 Experimental Methods for Assessing Herbivory Effects

Designing an experimental study so that the exotic species of interest is paired with a similar native species can provide a better understanding of the mechanisms responsible for invasion (Barrett and Richardson 1986; Schierenbeck et al. 1994; Mack 1996; Sakai et al. 2001; Keane and Crawley 2002; Daehler 2003). Ideally, the native and exotic species would be congeners, but many introduced species are taxonomically isolated, making pairing difficult. In such instances, an ecologically similar native species can be used for comparisons. In our studies, *Sapium sebiferum* is the only woody member of its genus in the region and there are no native Euphorbiaceae trees in Texas. By using both *Sapium* and a native tree species that shares multiple morphological, physiological and phenological characteristics, we can monitor similarities and differences between *Sapium* and native plants that are unrelated to their site of origin. This is especially important in grassland invasions because *Sapium* is a woody plant competing with herbaceous functional groups. As such, experiments that incorporate both the exotic tree and a native tree species will more likely reveal the mechanisms responsible for successful invasions and competitive dominance.

An alternative experimental method is to incorporate seedlings derived from seeds collected in both the native and introduced ranges. Such studies can provide valuable insights into genetic change as a potential mechanism causing invasiveness. Evolutionary change is increasingly being recognized as an important factor contributing to the success of exotic invaders (Blossey and Nötzold 1995; Thompson 1998; Mack et al. 2000; Keane and Crawley 2002; Mooney and Cleland 2001; Sakai et al. 2001; Lee 2002; Stockwell et al. 2003).

16.5.1 Common Garden/Reciprocal Transplant Studies

Reciprocal seedling transplants in common garden and greenhouse pot studies can be used to assess the effects of genetic change on invasive characteristics relative to the effects of phenotypic and developmental plasticity. Although the Enemy Release Hypothesis has been widely accepted to explain the invasive success of many exotic plant species, an alternative hypothesis, the Evolution of Increased Competitive Ability (EICA; Blossey and Nötzold 1995), proposes that invasive plants evolve reduced allocation to defence and increased allocation to growth and/or reproduction because they are seldom attacked by enemies (Thompson 1998; Willis et al. 1999; Willis et al. 2000). Because allocation to defence may be as costly as herbivore damage (Bazzaz et al. 1987; Simms 1992; Baldwin 1998; Strauss et al. 2002), plants that escape their enemies in an introduced range would gain a selective benefit from decreasing their defensive investment. While the Enemy Release Hypothesis predicts that both native and invasive genotypes would benefit from low levels of herbivore damage in the introduced range, the EICA hypothesis suggests that invasive genotypes would achieve an additional benefit derived from reduced allocation to energetically expensive defences. Some studies confirm differences in growth and competitive ability of invasive and native genotypes (Blossey and Nötzold 1995; Willis and Blossey 1999; Leger and Rice 2003), while others are inconclusive (Willis et al. 1999, 2000; Thebaud and Simberloff 2001).

Using long-term common garden and greenhouse experiments, we have recently shown genetic differences in growth and defence among native and introduced genotypes of *Sapium sebiferum* that likely contribute to its invasiveness (Siemann and Rogers 2001, 2003b, c). In a long-term common garden planted in east Texas, invasive genotypes of *Sapium* from North America had significantly higher growth rates, earlier and greater seed production but lower foliar tannin concentrations than native *Sapium* genotypes from Asia (Fig. 16.1). All genotypes had uniformly low amounts of leaf area removed by insect herbivores and damage was independent of genotype (Siemann and Rogers 2001). This outcome is unlikely to be explained by the Enemy Release Hypothesis since both native and invasive genotypes should have similarly displayed a plastic phenotypic reallocation from defence to growth in the

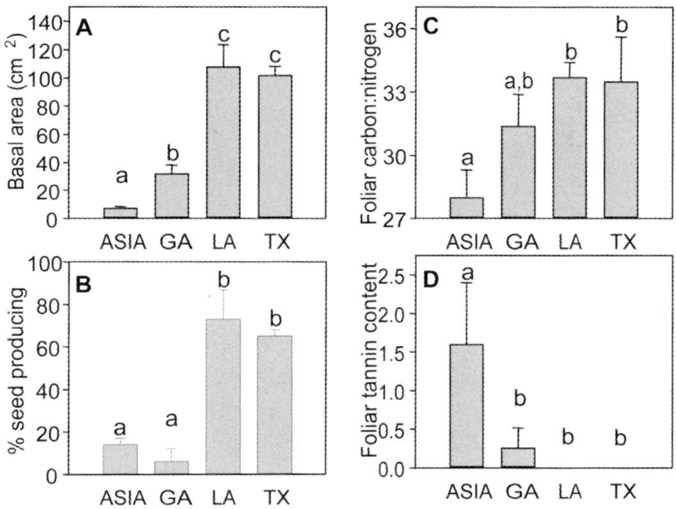

Fig. 16.1. Genotypes of *Sapium sebiferum* grown in a 14-year common garden in Texas differed in A growth, B seed production, C leaf chemistry and D defence chemicals [foliar tannin content measured as tannic acid equivalents (mg) per 100 mg tissue dry weight]. *ASIA* Native range; *GA* Georgia (site of North American introduction); *LA* Louisiana; *TX* Texas (areas colonized later). All trees had low levels of herbivore damage. *Different letters on bars* indicate significant statistical differences at $P<0.05$. (Modified from Siemann and Rogers 2001)

introduced range where herbivores are absent. Rather, the EICA hypothesis postulates an evolutionary mechanism for reallocation of resources from defence to growth in response to low herbivory and is consistent with these patterns for native and invasive genotypes of *Sapium* (Blossey and Nötzold 1995). In this scenario, there is little increase in the rate of herbivory on exotic plants with lower allocation to defence, whereas, for native plants, herbivory is expected to increase strongly as defences decrease. Reductions in defence likely lead to greater competitive ability only when the additional costs of herbivore damage do not exceed the reduced costs of defence (Coley et al. 1985; Bazzaz et al. 1987; Simms and Rausher 1987; Maschinski and Whitham 1989; Louda et al. 1990; Herms and Mattson 1992; Hunter and Price 1992; Mauricio 1998; Agrawal 2000). According to the EICA hypothesis, the discrepancy in growth rates between native plants and invasive exotics arises from the unique combination of low herbivory and low defence that native plants are unable to achieve.

We recently reinforced this interpretation by examining a companion long-term common garden study established in Hawaii that used *Sapium* seed collected from many of the same source trees as the east Texas common garden (Siemann and Rogers 2003 c). In Hawaii, the native Asian genotypes had less leaf damage and grew significantly larger than invasive Texas genotypes (Fig. 16.2A,B). This was contrary to our findings in the Texas common garden where invasive genotypes outperformed native genotypes. We believe this reversal of growth patterns for the different genotypes is due to Asian herbivores, which were inadvertently introduced to Hawaii, feeding more heavily on poorly defended Texas *Sapium* genotypes (Fig. 16.2). As a result, *Sapium* is currently not invasive on any of the Hawaiian islands despite being present for several decades.

Combined, these long-term common garden experiments in Texas and Hawaii with *Sapium* from its native range and areas where it is invasive suggest that post-introduction evolutionary change has occurred in response to an absence of herbivores. This may potentially explain *Sapium*'s current invasive status in Texas where Asian herbivores are non-existent and *Sapium*'s rel-

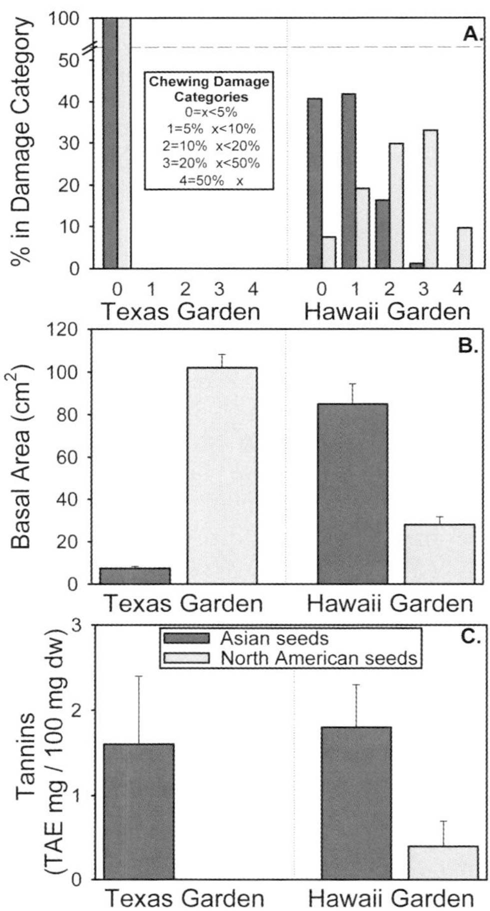

Fig. 16.2. Long-term common garden experiments in Texas and Hawaii, USA, with *Sapium* from its native range (Asia) and areas where it is invasive (North America) demonstrate that post-introduction evolutionary change promotes invasiveness in Texas, but not in Hawaii where Chinese herbivores are abundant. **A** In Texas, where *Sapium* is invasive, herbivory levels are low in the common garden. In Hawaii, where *Sapium* is not invasive, North American genotypes suffered greater herbivore damage than native Asian genotypes. **B** Invasive genotypes grew significantly larger than native genotypes in Texas, but the opposite occurred in Hawaii. **C** Foliar tannin content [measured as tannic acid equivalents (TAE, mg) per 100 mg tissue dry weight (dw)] was significantly greater in native genotypes compared to invasive genotypes regardless of common garden location. (Modified from Siemann and Rogers 2003 c)

ative scarcity in Hawaii where the herbivore *Adoretus sinicus* (Chinese rose beetle) is abundant (Siemann and Rogers 2003 c). In Texas, where *Sapium* is invasive, herbivores avoided feeding on all plants in the common garden. With uniformly low herbivory, fast-growing, poorly defended invasive genotypes outperformed slower-growing, better defended native genotypes. In Hawaii, where *Sapium* is not invasive, *A. sinicus* caused greater damage to Texas genotypes, which may be responsible for Asian genotypes being superior.

The reversal of growth patterns for native and invasive genotypes due to native herbivores being either present or absent emphasizes the importance of establishing common garden studies in both the introduced range where the species is invasive and in an area where native herbivores are present, preferably in the native range that contributed the original introduced source populations. Because insect herbivory pressures and plant resistance traits covaried across time and space, our insights into the genetic differences of invasive and native *Sapium* would have been greatly reduced without established gardens in both Texas and Hawaii. As a result, it is highly recommended that future common studies be concurrently established in the native and introduced range and include plant genotypes from both sources. We are currently pursuing studies that will establish common gardens of multiple *Sapium* genotypes at sites across a biogeographical gradient of invasion in several regions of the southeastern United States, sites in the native range of China, and sites in Hawaii where it has been introduced but is currently not invasive.

Another finding from these common garden studies is that *Sapium* derived from seed where it has been present longer as an introduced species is more similar in growth and defence to genotypes from the native Asian range than in North American areas where it has more recently invaded (Fig. 16.1). Within 300 years after introduction, invasive plants often support diverse insect communities similar to those on native plants (Strong et al. 1984). *Sapium* from Georgia, the site of original introduction, may more closely approximate the situation in Hawaii with Asian herbivores than that of Texas where herbivores are relatively inexperienced with the novel plant. The intermediate position of Georgia genotypes may reflect smaller genetic change compared to Texas genotypes. Alternatively, Georgia genotypes are potentially being recognized by native herbivores as an edible resource and are beginning to be selected for increased levels of defence as they accumulate a higher pest load and suffer greater amounts of damage. *Sapium* invasions present an ideal opportunity to conduct multiple common garden studies and various herbivore manipulations in order to examine native herbivores and an invasive plant in different stages of adjustment to a novel environment. The intriguing possibility that the ecological success of *Sapium* may be attributed to rapid post-introduction evolutionary change in competitive ability and defence against herbivores establishes a model system for investigations into the role of enemies in the success of other exotic plant species.

16.5.2 Reducing Herbivory on Target Plants Using Insecticide Sprays

There is a long history of debate in the ecological literature regarding top-down herbivore regulation of plant population dynamics, community structure and net primary productivity (e.g. Hairston et al. 1960; Strong et al. 1984; Crawley 1989; Brown and Gange 1990; Louda et al. 1990; Hunter and Price 1992; Schmitz, Chap. 14, this Vol.). In spite of this controversy, a central prediction of the Enemy Release Hypothesis is that if introduced plants suffer less damage than the native flora, removing herbivores should result in significantly greater damage reductions to natives than to exotic species. If differences in pest loads are responsible for the greater growth and lower mortality of introduced plants, removing herbivores should also minimize growth and survival differences between exotics and natives (Sakai et al. 2001; Keane and Crawley 2002).

Insect herbivores can be excluded by regular spraying of foliage on target plants with a variety of readily available insecticides, many of which have been successfully used for other ecological studies (reviewed by Siemann et al., Chap. 15, this Vol.). Plants not receiving insecticide should be sprayed with an equal amount of water. To avoid unintentional treatment of control plants with insecticide, sprays should be administered only on days when there is no wind.

Phytotoxic effects could potentially cause methodological artifacts. Companion studies in controlled environments should always be performed to discount the possibility of plant growth being directly affected by the chemical spray per se. Further, degree of toxicity to non-target organisms, residence time on vegetation and in the soil and nutrient levels (many insecticides contain trace amounts of nitrogen and phosphorus) need to be considered when designing studies involving the application of chemical sprays. While there is considerable utility in employing insecticides for disrupting insect herbivory on target plants, if concerns of toxicity or nutrient additions are warranted the more labour-intensive method of manually removing larger, sessile herbivores is also an effective manipulation (Karban and Strauss 1993; Agrawal 1998).

We have recently completed a 3-year test of the Enemy Release Hypothesis (Siemann and Rogers 2003a). Using chemical sprays, we suppressed insect herbivores on transplanted seedlings of *Sapium* and *Celtis laevigata*, a native tree, in forests and prairies in east Texas where *Sapium* is invading. Although not taxonomically related, pairing *Sapium* with the ecologically similar native tree *Celtis* established a control for woody establishment and encroachment in the absence of insect herbivory that is unrelated to the geographical origin of the species and should better reveal the mechanisms responsible for successful invasions.

As predicted by the Enemy Release Hypothesis, results from our study showed that insect herbivores caused greater damage to unsprayed native seedlings than unsprayed *Sapium* seedlings. However, contrary to predictions of the Enemy Release Hypothesis, suppression of insect herbivores caused significantly greater increases in survivorship and growth of *Sapium* compared to native seedlings (Siemann and Rogers 2003a). It was only due to common garden studies (mentioned above) and additional companion experiments that manipulated herbivore damage in alternative ways (see sections below) that we were able to explain these counterintuitive results and repeatedly obtain results consistent with the EICA hypothesis. We are currently pursuing studies using common garden plantings and factorial insecticide spray treatments on native plant species and native and invasive *Sapium* genotypes in a variety of biogeographical locations to further explore the role of genetic change in exotic species in response to the presence or absence of a herbivore load.

16.5.3 Reducing Herbivory on Community Assemblages Using Insecticide Sprays

Another method of assessing herbivore effects on native and exotic plants is to chemically treat the entire plant community in experimental plots. Other researchers have had dramatic success with this technique in native herbaceous communities (Brown et al. 1988; Carson and Root 1999, 2000). Again, if enemies strongly facilitate exotic plant invasions, removing herbivores from entire plant communities containing native and exotic species should reduce damage to native plants significantly more than exotics. The resultant competitive release caused by spraying should also reduce the growth of exotics relative to the growth of native plants in areas where herbivores are abundant. We recently completed a 3-year experiment chemically treating 2-m^2 prairie plots containing a mixture of native and invasive woody and herbaceous plant species. Surprisingly, no significant patterns in plant community composition or productivity were observed (Siemann and Rogers, unpubl. data). This type of study is likely most effective when implemented over longer time periods, particularly in instances where outbreaking insects occur (Carson and Root 1999, 2000).

The effect of belowground herbivores on exotic plant species and native plant community dynamics can also be manipulated by soaking insecticides into the soil of experimental plots. Belowground herbivory can have strong effects on plant community structure and the competitive environment of a plant often influences its response to belowground herbivory (Anderson 1987; Brown and Gange 1990; Mortimer et al. 1999; Rogers and Hartnett 2001; Verschoor et al. 2002). Several studies have found the impact of belowground insect herbivory to be greater when the host plant was competing with other plant species (e.g. Müller-Schärer 1991; Nötzold et al. 1998). Above- and belowground herbivores often damage plants simultaneously and complex interactions between different types of tissue damage frequently become manifest in varied growth responses (Seastedt et al. 1988; Moran and Whitham 1990; Müller-Schärer and Brown 1995; Houle and Simard 1996; Maron 1998; Masters et al. 2001; Masters, Chap. 5, this Vol.). Other studies have shown that root herbivory has a greater negative effect on plant growth and reproduction than foliar herbivory (Reichman and Smith 1991; Strong et al. 1995; Maron 1998; but see Moran and Whitham 1990; Houle and Simard 1996). Several root-feeding insects associated with *Sapium* have been identified in its native Asian range (Zhang and Lin 1994), but the effects of root damage on invasive North American genotypes and its effect on plant competition have not been previously examined. Despite belowground herbivores frequently having greater effects on plant community composition and productivity than aboveground herbivory, we are not aware of any studies to date that utilize insecticide manipulations to specifically examine the interactions between native and exotic plants with respect to root-damaging herbivores.

At the other end of a plant's life history, chemically removing insect herbivores from adult trees presents multiple logistic difficulties. Chemically fogging tree canopies has been successfully used in assessing insect species diversity in tropical rainforests (Erwin 1982; Basset 2001) and to control eruptive herbivores in European and North American forests (Perry 1994). Herbivore densities and effects on native and exotic adult trees could also be assessed by experimentally employing similar techniques in both the native and introduced range.

16.5.4 Factorial Manipulations of Herbivory, Resources and Competition

It is possible that herbivores exert their influence on exotic plant invasions by mediating resource competition. The negative effects of herbivory can be particularly pronounced with low nutrient availability and are frequently mitigated by an increased supply of limiting resources (Brown and Gange 1990; Louda et al. 1990; Maschinski and Whitham 1989; Steinger and Müller-Schärer 1992; Verschoor et al. 2002). Factorial experiments that simultaneously manipulate other environmental conditions such as soil resources, water, light availability and/or intensity of competition while concurrently manipulating herbivore damage will provide additional insights into the role of herbivores in facilitating exotic plant invasions. Recent reviews have emphasized the utility of making comparisons between native and invasive plants under multiple growing conditions and suggest context dependence for the invasiveness of many species (Alpert et al. 2000; Sakai et al. 2001; Keane and Crawley 2002; Daehler 2003). This strategy has been particularly fruitful in our studies with *Sapium* (Rogers et al. 2000, 2003; Rogers and Siemann 2002, 2003, 2004; Siemann and Rogers 2003d).

16.5.5 Simulating Herbivory Via Mechanical Leaf Damage

Realistic simulation of herbivory by mechanical means is problematic because many aspects of insect chewing cannot be accurately duplicated (Hendrix 1988; Karban and Baldwin 1997; Agrawal 1998; Hjältén, Chap. 12, this Vol.; Lehtilä and Boalt, Chap. 13, this Vol.). Artificial defoliation typically results in tissues being removed indiscriminately, whereas natural herbivory is frequently more selective. However, simulated damage can suitably represent the decreased leaf area and mass loss experienced by herbivore-damaged tree seedlings (Hendrix 1988; Marquis 1992; Stowe 1998; Tiffin and Inouye 2000). The advantages of simulated herbivory over other manipulations of herbivores include the ability to remove exact amounts of tissue, specify spatial and temporal patterns of removal and randomize controlled damage treatments with resource manipulations and competitive interactions.

Before initiating a mechanical defoliation study, it is useful to examine natural types and levels of herbivory. Preliminary examinations of insects on *Sapium* in Texas and Louisiana showed that the few herbivores observed on the foliage are generalists that also feed on a variety of native species in both forests and prairies (Johnson and Allain 1998; Hartley; Rogers and Siemann, unpubl. data). Although infrequent, small chewing holes are the most common form of leaf damage observed on naturally growing *Sapium* seedlings (Rogers et al. 2000). As a result, we used a steel paper hole punch to simulate the effects of leaf herbivory on *Sapium* while growing in various resource and competitive conditions (Rogers et al. 2000; Rogers and Siemann 2002, 2003). In the first study, seedlings were grown in pots and exposed to factorial combinations of three light treatments, three soil fertility treatments and three simulated herbivory treatments (control, moderate and high). Native *Celtis laevigata* tree seedlings were also subjected to the treatments for comparison. Hole punches were randomly and independently assigned to leaves twice during the growing season. New leaves near the top of each seedling were excluded to protect apical meristems and avoid affecting branching dynamics.

Focusing on early life-history stages has a greater capacity for revealing mechanisms that regulate community dynamics because young seedlings are frequently more susceptible to environmental stress than older plants (Fenner 1987; Meiners and Handel 2000). Regardless, partial herbivory rarely leads directly to the mortality of a seedling (Fenner 1987; Hendrix 1988). As a result, we have also manipulated the temporal patterns of leaf herbivory on *Sapium* (Rogers and Siemann 2003). We were able to concurrently assess the effects of casual herbivore consumption using a low-intensity, chronic defoliation treatment and the effects of an outbreaking insect using a high-intensity, acute defoliation treatment. Although the same number of leaf holes were punched for both simulated herbivory treatments in this study, the tempo and potential effect of the damage differed considerably. Again, the defoliation treatments were crossed factorially with light and soil fertility manipulations and the experiment was performed in both field and greenhouse settings (Rogers and Siemann 2003).

In another set of paired field and greenhouse studies simulating leaf herbivory on *Sapium* we used scissors to increase the damage severity by removing the front half of every full leaf blade twice during the growing season (Rogers et al. 2003). The scissors were sterilized with an alcohol wipe after defoliating each seedling to prevent the spread of disease or secondary allelochemicals. Clipped leaves that remained on seedlings were cut in half a second time as were all newly added leaves. In this and other studies, damaging the plants a second time allowed us to magnify the negative effects of artificial defoliation, particularly if the plants possess inducible defences that were activated by the initial leaf damage (Karban and Baldwin 1997; Rogers et al. 2003). Inducible defences can increase plant fitness in the presence of herbivores (Agrawal 1998), but can be costly if it does not deter future herbivore attacks. Unexpectedly, all of our studies involving simulated herbivory manipulations on invasive *Sapium* revealed that seedlings derived from seed collections obtained in the introduced range of east Texas were capable of rapidly compensating for all levels and types of tissue damage we imposed.

The success of *Sapium* as an invader is frequently attributed to an absence of pests (Bruce et al. 1995; Jubinsky and Anderson 1996) with the connotation that *Sapium* is resistant to native herbivores. Our results from these and other studies suggest that North American *Sapium* is a herbivory-tolerant plant that rapidly compensates for mass lost to defoliation. Consistent with the predictions of EICA, *Sapium*'s success as an invader may be that as a herbivory-tolerant species without an appreciable herbivore load, it is experiencing the benefits of a herbivore-resistant plant without incurring the associated costs of resistance (e.g. van der Meijden et al. 1988; Simms 1992; Rosenthal and Kotanen 1994; Strauss and Agrawal 1995; Stowe et al. 2001). In other words, invasive genotypes of *Sapium* are not experiencing a trade-off between herbivory resistance and tolerance like other native plant species because it has escaped the 'to grow or defend' dilemma of plants in its introduced range by allocating resources to growth rather than defence (Herms and Mattson 1992).

The predictions of the EICA hypothesis require that simulated herbivory be more costly to native genotypes of *Sapium* than to invasive genotypes. We conducted full-factorial, paired greenhouse and field experiments designed to assess the effects of soil fertility and simulated leaf herbivory using scissors on growth and survival of *Sapium* seedlings derived from seed collections obtained from the species' native range in China and introduced range along the Texas Gulf coast. Artificially defoliating *Sapium* significantly decreased the growth of Asian *Sapium* genotypes whereas Texas *Sapium* genotypes compensated for the leaf mass removed (Rogers and Siemann, unpublished data; Fig. 16.3). The negative effect of removing costly, defended leaves in native genotypes compared to substantial regrowth potential in poorly

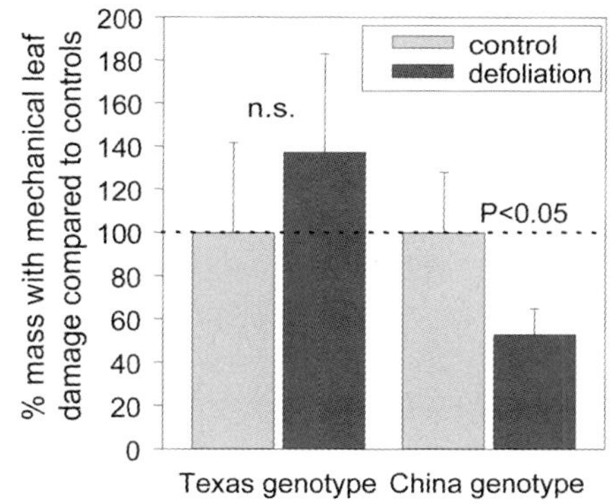

Fig. 16.3. Experimental defoliation of *Sapium* with scissors had a greater negative effect on native Chinese genotypes than on invasive Texas genotypes (Rogers and Siemann, unpubl. data). *n.s.* Not significant

defended invasive genotypes provides further support for EICA predictions that invasive genotypes of *Sapium* have undergone a post-introduction evolutionary change from a herbivore-resistant species to a fast-growing, herbivore-tolerant species that rapidly compensates for tissue damage.

16.5.6 Simulating Herbivory Via Mechanical Root Damage

Less common than herbivory manipulations involving mechanical defoliation are studies that simulate belowground root herbivory. While field studies can be performed (Reichman and Smith 1991; Rogers and Hartnett 2001), severing root tissue in pot experiments is considerably more tractable given the inaccessibility of the belowground environment (Detling et al. 1980; Schmid et al. 1990; Houle and Simard 1996). We conducted a full-factorial pot experiment designed to assess the effects of simulated root herbivory, soil fertility and competition on *Sapium* seedlings derived from seeds obtained in the ancestral Chinese range and introduced Texas range. Roots were severed using a sharp serrated steel blade inserted into a narrow opening cut in the plastic pot. Test pots were used prior to initiating root herbivory to ensure the effectiveness of this method. Belowground competition was achieved by adding annual ryegrass seed (*Lolium multiflorum* Lam.) to the pots. The results, again consistent with EICA predictions, reveal that Chinese genotypes were negatively affected by root damage, while Texas genotypes were able to compensate for root herbivory (Rogers and Siemann 2004). Increased soil fertility promoted growth of Chinese genotypes, but did not reduce the negative effects of root herbivory enough to allow the seedlings to completely compensate for damage. Grass competition increased the height growth rate of Chinese genotypes, but did not affect shoot or root mass. In competitive conditions, the shoot and root mass of Chinese genotypes was lower than undamaged controls in both fertilized and unfer-

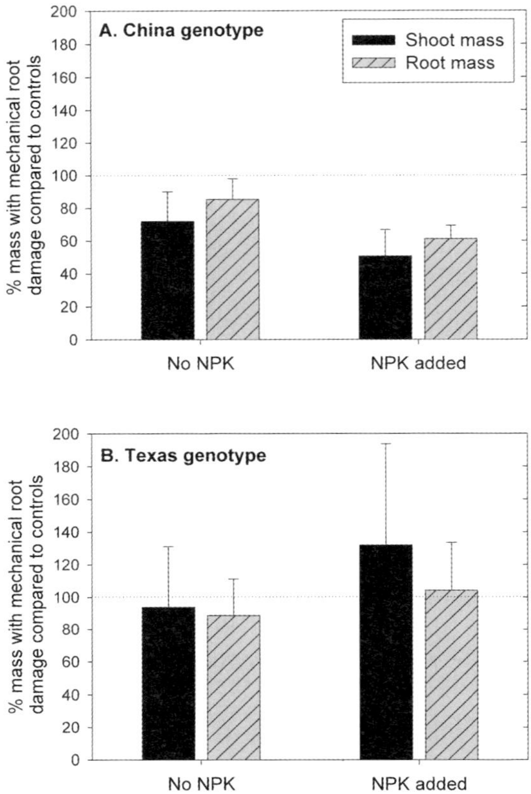

Fig. 16.4. Mass of potted *Sapium* seedlings grown in competition with ryegrass and receiving factorial combinations of fertilizer (*NPK*) and simulated root herbivory. Per cent *Sapium* shoot and root mass (mean +1 SE) of mechanically damaged seedlings relative to control seedlings (100 %) reveals that China genotype seedlings (**A**) were markedly less likely to compensate for simulated root herbivory than Texas genotype seedlings (**B**), especially with the addition of soil nutrients. (Modified from Rogers and Siemann, 2004)

tilized conditions (Fig. 16.4A). By contrast, the shoot and root mass of Texas genotypes compensated for simulated root herbivory relative to undamaged plants, particularly in fertilized conditions (Fig. 16.4B). These results provide additional support for previous studies indicating that invasive Texas *Sapium* has undergone a genetic shift away from possessing costly herbivore-defended tissues to producing relatively inexpensive tissues that are capable of rapidly compensating for damage.

16.5.7 Simulating Herbivory Using Herbicide Sprays

Species removal studies have a long history in ecological experiments (Connell 1961; Paine 1966). Many studies have used herbicides to selectively remove particular plant species and examine the subsequent community responses to altered competitive interactions (see reviews by Aarssen and Epp 1990; Goldberg and Barton 1992). These experiments are typically performed on small scales in herbaceous ecosystems and involve the removal of dominant native species (McLellan et al. 1997; D'Antonio et al 1998; Smith et al. 1999; Cabin et al. 2002). Simulating insect herbivory via experimental removal of designated plants provides unique opportunities to investigate the role of a specific plant population on competitive interactions, community assembly dynamics and ecosystem function.

In order to assess whether *Sapium* invasions can be controlled by eliminating local seed production, we killed all mature *Sapium* trees in eight 16-ha plots in an east Texas bottomland floodplain forest. Eight 16-ha control plots were also established. This study was designed to represent a highly effective seed predator or devastating insect outbreak that causes tree mortality (Perry 1994). In the removal plots, all *Sapium* trees with a diameter at breast height (dbh) greater than 4.5 cm were killed by girdling and basal bark applications of Garlon herbicide in an oil base. In total, seed recruitment from nearly 14,000 mature *Sapium* trees was eliminated. Our preliminary findings suggest that killing seed-producing *Sapium* trees results in *Sapium* seedling density being reduced while seedling density of native trees is increased (Fig. 16.5A). Although these trends were not statistically significant, 4 years after initiating the herbicide treatments the ratio of native to *Sapium* seedling density was significantly greater in plots where mature *Sapium* trees had been killed (Fig. 16.5B). This shift could potentially alter the competitive balance back in favour of the native species. Mature native tree growth and native sapling growth also increased with *Sapium* removal (Siemann and Rogers, unpublished data). Because *Sapium* forms a short-statured, short-lived forest

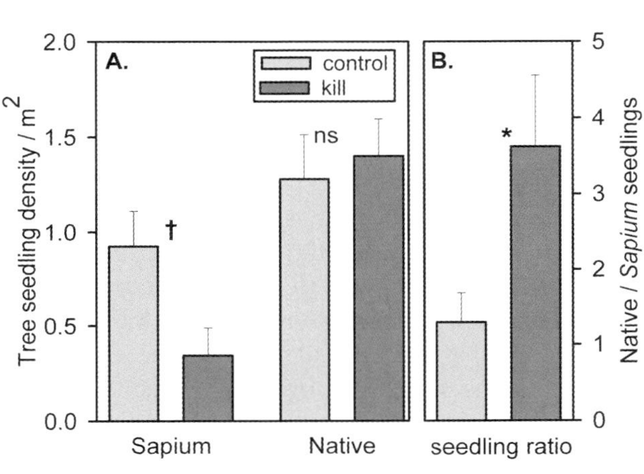

Fig. 16.5. A Density of *Sapium* seedlings and all native tree seedlings in control plots of undisturbed floodplain forest and plots where all mature *Sapium* trees were killed with herbicide. **B** Ratio of native tree seedling density to *Sapium* seedling density. *Sapium* kill plots represent a 20-fold reduction in adult *Sapium* trees (Siemann and Rogers, unpubl. data). *Cross* $P<0.01$; *asterisk* $P<0.05$; *ns* not significant

canopy compared to the native floodplain forest trees it displaces, there are also likely to be important differences in ecosystem processes, such as nutrient cycling and carbon sequestration, of forests dominated by exotic trees. Long-term monitoring of forest regeneration dynamics will be necessary to fully assess the effectiveness of this experiment, but these early results suggest that killing mature *Sapium* trees and reducing local seed supply, either through an outbreaking insect or field-worker efforts, would be a worthy strategy for conserving native forest communities imperiled by *Sapium* invasions. To our knowledge, other replicated experimental removals of a dominant invasive species have not been previously conducted on these large spatial and temporal scales.

16.5.8 Assessing Herbivore Damage Using Exclosures and Enclosures

Experimental cages are a useful tool for comparing insect herbivore impacts on exotic plants and native plants. Cages can be constructed to serve as either exclosures to prevent herbivore access to target plants or they can be used as enclosures to ensure exposure of target plants to predetermined species and densities of insect herbivores.

Cages and fences used to prevent herbivore access are most effective at excluding vertebrate herbivores (Brown and Heske 1990; Jefferies et al. 1994; McNaughton et al. 1996; Ritchie et al. 1998; Knapp et al. 1999), but could be used to prevent access of insects to target plants or community plots (Schmitz, Chap. 14, this Vol.). Because of potentially strong microclimatic effects of cages on plant growth, control cages that contain large enough holes to provide access by insects should be also erected. Cages can also be used to examine trophic interactions by adding or excluding predators that feed on herbivorous insects (Price et al. 1980; Marquis and Whelan 1996; Schmitz 1998; Schmitz, Chap. 14, this Vol.).

For insect herbivory studies, experimental cages have been most successfully used by enclosing stocked herbivores on a single leaf, individual plant or assemblages of multiple species and individuals both in pots and under field conditions (Belovsky 1986; Ritchie and Tilman 1992; Schmitz 1993; Agrawal 1998; Lill and Marquis 2001). Cage enclosures are particularly useful in experimental manipulations with potted plants. Cages can easily be erected around aboveground plant tissues, allowing the investigator to control herbivore species, herbivore density and time of exposure to herbivory. Additionally, the pot itself can serve as an effective enclosure for stocking and manipulating belowground herbivores (Steinger and Müller-Schärer 1992; Blossey 1993; Nötzold et al. 1998). Other variables like resource availability and plant competition can be simultaneously manipulated in pot experiments with caged herbivores to examine how top-down and bottom-up interactions are affected by biotic and abiotic conditions. Pot experiments provide simple access to multiple morphological and physiological measures of plant responses to damage including stem height and diameter, branch and leaf numbers, leaf area damage, water potential and photosynthesis rates. Potted plants are useful for obtaining data on measures that are frequently difficult to collect in field conditions such as both above- and belowground productivity at the termination of the experiment. Herbivore survival and performance are also easier to measure in controlled enclosure environments. In order to simultaneously examine the effects of plant competition, herbivore choice and community dynamics in controlled environments, larger foraging arenas using container mesocosms containing more diverse assemblages of plant species can be erected and stocked with herbivores.

Realism can be increased while only modestly sacrificing precision by erecting cages over transplants and/or existing vegetation in field settings and similarly stocked with herbivores. The focus can be a single target plant or a diverse community assemblage. Likewise, field studies with caged herbivores can be designed with factorial manipulations of resources and competitive interactions. Important additional insights regarding herbivory effects can be gained by conducting these bioassays with both generalist and specialist herbivores (Bernays and Chapman 1994; Marcel et al. 2002). Specialist herbivores are typically absent from exotic plant species because they have unique behavioural and physiological adaptations to their host species. Using exclosure bioassays to determine and eventually re-establish feeding relationships between exotic plants and their specialist herbivores is central to biological control efforts (McFayden 1998; Louda et al. 2003). Generalist herbivores lack specific adaptations to particular host plants and may avoid exotic plant species because they possess unusually toxic novel defences to which the herbivore is unaccustomed. Alternatively, the exotic plant may be suitable to native generalist herbivores, but the herbivores might lack behavioural adaptations necessary to recognize and utilize a new food source. Dietary experimentation is generally selected against because the risks of selecting a toxic plant often exceed the benefits of gaining an additional food source (Feeny 1975; Abrahamson and Weis 1997). Insects with many potential hosts are less efficient in their decisions and therefore suffer increased vulnerability to natural enemies. Thus, insects have evolved to quickly recognize specific chemical cues associated with suitable hosts and ignore or avoid plants that lack these cues (Bernays and Chapman 1994; Bernays 2001).

We have conducted several bioassay experiments using North American acridid grasshoppers exposed to native vegetation and exotic *Sapium* seedlings in both pot and field exclosures. In all of our studies we have found that, despite negligible herbivory damage on *Sapium* in natural grassland and forest conditions, when grasshoppers are placed in enclosures with *Sapium* they readily feed on its foliage and show strong preferences for it over native trees, forbs and grasses (Lankau et al. 2004; Fig. 16.6). We believe this indicates a behavioural barrier rather than a biochemical deterrent to utilization.

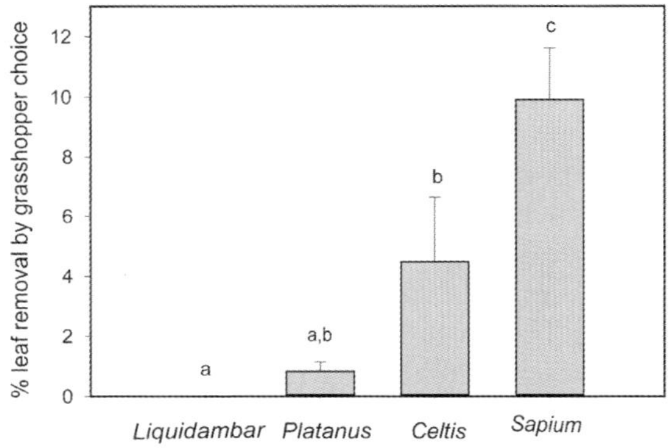

Fig. 16.6. Grasshopper (*Melanaplus angustipennis*) herbivory preferences determined by feeding trials in enclosures containing leaves of native (*Liquidambar styracifula*, *Platanus occidentalis* and *Celtis laevigata*) and an exotic (*Sapium sebiferum*) species. Grasshoppers consumed (estimated as mean percent leaf removal +1 SE) significantly more *Sapium* foliage than the native foliage in enclosures, suggesting a behavioural avoidance of the invasive species under field conditions. *Different letters on bars* indicate significant statistical differences at $P<0.05$. (Modified from Lankau et al. 2004)

Sapium is a potentially suitable food choice that is avoided because there is strong selection against host range expansion when new host plants may be toxic (Chew and Courtney 1991), temporally or spatially uncommon (Chew and Courtney 1991; Beccaloni and Symons 2000) or of limited use due to natural enemy influences (Camara 1997). We have also found that grasshoppers that were first conditioned on *Sapium* foliage in small cages fed more on the exotic tree after being introduced to multiple species mesocosms than grasshoppers first conditioned on native trees (Lankau et al. 2004). Together, these results suggest that behavioural constraints, rather than toxic novel defences, prevent generalist herbivores from more fully utilizing this abundant plant species.

Using grasshopper bioassays with potted *Sapium* seedlings derived from native China and invasive Texas seed, we have also shown significant herbivore preferences for invasive Texas foliage when offered a choice between native and invasive *Sapium* (Siemann and Rogers 2003b; Fig. 16.7A). The higher levels of consumption on native foliage caused significant decreases in the growth of China seedlings compared to the growth of Texas seedlings (Fig. 16.7B). Conversely, using the same seed sources, Texas *Sapium* seedlings grew 40 % faster than Asia *Sapium* seedlings when grown in unmanipulated field conditions where herbivores remove less than 1 % leaf area of both genotypes (Siemann and Rogers 2003b). When grasshoppers were stocked in cages with potted seedlings from the same continent, herbivory damage and *Sapium* growth rates were indistinguishable between the different genotypes

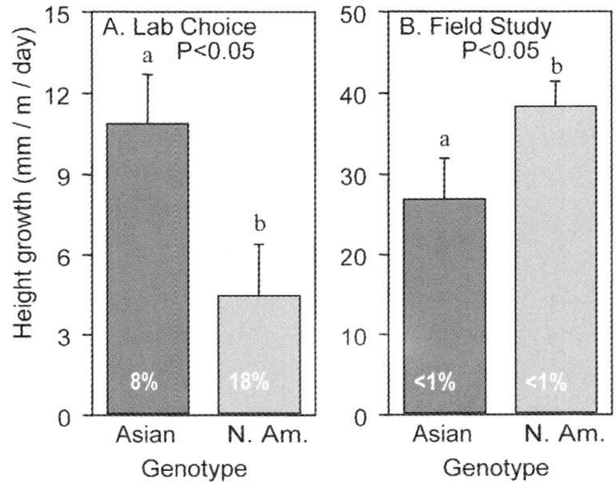

Fig. 16.7. A Asian genotypes of *Sapium* outgrew North American (*N. Am.*) genotypes of *Sapium* in direct competition in laboratory bioassays containing grasshoppers. Grasshoppers removed less leaf area of native Asian genotypes than invasive North American genotypes (percentage indicated in *bars*). B Conversely, a field study in east Texas with the same seed sources revealed that herbivory was uniformly low (percentage indicated in *bars*) and North American genotypes outgrew Asian genotypes. *Different letters on bars* indicate significant statistical differences at $P<0.05$. (Modified from Siemann and Rogers 2003b)

(Fig. 16.7). In a companion study, we allowed the *Sapium* seedlings 8 weeks to regrow after grasshoppers were removed. Texas genotypes were able to compensate for herbivory damage such that there was no statistical difference between the growth of damaged and undamaged trees (Rogers and Siemann, unpubl. data). By contrast, China genotypes exposed to grasshoppers in the same study had significantly reduced growth compared to undamaged plants despite 8 weeks of regrowth following herbivory (Rogers and Siemann, unpublished data).

These findings are further support for predictions of the EICA hypothesis. If the Enemy Release Hypothesis were correct, grasshoppers should have consumed or avoided seedlings from both regions similarly and growth rates for the different *Sapium* genotypes should have been indistinguishable within the same environmental conditions. In comparison, the EICA hypothesis predicts that even though herbivores in the introduced range avoid feeding on invasive plants in field settings, in controlled feeding trials they should overcome behavioural barriers and prefer fast-growing, less-defended invasive genotypes over slow-growing, better defended native genotypes.

16.6 Implications and Potential Significance

Invasions by exotic plant species are a large and growing environmental problem with tremendous societal costs. There is a pressing need to better understand the mechanisms responsible for exotic plant invasions and to develop effective management strategies to lessen their effects on a variety of threatened species and imperiled ecosystems. In this chapter we have described a unique, complementary suite of experimental field and greenhouse studies examining the manner in which enhancing or reducing insect herbivory influences the success of an exotic plant species, *Sapium sebiferum*, which is aggressively invading many habitats throughout the southeastern United States. Because invasion is a key step in community assembly, new studies like these with other problematic invasive species will provide valuable insights into the factors influencing successional dynamics, community structure and ecosystem stability and integrity.

Many of these studies were designed to experimentally test between two leading hypotheses, the Enemy Release Hypothesis and the Evolution of Increased Competitive Ability (EICA) hypothesis. The various experimental designs described above provide suggestions for differentiating between these hypotheses despite the differences in spatial and temporal scales of investigation. Experimental tests of the Enemy Release Hypothesis involve disrupting local patterns of insect herbivory and manipulating amounts of damage on different plants, whereas direct experimental tests of the EICA hypothesis involve disrupting geographical patterns of herbivory and manipulating evolutionary selection pressures. Nevertheless, it is important to note that the central premise of these hypotheses shares a similar origin. Evolution of invasiveness can only occur if exotic plants first experience an ecological release from enemies in their introduced range that strongly alters selection pressures and leads to genetic shifts away from defence allocation and toward greater growth and reproduction. The EICA hypothesis predicts that while genotypic changes in introduced species may contribute to their ecological success, it may also increase their susceptibility to herbivores introduced from their native range. In fact, it is possible that this phenomenon may explain the striking success of certain biological control efforts.

Successful biological control may be not only due to a re-establishment of feeding relationships with native herbivores, but also partially due to a host plant that has become unusually susceptible to its native herbivores because of a genetic shift away from chemical deterrents (Sakai et al. 2001). Invasive plants begin to support diverse insect communities similar to those on native plants within 300 years after introduction (Strong et al. 1977, 1984). If an invasive plant has evolved a reduction in herbivore deterrents, local herbivores in the introduced range may more likely begin to recognize and utilize the exotic species as a viable food alternative. While the introduction of an exotic species will by itself have profound effects on plant community composition and ecosystem processes like primary productivity, carbon sequestration and nutrient cycling, a shift in feeding preferences of native insect herbivores to the invasive exotic plant species will likely have equally dramatic consequences for community dynamics, trophic interactions and ecosystem function.

Although immediate action is required to suppress certain aggressive invaders (Simberloff 2003), a management strategy that encourages local herbivore recognition may be warranted considering the substantial risks associated with introducing biological control agents (Louda et al. 2003). Our results supporting the EICA hypothesis also suggest that commonly observed time lags from introduction to emergence as a problem invasive species may reflect a genetic adjustment period by the exotic plant and not merely demographic delays in recruitment and migration. This will greatly complicate the ability of horticulturists and land managers to identify and predict problem species a priori because the initial status of an introduced species may be a poor indicator of its future ecological success if the evolution of invasiveness is commonplace.

Acknowledgements. Funding for this research was made possible through grants provided by the United States Department of Agriculture (NRI-35320-13498), National Science Foundation (DEB-981654 and DEB-0315796), Environmental Protection Agency (STAR-R828903), National Park Service, Big Thicket National Preserve and Rice University.

References

Aarssen LW, Epp GA (1990) Neighbor manipulations in natural vegetation: a review. J Veg Sci 1:13–30

Abrahamson WG, Weis AE (1997) Evolutionary ecology across three trophic levels: goldenrods, gallmakers, and natural enemies. Princeton University Press, Princeton, NJ

Agrawal AA (1998) Induced responses to herbivory and increased plant performance. Science 279:1201–1202

Agrawal AA (2000). Benefits and costs of induced plant defense for *Lepidium virginicum* (Brassicaceae). Ecology 81:1804–1813

Alpert P, Bone E, Holzapfel C (2000) Invasiveness, invasibility and the role of environmental stress in the spread of non-native plants. Perspect Plant Ecol Evol Syst 3:52–66

Andersen DC (1987) Below-ground herbivory in natural communities: a review emphasizing fossorial animals. Q Rev Biol 62:261–286

Baldwin IT (1990) Herbivory simulations in ecological research. Trends Ecol Evol 5:91–93

Baldwin IT (1998) Jasmonate-induced responses are costly but benefit plants under attack in native populations. Proc Natl Acad Sci USA 95:8113–8118

Barrett SCH, Richardson BJ (1986) Genetic attributes of invading species. In: Groves RH, Burdon JJ (eds) Ecology of biological invasions. Cambridge University Press, Melbourne, Australia, pp 21–33

Basset Y (2001) Invertebrates in the canopy of tropical rain forests: how much do we really know? Plant Ecol 153:87–107

Bazzaz FA, Chiariello NR, Coley PD, Pitelka LF (1987) Allocating resources to reproduction and defense. BioScience 37:58–67

Beccaloni GW, Symons FB (2000) Variation of butterfly diet breadth in relation to host-plant predictability: results from two faunas. Oikos 90:50–66

Belovsky GE (1986) Optimal foraging and community structure: implications for a guild of generalist grassland herbivores. Oecologia 70:35–52
Bernays EA (2001) Neural limitations in phytophagous insects: implications for diet breadth and evolution of host affiliation. Annu Rev Entomol 46:703–727
Bernays EA, Chapman RF (1994) Host-plant selection by phytophagous insects. Chapman and Hall, New York
Blossey B (1993) Herbivory below ground and biological weed control: life history of a root-boring weevil on purple loosestrife. Oecologia 94:380–387
Blossey B, Nötzold R (1995) Evolution of increased competitive ability in invasive non-indigenous plants: a hypothesis. J Ecol 83:887–889
Brown JH, Heske EJ (1990) Control of a desert grassland transition by a keystone rodent guild. Science 250:1705–1708
Brown VK, Gange AC (1990) Insect herbivory below ground. Adv Ecol Res 20:1–58
Brown VK, Jepsen M, Gibson CWD (1988) Insect herbivory: effects on early oldfield succession demonstrated by chemical exclusion methods. Oikos 52:293–302
Bruce KA, Cameron GN, Harcombe PA (1995) Initiation of a new woodland type on the Texas Coastal Prairie by the Chinese tallow tree (*Sapium sebiferum* (L.) Roxb.). Bull Torr Bot Club 122:215–225
Bruce KA, Cameron GN, Harcombe PA, Jubinsky G (1997) Introduction, impact on native habitats, and management of a woody invader, the Chinese tallow tree, *Sapium sebiferum* (L.). Natl Areas J 17:255–260
Cabin RJ, Weller SG, Lorence DH, Cordell S, Hadway LJ, Montgomery R., Goo D, Urakami A (2002) Effects of light, alien grass, and native species additions on Hawaiian dry forest restoration. Ecol Appl 12:1595–1610
Camara MD (1997) A recent host range expansion in *Junonia coenia* Hubner (Nymphalidae): oviposition preference, survival, growth, and chemical defense. Evolution 51:873–884
Carson WP, Root RB (1999) Top-down effects of insect herbivores during early succession: influence on biomass and plant dominance. Oecologia 121:260–272
Carson WP, Root RB (2000) Herbivory and plant species coexistence: community regulation by an outbreaking phytophagous insect. Ecol Monogr 70:73–99
Chapin FS, Zavaleta ES, Eviner VT, Naylor RL, Vitousek PM, Reynolds HL, Hooper DU, Lavorel S, Sala OE, Hobbie SE, Mack MC, Diaz S (2000) Consequences of changing biodiversity. Nature 405:234–242
Chew FS, Courtney SP (1991) Plant apparency and evolutionary escape from insect herbivory. Am Nat 138:729–750
Coblentz BE (1990) Exotic organisms: a dilemma for conservation biology. Conserv Biol 4:261–265
Coley PD, Bryant JP, Chapin FS (1985) Resource availability and plant antiherbivore defense. Science 230:895–899
Connell JH (1961) The influence of interspecific competition and other factors on the distribution of the barnacle *Chthamalus stellatus*. Ecology 42:710–723
Crawley MJ (1989) Insect herbivores and plant population dynamics. Annu Rev Entomol 34:531–564
Crawley MJ (1997) Plant–herbivore dynamics. In: Crawley MJ (ed) Plant ecology, 2nd edn. Blackwell, Oxford, pp 401–474
Daehler CC (2003) Performance comparisons of co-occurring native and alien invasive plants: implications for conservation and restoration. Annu Rev Ecol Syst 34:183–211
D'Antonio CD, Hughes RF, Mack M, Hitchcock D, Vitousek PM (1998) The response of native species removal of invasive exotic grasses in a seasonally dry Hawaiian woodland. J Veg Sci 9:699–712

Detling JD, Winn T, Procter-Gregg C, Painter EL (1980) Effects of simulated grazing by below-ground herbivores on growth, CO_2 exchange, and carbon allocation patterns of *Bouteloua gracilis*. J Appl Ecol 17:771–778
DeWalt SJ, Denslow JS, Ickes K (2004) Natural enemy release facilitates habitat expansion of the invasive tropical shrub *Clidemia hirta*. Ecology 85:471–483
Elton CS (1958) The ecology of invasion by plants and animals. Chapman and Hall, London
Erwin TL (1982) Tropical forests: their richness in Coleoptera and other arthropod species. Coleopt Bull 36:74–75
Feeny P (1975) Biochemical coevolution between plants and their insect herbivores. In: Gilbert LE, Raven PH (eds) Coevolution of animals and plants. University of Texas Press, Austin, Texas, pp 3–19
Fenner M (1987) Seedlings. Can J Bot 106:35–47
Goldberg DE, Barton AM (1992) Patterns and consequences of interspecific competition in natural communities: a review of field experiments with plants. Am Nat 139:771–801
Grace JB (1998) Can prescribed fire save the endangered coastal prairie ecosystem from Chinese tallow invasion? Endangered Species Update 15:70–76
Grover JP (1994) Assembly rules for communities of nutrient-limited plants and specialist herbivores. Am Nat 143:258–282
Groves RH (1989) Ecological control of invasive terrestrial plants. In: Drake JA, Mooney HA, di Castri F, Groves RH, Kruger FJ, Rejmanek M, Williamson MW (eds) Biological invasions. Wiley, New York, pp 437–462
Hairston NG, Smith FE, Slobotkin LB (1960) Community structure, population control and competition. Am Nat 94:421–425
Hendrix SD (1988) Herbivory and its impact on plant reproduction. In: Lovett-Doust J, Lovett-Doust L (eds) Plant reproductive ecology: patterns and strategies. Oxford University Press, New York, pp 246–263
Herms DA, Mattson WJ (1992) The dilemma of plants: to grow or to defend. Q Rev Biol 67:283–335
Houle G, Simard G (1996) Additive effects of genotype, nutrient availability and type of tissue damage on the compensatory response of *Salix planifolia* ssp. *planifolia* to simulated herbivory. Oecologia 107:373–378
Hunter MD, Price PW (1992) Playing chutes and ladders: heterogeneity and the relative roles of bottom-up and top-down forces in natural communities. Ecology 73:724–732
Jefferies RL, Klein DR, Shaver GR (1994) Vertebrate herbivores and northern plant-communities: reciprocal influences and responses. Oikos 71:193–206
Johnson SR, Allain LK (1998) Observations on insect use of Chinese tallow (*Sapium sebiferum* (L.) Roxb) in Louisiana and Texas. Castanea 63:188–189
Jubinsky G, Anderson LC (1996) The invasive potential of Chinese tallow-tree (*Sapium sebiferum* Roxb.) in the southeast. Castanea 61:226–231
Karban R, Baldwin IT (1997) Induced responses to herbivory. University of Chicago Press, Chicago, Illinois
Karban R, Strauss SY (1993) Effects of herbivores on growth and reproduction of their perennial host, *Erigeron glaucus*. Ecology 74:39–46
Keane RM, Crawley MJ (2002) Exotic plant invasions and the enemy release hypothesis. Trends Ecol Evol 17:164–170
Knapp AK, Blair JB, Briggs JM, Collins SC, Hartnett DC, Johnson L, Towne G (1999) The keystone role of bison in North American tallgrass prairie. BioScience 49:39–50
Lankau RA, Siemann E, Rogers WE (2004) Constraints on the utilisation of invasive Chinese tallow tree (*Sapium sebiferum*) by generalist native herbivores in coastal prairies. Ecol Entomol 29:66–75

Lee CE (2002) Evolutionary genetics of invasive species. Trends Ecol Evol 17:386–391
Leger EA, Rice K (2003) Invasive California poppies *(Eschscholzia californica* Cham.) grow larger than native individuals under reduced competition. Ecol Lett 6:257–264
Leibold MA (1996) A graphical model of keystone predators in food webs: trophic regulation of abundance, incidence, and diversity patterns in communities. Am Nat 147:784–812
Lill JT, Marquis RJ (2001) The effects of leaf quality on herbivore performance and attack from natural enemies. Oecologia 126:418–428
Louda SM, Keeler KH, Holt RD (1990) Herbivore influence on plant performance and competitive interactions. In: Grace JB, Tilman D (eds) Perspectives on plant competition. Academic Press, New York, pp 413–444
Louda SM, Arnett AE, Rand A, Russell FL (2003) Invasiveness of some biological control insects and adequacy of their ecological risk assessment and regulation. Conserv Biol 17:73–82
Mack RN (1996) Predicting the identity and fate of plant invaders: emergent and emerging approaches. Biol Conserv 78:107–121
Mack RN, Simberloff D, Lonsdale WM, Evans H, Clout M, Bazzaz FA (2000) Biotic invasions: causes, epidemiology, global consequences, and control. Ecol Appl 10:689–710
Marcel M, Klinkhamer PGL, Vrieling K, van der Meijden E (2002) Diversity of pyrrolizidine alkaloids in *Senecio* species does not affect the specialist herbivore *Tyria jacobaeae*. Oecologia 133:541–550
Maron JL (1998) Insect herbivory above- and belowground: individual and joint effects on plant fitness. Ecology 79:1281–1293
Maron JL, Vila M (2001) When do herbivores affect plant invasion? Evidence for the natural enemies and biotic resistance hypotheses. Oikos 95:361–373
Marquis RJ (1992) The selective impact of herbivores. In: Fritz RS, Simms EL (eds) Plant resistance to herbivores and pathogens: ecology, evolution, and genetics. University of Chicago Press, Chicago, pp 301–325
Marquis RJ, Whelan C (1996) Plant morphology and recruitment of the third trophic level: subtle and little-recognized defenses? Oikos 75:330–334
Maschinski J, Whitham T (1989) The continuum of plant responses to herbivory: the influence of plant association, nutrient availability and timing. Am Nat 134:1–19
Masters GJ, Jones TH, Rogers M (2001) Host-plant mediated effects of root herbivory on insect seed predators and their parasitoids. Oecologia 127:246–250
Mauricio R (1998) Costs of resistance to natural enemies in field populations of the annual plant *Arabidopsis thaliana*. Am Nat 151:20–28
McFayden RE (1998) Biological control of weeds. Annu Rev Entomol 43:369–393
McLellan AJ, Law R, Fitter AH (1997) Response of calcareous grassland plant species to diffuse competition: results from a removal experiment. J Ecol 85:479–490
McNaughton SJ, Milchunas DG, Frank DA, Williams KJ (1996) How can net primary productivity be measured in grazing ecosystems? Ecology 77:974–977
Meiners SJ, Handel SN (2000) Additive and nonadditive effects of herbivory and competition on tree seedling mortality, growth, and allocation. Am J Bot 87:1821–1826
Mooney HA, Cleland EE (2001) The evolutionary impact of invasive species. Proc Natl Acad Sci USA 98:5446–5451
Mooney HA, Drake JA (1989) Biological invasions: a SCOPE program overview. In: Drake JA, Mooney HA, di Castri F, Groves RH, Kruger F, Rejmanek M, Williamson MW (eds) Biological invasions. Wiley, New York, pp 491–508
Moran NA, Whitham TG (1990) Interspecific competition between root-feeding and leaf-galling aphids mediated by host-plant resistance. Ecology 71:1050–1058
Mortimer SR, van der Putten WH, Brown VK (1999) Insect and nematode herbivory below ground: interactions and role in vegetation succession. In: Olff H, Brown V

Drent R (eds) Herbivores: between plants and predators. Blackwell, Cambridge, pp 205–238

Müller-Schärer H (1991) The impact of root herbivory as a function of plant density and competition: survival, growth and fecundity of Centaurea maculosa in field plots. J Appl Ecol 28:759–776

Müller-Schärer H, Brown VK (1995) Direct and indirect effects of above- and below-ground insect herbivory on plant density and performance of *Tripleurospermum perforatum* during early plant succession. Oikos 72:36–41

Nötzold R, Blossey B, Newton E (1998) The influence of below ground herbivory and plant competition on growth and biomass allocation of purple loosestrife. Oecologia 113:82–93

Osterheld M, McNaughton SJ (2000) Herbivory in terrestrial ecosystems. In: Sala OE, Jackson RB, Mooney HA, Howarth RW (eds) Methods in ecosystem science. Springer, Berlin Heidelberg New York, pp 151–158

Paine RT (1966) Food web complexity and species diversity. Am Nat 100:65–75

Perry DA (1994) Forest ecosystems. John Hopkins University Press, Baltimore

Pimentel D, Lach L, Zuniga R, Morrison D (2000) Environmental and economic costs of nonindigenous species in the United States. BioScience 50:53–65

Price PW, Bouton CE, Gross P, McPheron BA, Thompson JN, Weis AE (1980) Interactions among three trophic levels: influence of plants on interactions between insect herbivores and natural enemies. Annu Rev Ecol Syst 11:41–65

Reichman OJ, Smith CC (1991) Responses to simulated leaf and root herbivory by a biennial, Tragopogon dubius. Ecology 72:116–124

Ritchie ME, Tilman D (1992) Interspecific competition among grasshoppers and their effect on plant abundance in experimental field environments. Oecologia 89:524–532

Ritchie ME, Tilman D, Knops JMH (1998) Herbivore effects on plant and nitrogen dynamics in oak savanna. Ecology 79:165–177

Rogers WE, Hartnett DC (2001) Vegetation responses to different spatial patterns of soil disturbance in burned and unburned tallgrass prairie. Plant Ecol 155:99–109

Rogers WE, Siemann E (2002) Effects of simulated herbivory and resource availability on native and invasive exotic tree seedlings. Basic Appl Ecol 4:297–307

Rogers WE, Siemann E (2003) Effects of simulated herbivory and resources on Chinese tallow tree (*Sapium sebiferum*, Euphorbiaceae) invasion of native coastal prairie. Am J Bot 90:241–247

Rogers WE, Siemann E (2004) Invasive ecotypes tolerate herbivory more effectively than native ecotypes of the Chinese tallow tree. J Appl Ecol 41:561–570

Rogers WE, Nijjer S, Smith C, Siemann E (2000) Effects of resources and herbivory on leaf morphology and physiology of Chinese tallow (*Sapium sebiferum*) tree seedlings. Texas J Sci 52S:43–56

Rogers WE, Siemann E, Lankau RA (2003) Damage induced production of extrafloral nectaries in native and introduced seedlings of Chinese tallow tree (*Sapium sebiferum*). Am Midl Nat 149:413–417

Rosenthal JP, Kotanen PM (1994) Terrestrial plant tolerance to herbivory. Trends Ecol Evol 9:145–148

Sakai AK, Allendorf FW, Holt JS, Lodge DM, Molofsky J, With KA, Baughman S, Cabin RJ, Cohen JE, Ellstrand NC, McCauley DE, O'Neil P, Parker IM, Thompson JN, Weller SG (2001) The population biology of invasive species. Annu Rev Ecol Syst 32:305–332

Schierenbeck KA, Mack RN, Sharitz RR (1994) Effects of herbivory on growth and biomass allocation in native and introduced species of *Lonicera*. Ecology 75:1661–1672

Schlichting CD, Smith H (2002) Phenotypic plasticity: linking molecular mechanisms with evolutionary outcomes. Evol Ecol 16:189–211

Schmid B, Miao SL, Bazazz FA (1990) Effects of simulated root herbivory and fertilizer application on growth and biomass allocation in the clonal perennial *Solidago canadensis*. Oecologia 84:9–15

Schmitz OJ (1993) Trophic exploitation in grassland food chains: simple models and a field experiment. Oecologia 93:327–335

Schmitz OJ (1998) Direct and indirect effects of predation and predation risk in old-field interaction webs. Am Nat 151:327–342

Seastedt TR, Ramundo R, Hayes D (1988) Maximization of densities of soil animals by foliage herbivory: empirical evidence, graphical and conceptual models. Oikos 51:243–248

Siemann E, Rogers WE (2001) Genetic differences in growth of an invasive tree species. Ecol Lett 4:514–518

Siemann E, Rogers WE (2003a) Herbivory, disease, recruitment limitation and the success of an alien tree species. Ecology 84:1489–1505

Siemann E, Rogers WE (2003b) Reduced resistance of invasive varieties of the alien tree *Sapium sebiferum* to a generalist herbivore. Oecologia 135:451–457

Siemann E, Rogers WE (2003 c) Increased competitive ability of an invasive tree may be limited by an invasive beetle. Ecol Appl 13:1503–1507

Siemann E, Rogers WE (2003d) Changes in light and nitrogen availability under pioneer trees may indirectly facilitate tree invasions of grasslands. J Ecol 91:923–931

Simberloff D (2003) How much information on population biology is needed to manage introduced species? Conserv Biol 17:83–92

Simms EL (1992) Costs of plant resistance to herbivory. In: Fritz RS, Simms EL (eds) Plant resistance to herbivores and pathogens. University of Chicago Press, Chicago, pp 392–425

Simms EL, Rausher M (1987) Costs and benefits of plant resistance to herbivory. Am Nat 130:570–581

Smith MD, Hartnett DC, Wilson GW (1999) Interacting influence of mycorrhizal symbiosis and competition on plant diversity in tallgrass prairie. Oecologia 121:574–582

Soule ME (1990) The onslaught of alien species, and other challenges in the coming decades. Conserv Biol 4:233–239

Steinger T, Müller-Schärer H (1992) Physiological and growth responses of *Centaurea maculosa* (Asteraceae) to root herbivory under varying levels of interspecific plant competition and soil nitrogen availability. Oecologia 91:141–149

Stockwell CA, Hendry AP, Kinnison MT (2003) Contemporary evolution meets conservation biology. Trends Ecol Evol 18:94–101

Stowe KA (1998) Experimental evolution of resistance in *Brassica rapa*: correlated response of tolerance in lines selected for glucosinolate content. Evolution 52:703–712

Stowe KA, Marquis RJ, Hochwender CG, Simms EL (2001) The evolutionary ecology of tolerance to consumer damage. Annu Rev Ecol Syst 31:565–595

Strauss SY, Agrawal AA (1995) The ecology and evolution of plant tolerance to herbivory. Trends Ecol Evol 14:179–185

Strauss SY, Rudgers JA, Lau JA, Irwin RE (2002) Direct and ecological costs of resistance to herbivory. Trends Ecol Evol 17:278–285

Strong DR, McCoy ED, Rey JR (1977) Time and the number of herbivore species: the pests of sugarcane. Ecology 58:167–175

Strong DR, Lawton JH, Southwood TRE (1984) Insects on plants. Harvard University Press, Cambridge

Strong DR, Maron JL, Connors PG, Whipple A, Harrison S, Jefferies RL (1995) High mortality, fluctuation in numbers, and heavy subterranean insect herbivory in bush lupine, *Lupinus arboreus*. Oecologia 104:85–92

Thebaud C, Simberloff D (2001) Are plants really larger in their introduced ranges? Am Nat 157:231–236
Thompson JN (1998) Rapid evolution as an ecological process. Trends Ecol Evol 13:329–332
Tiffin P, Inouye BD (2000) Measuring tolerance to herbivory: accuracy and precision of estimates made using natural versus imposed damage. Evolution 54:1024–1029
Tilman D (1999) The ecological consequences of changes in biodiversity: a search for general principles. Ecology 80:1455–1474
Tucker KC, Richardson DM (1995) An expert system for screening potentially invasive alien plants in South African fynbos. J Environ Manage 44:309–338
van der Meijden E, Wijn M, Verkaar HJ (1988) Defense and regrowth, alternative strategies in the struggle against herbivores. Oikos 51:355–363
Verschoor BC, Pronk TE, de Goede RGM, Brussaard L (2002) Could plant-feeding nematodes affect the competition between grass species during succession in grasslands under restoration management? J Ecol 90:753–761
Williamson M (1996) Biological invasions. Chapman and Hall, London
Willis AJ, Blossey B (1999) Benign environments do not explain the increased vigour of non-indigenous plants: a cross-continental transplant experiment. Biocontrol Sci Tech 9:567–577
Willis AJ, Thomas M, Lawton JH (1999) Is the increased vigour of invasive weeds explained by a trade-off between growth and herbivore resistance? Oecologia 120:632–640
Willis AJ, Memmott J, Forrester RI (2000) Is there evidence for the post-invasion evolution of increased size among invasive plant species? Ecol Lett 3:275–283
Wolfe LM (2002) Why alien invaders succeed: support for the escape from enemy hypothesis. Am Nat 160:705–711
Yela JL, Lawton JH (1997) Insect herbivore loads on native and introduced plants; a preliminary study. Entomol Exp Appl 85:275–279
Zhang K, Lin Y (1994) Chinese tallow. China Forestry Press, Beijing

17 Herbivore-Specific Transcriptional Responses and Their Research Potential for Ecosystem Studies

C. Voelckel and I.T. Baldwin

17.1 Summary

While simulations of herbivory with mechanical clipping provide many experimental advantages over true herbivory for ecological research, faithful mimicking can be onerous. Not only do herbivores differ in how and what they remove from a plant, but also they differ in saliva and regurgitate composition, microbial commensalists, pathogen vectoring, feeding phenology and tritrophic interactions, all of which can dramatically alter a plant's response. These differences in response emerge from alterations in primary and secondary metabolism that are activated by specific signalling pathways and signal recognition systems. Frequently, these responses are under transcriptional control and affect genes involved in hormone biosynthesis and perception, volatile organic compound and secondary metabolite synthesis, photosynthesis and transcriptional and translational processes, in other words, in metabolism *sensu lato*. We review studies that have identified herbivore-specific transcriptional responses, introduce the molecular techniques used to measure these changes, and argue that research into molecular mechanisms provides ecologists with tools to monitor and manipulate the subtle effects that insects have on ecosystem function.

17.2 The Subtle Effects of Insects on Ecosystem Function

When insects eat leaves, suck sap, shred litter and pollinate plants, they consume net primary productivity (NPP), recycle nutrients and influence ecosystem function directly through their activities. In addition to these direct effects, the activities of insects are known to alter plant phenotypes, which can indirectly influence ecosystem function. These phenotypic alterations have primarily been described as changes in the concentrations of secondary metabolites (Karban and Baldwin 1997) which, in turn, can influence the

Ecological Studies, Vol. 173
W.W. Weisser and E. Siemann (Eds.)
Insects and Ecosystem Function

'afterlife' of plant parts and thereby the cycling of nutrients in the ecosystem (Driebe and Whitham 2000; Hättenschwiler and Vitousek 2000; Hartley and Jones, Chap. 2, this Vol.). More subtle are the effects of insect attack on plant metabolism. In order to meet the metabolic demands of the large investments in secondary metabolism and/or structural defences (thorns, spines), plants reconfigure their metabolism, which frequently has consequences for plant growth and the outcome of competitive interactions (Baldwin 1998; Heil and Baldwin 2002; Glawe et al. 2003). These indirect effects of insect attack on NPP (the 'costs' of resistance) can be large (for a review on cost estimates see Strauss et al. 2002) and add to the direct effects of NPP consumption by insects (compare 10 % annual removal of NPP by herbivores, references in Coley et al. 1985). Since plants play a fundamental role in most ecosystems by providing the energy, nutrient and material input for all other trophic levels, processes that alter plant metabolic function will, in turn, influence ecosystem function. In this chapter, we explore the mechanisms by which insect attack influences plant gene expression and how an understanding of these mechanisms can be used by ecologists to understand ecosystem function.

17.3 Transcriptional Regulation of Plant Responses

All cellular biological processes, including maintenance of metabolic and physiological balance (homeostasis), and responses to the environment are controlled at the level of metabolite production/concentration, enzyme activity or gene expression. The latter is realized through a cascade comprising transcriptional and post-transcriptional, translational and post-translational regulation, as well as regulation through protein degradation (Libbert 1993). Since plants are largely immobile, they have evolved a large degree of physiological plasticity to cope with fluctuating environments. For example, *Arabidopsis* dedicates 5.9 % of its genome to the production of more than 1,500 transcription factors, which is considerably more than the amount dedicated by the genomes of *Caenorhabditis elegans* (3.5 %) and *Drosophila melanogaster* (4.5 %), respectively (Riechmann et al. 2000). Transcription factors can regulate the expression of many genes, and are themselves activated by complex signalling pathways, which, in turn, are triggered by various internal and external stimuli. Through such networks of signal transduction cascades, environmental stimuli can alter a plant's 'transcriptome' (the expressed portion of the genome) by both fine-tuning responses of individual genes to specific challenges and activating specific sets of genes. Since changes in the transcriptome may not result in altered protein or metabolite phenotypes, studying these transcriptional changes gives researchers the ability to monitor the stimuli that a plant perceives but chooses to ignore (not respond to). In short, transcriptional regulation in all its complexity confers the means to specifically tailor responses to the environment and reveals new insights into the behavioural repertoire of a plant. In the following, we provide a short primer (in the terminological turgidity that characterizes this literature) on transcriptional regulation and an example of a signal transduction cascade that mediates the transcriptional regulation of wound-induced alkaloid production.

A typical eukaryotic gene is composed of several parts. Its transcribed region serves as a template for RNA and protein synthesis and is interspersed with non-coding regions (introns), which are eliminated before translation of the coding regions (exons) (Fig. 17.1A). The transcribed region is flanked on either side by non-coding sequences that can play a role in the regulation of the gene. The first 1 kb or so of the 5′ flanking region is referred to as the gene promoter and contains sequence motifs (*cis*-acting elements, e.g. TATA) that recruit proteins (*trans*-activating factors) that modulate the rate of initiation of mRNA synthesis by the RNA polymerase II complex. *cis*-Acting elements outside the promoter region can either enhance or suppress transcription (Buchanan et al. 2000). See Fig. 17.1 for additional details.

Jasmonic acid (JA), the plant analogue of prostaglandin in animals, is a member of the oxylipin family of signalling molecules that mediate increases in the synthesis of defensive proteins, such as proteinase inhibitors (Farmer et al. 1992), and protective metabolites, such as furanocoumarins (Miksch and Boland 1996), terpenoids (Martin et al. 2002) and alkaloids (nicotine, Baldwin 1999; terpenoid indole alkaloids, TIAs, Aerts et al. 1994) after herbivore attack or wounding. In *Catharanthus roseus*, JA increases the expression of genes involved in TIA synthesis and TIA-precursor formation via *trans*-activating factors called ORCAs (octadecanoid-responsive *Catharanthus* AP2-domain proteins). ORCAs bind to a *cis*-acting element called JERE (jasmonate and elicitor response element) in the promoter region of JA-inducible genes, such as strictosidine synthase (*str*), which catalyzes the initial step in TIA biosynthesis. However, at least three more types of transcription factors interacting with *cis*-elements other than JERE have been described to regulate the expression of the *str* gene (Gantet and Memelink 2002 and references therein; for details see Fig. 17.1).

While the expression of a gene is controlled by several transcriptional regulators, the activity of the transcriptional regulators themselves can be controlled by several mechanisms. For example, the peaks of JA-induced ORCA expression precede maximal induction of target genes such as *str*, suggesting the existence of a transcriptional cascade in which a putative transcription activating factor (TAF) promotes ORCA expression (Fig. 17.1C, 1). However, JA-induced *str* expression is not susceptible to protein synthesis inhibitors, indicating that JA does not induce TIA gene expression simply by increasing ORCA protein abundance, but rather activates pre-existing ORCA protein, which, once activated, binds to the promoters of TIA genes and the ORCA gene itself (Fig. 17.1C, 2). Protein activity can be regulated through post-

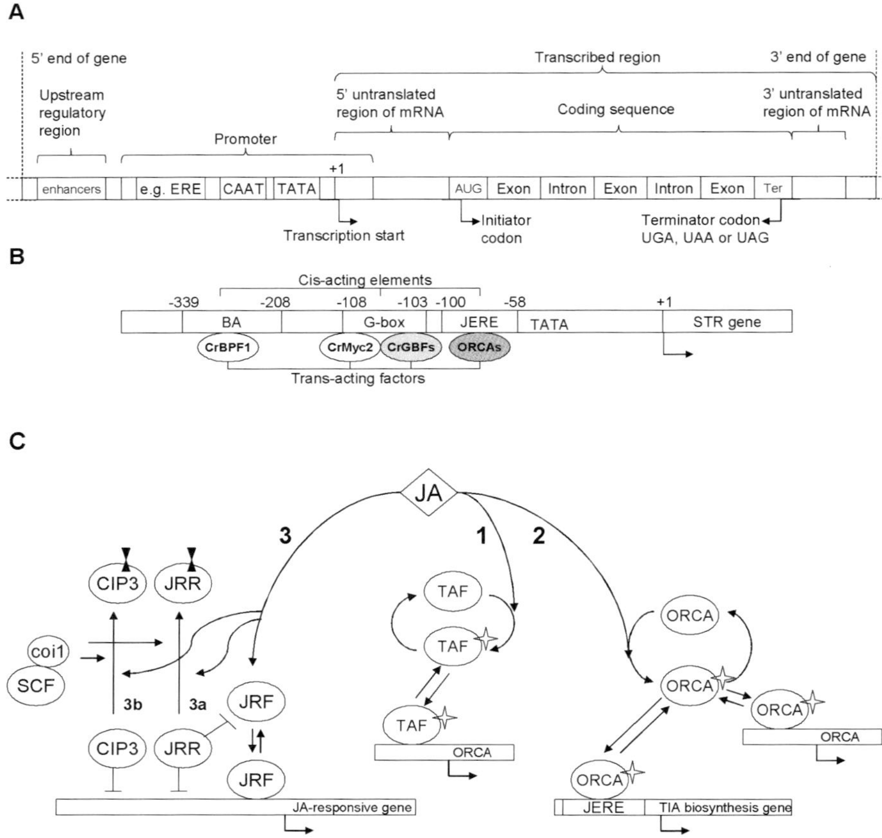

Fig. 17.1. **A** Structure and organization of a eukaryotic gene (after Buchanan et al. 2000). Basic promoter sequence motifs such as TATA and CAAT, additional promoter elements such as ERE (ethylene response element) and up- or downstream regulatory regions on the same strand as the coding region are called *cis*-elements. Before the RNA transcript (mRNA) serves as a template for protein biosynthesis, non-coding sequences (introns) are eliminated, coding sequences (exons) are fused (referred to as 'splicing') and the 5′ and 3′ untranslated regions are post-transcriptionally modified. Open reading frames (ORFs) that are translated into a protein always start with the initiator codon AUG and end with one of the terminator codons UGA, UAA or UAG. **B** The promoter region of the jasmonate (JA)-inducible strictosidine synthase (*str*) gene after Gantet and Memelink (2002). JA- and elicitor-responsive ORCAs (octadecanoid-responsive *Catharanthus* AP2-domain proteins) bind to JERE (jasmonate- and elicitor-responsive element) leading to *str* expression; CrGBFs (*C. roseus* G-box-binding factors) bind to the G-Box (5′-CACGTG-3′) leading to *str* repression. JA-responsive CrMYC2 (*C. roseus* MYC-type basic helix-loop-helix transcription factor) and elicitor-responsive CrBPF1 (*C. roseus* box P-binding factor 1 homologue) bind to the G-box and the BA-box, respectively, but their transcriptional activity remains to be discovered. **C** Putative mechanisms regulating transcription factors after Devoto et al. (2002) and Vom Endt et al. (2002). *1* Regulation of factor abundance by adjusting production of the encoding mRNA through transcriptional cascades involving transcription activating factors (TAFs). *2* Regulation of ORCA activity by post-translational modification (✧) of pre-existing transcription factor pro-

translational modifications (e.g. phosphorylation, acetylation, glycosylation and others) and/or interactions with other proteins. In the case of ORCA, JA-responsive *str* expression is sensitive to protein kinase inhibitors, suggesting that ORCA phosphorylation is required for ORCA activation and TIA gene expression (Vom Endt et al. 2002).

In addition to regulating transcription factor production, changes in transcription factor abundance can stem from modifications of stability and turnover of these proteins. In *Arabidopsis*, JA-induced biosynthesis of indole glucosinolates requires a functional COI1 (*coronatine-insensitive1*) protein, which is part of the SCF (Skp1/Cul1/F-box protein) complexes that specifically recognize and target proteins for ubiquitin-mediated proteolysis. By analogy to what is known about auxin responses, Vom Endt et al. (2002) proposed that JA promotes the binding of JA response repressors (JRR) to the SCF^{COI1} ubiquitin ligase, which leads to their ubiquitination and subsequent proteasome-mediated degradation. This results in the activation of JA response factors (JRFs) and the de-repression of JA-responsive genes (Fig. 17.1C, 3a). Moreover, Devoto et al. (2002) characterized COI1-interacting proteins and recovered a histone deacetylase (RPD3b) co-immunoprecipitating with COI1. Histone deacetylation is believed to decrease the accessibility of chromatin to the transcriptional machinery, thereby repressing transcription. It is now examined whether COI1 regulates the ubiquitination and proteolytic destruction of RPD3b, representing another mechanism of JA-mediated de-repression of JA-responsive genes (Fig. 17.1C, 3b).

In summary, all steps involved in converting the information content of a gene into a protein appear to be involved in transcriptional control. Hence, a deep knowledge of the mechanisms underlying transcriptional regulation is required to understand how specific environmental stimuli, such as attack from different herbivore species or feeding guilds, are 'recognized' by a plant. Furthermore, this mechanistic understanding enables molecular ecologists to identify targets for the manipulation of herbivore-induced phenotypes, which in turn provide the tools required to elucidate the function and significance of herbivore-induced traits and their significance for ecosystem function.

---

tein, which leads to interaction with cognate elements of responsive genes and/or genes coding for the factors themselves (auto-regulation). *3* Regulation of factor abundance by adjusting protein turnover rate through COI1-mediated ubiquitination (⧗) and subsequent proteolysis of jasmonate response repressors (JRRs, *3a*) or a histone deacetylase (RPD3b=COI1-interacting protein3, CIP3, *3b*), resulting in activation of jasmonate response factors (JRFs) and/or increased access of JRFs or other regulators to the regulatory regions, respectively

17.4 Insect-Induced Transcriptional Changes

From a plant's perspective, herbivorous insects differ in their feeding apparatus, saliva composition, the type of plant tissue they attack, the amount of tissue they remove or destroy and feeding rhythms; in short, they can be assigned to different feeding guilds (Root 1973), whose damage has different fitness consequences for the plant. Moreover, the plant–insect interaction is not a binary interaction, since other trophic levels can considerably influence its outcome. Herbivorous insects may or may not vector diseases and they may or may not be negatively affected by predacious insects, parasitoids or intra- and inter-specific competitors. These ancillary interactions can all be influenced by plant traits, and as a consequence the plant–insect interaction is frequently played out on a spatial scale larger than the plant itself and includes components of the plant's community. Consequently, plants may fend off insect attack by eliciting direct and indirect defences that influence herbivore performance and survival through bottom-up or top-down control, respectively.

Herbivore attack elicits a myriad of plant responses (Karban and Baldwin 1997; Walling 2000) and some of these responses may be tailored to the particular attacker, as has been demonstrated on the level of signal molecules (e.g. JA: Schittko et al. 2000; Ziegler et al. 2001; or ethylene: Kahl et al. 2000), secondary metabolites (e.g. nicotine: Winz and Baldwin 2001), protein amounts and activities (Stout et al. 1994; Tamayo et al. 2000), as well as volatile organic compound (VOC) emission (Halitschke et al. 2001; Kahl et al. 2000). Here we review studies that examine insect-induced changes at an earlier stage of phenotypic expression, namely in alterations in transcript abundance. Insect-induced changes in transcript abundance have been measured using the methods summarized in Fig. 17.2, either singly or in combinations. Real-time quantitative polymerase chain reaction (RTQ-PCR) and reverse transcription PCR (RT-PCR) are based on the ability of a PCR to exponentially amplify initial differences in transcript number even when the transcripts are present in trace quantities. The amplified products are visualized either in real time during the PCR (RTQ-PCR, Fig. 17.2B) or after (RT-PCR, Fig. 17.2A) the reaction. In Northern blot analyses (NA), transcript-specific radioactive probes are used to identify (by sequence-specific hybridization) a target mRNA species within an immobilized RNA sample. The generated signal is proportional to the amount of target mRNA in the sample (Fig. 17.2, C). In a microarray analysis, the hybridization procedure is reversed: many (e.g. frequently more than 8,000 genes) transcript-specific, unlabelled probes are arrayed on glass slides and competitively hybridized to two RNA pools, which originate from samples of a binary comparison (e.g. control and insect-attacked plants) and are labelled with two different fluorescent dyes. The signal intensities stemming from the two dyes are normalized and expressed as a ratio for each gene. Genes with expression ratios that deviate significantly, either positively or negatively, from 1 indicate up- or down-regulation of a gene, respectively (Fig. 17.2D). Transcript accumulation in response to various environmental stimuli can also be measured indirectly by transforming plants with promoter:glucuronidase (GUS) constructs (fusions of a promoter of a gene of interest with a reporter gene). In comparison to measuring transcript abundance directly by RT-PCR, RTQ-PCR, NA or microarray procedures, the activity of the reporter gene, GUS, is measured histochemically. In many cases, the reporter gene activity has faithfully mimicked the expression patterns of the endogenous gene from which the promoter was derived. While these 'reporter plants' are only able to report the transcriptional activity of a single gene, they allow for detailed spatial and kinetic analyses of transcript accumulation (Fig. 17.2E).

Since insect herbivory is inevitably accompanied by wounding, many transcriptional studies compared responses to herbivore attack with those to mechanical damage. These studies confirmed the central conclusion from secondary metabolite analyses, namely that herbivore attack may modify a plant's wound response. Frequently, a stronger and faster induction of a gene is observed after insect herbivory and simulations thereof than after mechanical wounding (Korth and Dixon 1997; Shen et al. 2000; Berger et al. 2002; Table 17.1). Moreover, wound-inducible genes may only slightly or not at all be induced after insect herbivory (Reymond et al. 2000; Table 17.1) and there are genes only (Reymond et al. 2000; Table 17.1) or preferentially (Berger et al. 2002; Table 17.1) induced by insect feeding as compared to mechanical damage. The wound response of wild tobacco, *Nicotiana attenuata*, is altered in three different ways when challenged by tobacco hornworm (*Manduca sexta*) larvae. Application of *M. sexta*-regurgitant to wounded leaves, which simulates the responses elicited when *M. sexta* larvae attack plants, reduced the accumulation of wound-induced transcripts (e.g. threonine deaminase and two putrescine-methyltransferase genes) or increased both the wound-induced suppression (e.g. a subunit of light harvesting complex II) and the wound-induced amplification (e.g. pathogen-induced oxygenase) of many transcripts (Schittko et al. 2001; Winz and Baldwin 2001; Table 17.1).

Herbivory is associated with a continuum of wounding intensities, which in turn is associated with differential gene induction. While leaf chewers such as lepidopteran larvae or adult beetles severely wound and remove plant tissues, cell-content feeders such as mites, thrips or phytophagous bugs, by piercing and sucking out mesophyll cells, cause considerable damage, but do not substantially decrease a plant's leaf area. Phloem feeders such as aphids and whiteflies must be able to delicately tap into phloem cells without eliciting wound-induced termination of phloem flow, and hence feed stealthily with the least amount of damage of all insect herbivores. With their flexible stylets, they move interstitially (fungus-like) before penetrating phloem cells and appear to suppress a plant's wound recognition system.

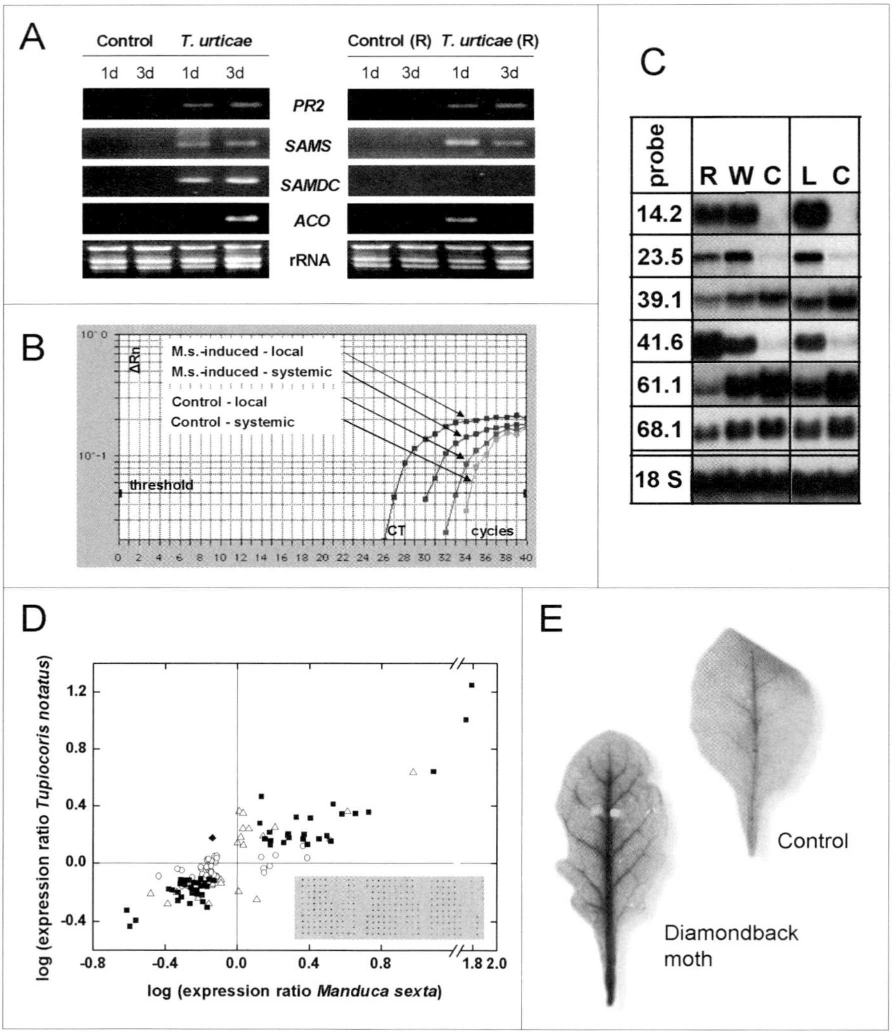

Fig. 17.2. Examples of commonly applied procedures in gene-expression profiling in plant–insect interactions. **A** Reverse transcription-polymerase chain reaction (RT-PCR) analysis of pathogenesis-related protein 2 (*PR2*), S-adenosylmethionine synthetase (*SAMS*), S-adenosylmethionine decarboxylase (*SAMDC*) and 1-aminocyclopropane-1-carboxylic acid oxidase (*ACO*) transcripts in unattacked lima bean leaves, leaves infested with the spider mite *Tetranychus urticae*, (receiver-)leaves exposed to volatiles emitted by control leaves (control, *R*) and (receiver-)leaves exposed to volatiles emitted by *T. urticae*-infested leaves (*T. urticae*, *R*). (After Arimura et al. 2002) **B** Amplification plot, taken from a real-time quantitative PCR analysis of trypsin inhibitor (TI) transcripts in *Nicotiana attenuata* plants attacked by *Manduca sexta* larvae, depicting local and systemic induction of TIs. The cycle threshold (*CT*) indicates the number of cycles necessary for an increase of the reporter signal (*Rn*) above the background signal (indicated by threshold line) and is inversely proportional to the initial copy number of the tran-

The elicitation of wound-inducible proteinase inhibitors (PI), proteins that interfere with herbivore digestion, reflects this pattern: while PI transcripts increased after herbivory from corn earworm larvae (Fidantsef et al. 1999; Stout et al. 1999), tomato hornworm larvae (Voelckel and Baldwin 2003), two-spotted spiders (Li et al. 2002) and mirids (Voelckel and Baldwin 2003), they do not after whitefly (Walling 2000) and aphid attack (Fidantsef et al. 1999; Table 17.1). Interestingly, these stealthy herbivores appear to elicit transcriptional responses characteristic of those elicited by pathogen attack. For example, transcripts of pathogenesis-related (PR) proteins, which are usually involved in conferring systemic acquired resistance (SAR) after pathogen infestation, are induced after feeding of *Macrosiphum euphorbiae*, *Myzus persicae* and *Brevicoryne brassicae* aphids (Fidantsef et al. 1999; Moran and Thompson 2001; Moran et al. 2002), *Bemisia argentifolii* and *Trialeurodes vaporariorum* whiteflies (Walling 2000) and *Tetranychus urticae* mites (Arimura et al. 2000a, b). In contrast, in response to attack from grazing herbivores such as *Helicoverpa zea* and *Pieris rapae* larvae, PR transcripts are only weakly elicited (Fidantsef et al. 1999; Reymond et al. 2000; Table 17.1). Whether this difference in response is due to the more localized and pathogen-like damage resulting from attack by sucking as opposed to chewing insects, or a higher probability of stylet feeders to vector pathogens, remains to be determined.

Evidence is accumulating that plants can distinguish attack from closely related species and even from different stages of the same species. Species-specific changes in transcript accumulation are best exemplified in white-

---

script. (J. Zavala and I.T. Baldwin, unpubl. results.) **C** Northern blot analysis of threonine deaminase (14.2), α-dioxygenase (41.6), a light-harvesting complex II subunit (61.1) and three unknown (23.5, 39.1, 68.1) transcripts comparing expression in wounded+(*W*)ater-treated and wounded+(*R*)egurgitate-treated and *M. sexta* second and third instar (*L*)arvae-attacked tissue with expression in (*C*)ontrol tissue. Hybridization with an 18 S rRNA probe demonstrates equal loading of samples. (After Schittko et al. 2001) **D** Microarray analysis with a cDNA microarray printed with 240 herbivory-related genes comparing changes in *N. attenuata*'s transcriptome after infestation with leaf chewers (*M. sexta*) and cell-content feeders (*Tupiocoris notatus*). Data evaluation based on statistical significance tests and arbitrary threshold revealed 123 genes to be repressed or induced: 59 were regulated by both species, either similarly (*squares*, 58) or inversely (*diamond*, 1), while 40 were regulated only after attack by *Manduca* (*circles*) and 24 only after attack by *Tupiocoris* (*triangles*). Unresponsive transcripts (117) were omitted; *insert* depicts a fluorescence image of the array. (C. Voelckel and I.T. Baldwin, unpubl. results.) **E** Expression of a vegetative storage protein (*VSP*) as reported by the activity of β-glucuronidase (GUS), which was fused to a *vspB* promoter from soybean and transformed into *Arabidopsis*. Note enhanced GUS expression in the midribs after diamondback moth feeding as compared to control leaves. VSPs are assumed to sequester plant resources during periods of insect attack that will be remobilized when conditions become more favourable. (After Berger et al. 2002)

Table 17.1. Summary of studies examining insect-induced transcriptional changes; plant system, herbivorous elicitors (=insects), investigated genes and profiling tools are listed for each reference. According to the standards adopted by the *Arabidopsis* community, gene symbols are *italicized* (*lowercase letters* for mutant genes), while protein products of genes are written in *uppercase letters without italics* (http://mutant.lse.okstate.edu/genepage/namerule.html; Meinke and Koornneef 1997). All gene names appear as in the original reference; references are organized by plant taxa. Since characterization of many of these genes is still in its infancy, classification into functional groups is at present not possible

Plant	Herbivore/treatment	Induced transcripts	Method	Reference
Phaseolus lunatus (lima bean)	*Tetranychus urticae* (two-spotted spider mite) and *T. urticae*-induced volatiles	β-1,3-glucanase (PR-2), chitinases (PR-3, PR-4), lipoxygenase (LOX), phenylalanine ammonia-lyase (PAL), farnesyl-pyrophosphate synthetase (FPS), S-adenosylmethionine (SAM) synthetase (SAMS), 1-aminocyclo-propane-1-carboxylic acid (ACC) oxidase (ACO), SAM decarboxylase (SAMDC)	Polymerase chain reaction after reverse transcription of RNA (RT-PCR)	Arimura et al. (2000a, 2002)
P. lunatus	*T. urticae* and *T. urticae*-induced volatiles	Ca. 80 genes related to: pathogenesis/wounding (52 %), hormones (3.6 %), ethylene biosynthesis (7.7 %), flavonoid biosynthesis (1 %), transcriptional modifications (2 %), translations (13.3 %), chaperons (2 %), secondary signaling messengers (1.5 %), membrane transports 3.6 %), protein/peptide degradations (2 %) and photosynthesis (1.5 %)	Microarray with 2,032 cDNAs derived from a cDNA library of *T. urticae* and *Spodoptera exigua*-treated *P. lunatus* plants	Arimura et al. (2000b)
Lycopersicon esculentum (tomato)	*T. urticae*	Serin PI-II, Serin PI-I, cathepsin D inhibitor (strong increase), LoxD, AOS1 (modest and gradual increase)	Northern blot analysis (NA)	Li et al. (2002)
L. esculentum	*Macrosiphum euphorbiae* (potato aphid)/*Myzus persicae* (green peach aphid)	LOX (modest), proteinase inhibitor II (PINII) (no effect), PR-4 (strong)	NA	Fidantsef et al. (1999)
L. esculentum	Whitefly and pink potato aphid	Wfi (NADPH oxidase subunit) RNAs accumulate locally and systemically by whitefly nymphs but not by pink potato aphids or whitefly adults	?	Walling (2000)
L. esculentum	*Bemisia argentifolii* (silverleaf whitefly) and *Trialeurodes vaporariorum* (greenhouse whitefly)	No accumulation of wound-inducible leucine aminopeptidase (LapA) and pin2, PR (pathogenesis-related) transcripts regulated by JA and/or ethylene (basic β-1,3-glucanase, basic chitinase, PR-1) accumulate to higher levels than SA-regulated PR mRNAs (acidic chitinase, acidic β-1,3-glucanase)	?	Walling (2000)

L. esculentum	*Helicoverpa zea* (corn earworm)	LOX (weak), proteinase inhibitor II (PINII) (strong, locally and systemically), PR-4 (weak, locally and systemically)	NA	Fidantsef et al. (1999); Stout et al. (1999)
L. esculentum	*Manduca sexta* (tobacco hornworm)	Earlier and higher accumulation of prosystemin mRNA levels in wild-type than in prosystemin antisense plants after herbivory	NA	Orozco-Cardenas et al. (1993)
L. esculentum	*Spodoptera littoralis*	Increase in LAP transcripts	NA	Pautot et al. (1993)
Solanum tuberosum (potato)	*M. sexta*	PI-II- and 3-hydroxy-3-methylglutaryl-coenzyme A reductase (HMGR) transcripts accumulate faster after simulated and real herbivory than after mechanical wounding	NA	Korth and Dixon (1997)
S. tuberosum	*Leptinotarsa decemlineata* (Colorado potato beetle) and *Spodoptera exigua* (beet armyworm)	Both insects induce isoforms of lipoxygenase (H1, H3) and proteinase inhibitor 2 transcripts while feeding on wild-type but not feeding on LOX-H3-depleted plants	NA	Royo et al. (1999)
Zea mays (maize)	*S. exigua*	Stronger and earlier stc1 (sesquiterpene cyclase gene 1) transcript induction after real and simulated herbivory than after mechanical damage	NA	Shen et al. (2000)
Cucurbita pepo cv. *Chefini* (squash)	*B. argentifolii* and *Bemisia tabaci* (sweet potato whitefly)	M20B metallo-peptidase-like protein (SLW1) and β-glucosidase (SLW3) transcripts were expressed locally in response to both species, only after silverleaf whitefly feeding in apical leaves, and only after nymph but not adult feeding	DDRT-PCR, NA	Van de Ven et al. (2000)
Arabidopsis thaliana	*Pieris rapae* (cabbage butterfly) (*Pieris brassicae*)	Many genes strongly induced by mechanical damage were less or not at all induced after herbivore attack [e.g. PR-2, phenylalanine ammonia lyase (PAL), cinnamoyl-coA reductase (CCR), o-methyltransferase (COMT)]; hevein-like protein gene (HEL) was induced by *P. rapae* larvae but not by mechanical wounding; no preferential activation of either jasmonate-dependent or jasmonate-independent genes by *P. rapae*; no activation of water-stress-inducible genes by *P. rapae*	cDNA microarray consisting of 150 expressed sequence tags (ESTs) of defence-related genes	Reymond et al. (2000)
A. thaliana	*Plutella xylostella* (diamondback moth)	Increased levels of LOX2, a vegetative storage protein (VSP), β-glucosidase 1 (BGL1), glutathione S-transferase 2 (GST2), GST6, a putative calcium-binding elongation factor hand protein (CaEF) in infested rosette tissue	Differential display (DDRT-PCR), NA	Stotz et al. (2000)

Table 17.1. (*Continued*)

Plant	Herbivore/treatment	Induced transcripts	Method	Reference
A. thaliana	*P. xylostella* and *Spodoptera littoralis* (Egyptian cotton worm)	*S. littoralis* regurgitate accelerates the accumulation of *vsp*1 and *vsp* 2, preferential induction of *vsp*1 relative to *vsp*2 by both insect species in contrast to wounding, apparent suppression of systemic *vsp* expression after *S. littoralis* feeding relative to *P. xylostella* herbivory	Quantitative real-time PCR, *vspB*:GUS fusion assays	Berger et. al. (2002)
A. thaliana	*M. persicae*	PR-1, an apoplastic form of β-1,3-glucanase (BGL2), a defensin (PDF1.2), LOX2, PAL1, and a sugar transport protein gene (STP4) are induced in aphid-infested rosette leaves and BGL2 also in inflorescence bolts	NA	Moran and Thompson (2001)
A. thaliana	*M. persicae* and *Brevicoryne brassicae* (cabbage aphid)	Response profiles at 72 h (microarray analysis) and 96 h (macroarray analysis) contain induced and repressed genes related to oxidative stress, Ca^{2+}/calmodulin signaling, tryptophan-, ethylene- and aromatic biosynthesis, a hevein-like protein, α-dioxygenase and endo-transglycosylase, increased levels of PR-1, BGL2, PDF1.2 and STP4 transcripts in infested leaves	cDNA microarray consisting of 105 ESTs of defence-related genes, macroarray consisting of differential ESTs from the preceding microarray, NA	Moran et al. (2002)
Nicotiana attenuata	*M. sexta*	Up- and downregulation of 27 genes related to photosynthesis (e.g. chelatase), electron transport (cytochrome c-type protein), cytoskeleton (β-tubulin), carbon- (threonine deaminase) and nitrogen- (ferredoxin-dependent Glu synthase) metabolism, signalling (importin α, kinase cofactor GAL 83) and stress, wounding or pathogen invasion (e.g. Sn-1 gene, dehydration-responsive protein, germin, retrotransposon, α-dioxygenase, luminal binding protein)	DDRT-PCR, NA	Hermsmeier et al. (2001); Schittko et al. (2001)
N. attenuata	*M. sexta*	Increase in putrescine N-methyltransferase I and II transcripts after pretreatments with ethylene perception inhibitor 1-MCP	NA	Winz and Baldwin (2001)
N. attenuata	*M. sexta*	Increase in allene oxide synthase (AOS) levels in attacked leaves	NA	Ziegler et al. (2001)

N. attenuata	*M. sexta*	73 Differentially expressed transcripts: decrease in photosynthesis-related transcripts, increase in transcripts associated with protein and nucleotide turnover and carbohydrate metabolism, altered transcript levels for RNA binding proteins, putative ADP/ATP translocators, chaperonins, histones, water channel proteins, terpenoid-, cell-wall-related and oxylipin-related transcripts	DDRT-PCR, SHMB, c-DNA microarray with genes related to real and simulated *M. sexta* herbivory	Hui et al. (2003)
N. attenuata	*M. sexta*/*Tupiocoris notatus* (suckfly)	Altered expression of photosynthesis-related transcripts (small subunit of RuBPCase, RuBPCase activase, cytochrome f, phosphoglycerate kinase, photosystem IE, plastidic aldolase), rhamnosyl-transferase, thionin, SAM-decarboxylase, phospholipase C, trypsin inhibitors and others after mirid attack	DDRT-PCR, SHMB, NA	Voelckel and Baldwin (2003)
Saccharum officinarum (sugarcane)	Potential insect pests of sugarcane	Transcripts related to signal perception, signalling pathways, defence and proteolysis	*In silico* gene expression analysis from cDNA libraries of different sugarcane tissues	Falco et al. (2001)

fly–squash and whitefly–tomato interactions. Transcripts coding for an M20B metallopeptidase-like protein were elicited to a much higher level in silverleaf whitefly-infested than in sweet potato whitefly-infested leaves, and for transcripts of a β-glucosidase-like protein, which are induced by both whitefly species in infested leaves, systemic induction was only observed after silverleaf whitefly attack (van de Ven et al. 2000). Transcripts for a subunit of NADPH oxidase (Wfi1) accumulate in local and systemic tomato leaves only after whitefly- but not pink potato aphid feeding (Walling 2000). Interestingly, only whitefly nymphs, but not adults, are responsible for these changes in transcript accumulation (van de Ven et al. 2000; Walling 2000). In contrast, plants appear not to distinguish attack from different instars of the same lepidopteran species by differences in larval oral secretion chemistry, because R collected from 3rd, 4th and 5th instars of *M. sexta* was equally active in causing R-induced changes in gene expression (Schittko et al. 2001; Table 17.1).

By monitoring changes in gene expression for hundreds of genes simultaneously, microarray techniques (Fig. 17.2D) have allowed plant–insect interactions to be explored at the level of the plant's transcriptome, which has led to important new insights. First, the transcriptional signatures of various stressors may overlap substantially. To paraphrase Shakespeare, stresses in nature, not as single spies but in battalions come (e.g. wounding is often associated with water stress; pathogen attack leads to oxidative stress). As a consequence, plants have not been under selection to perceive well-defined stresses individually. For example, Moran et al. (2002) found some oxidative stress genes, which are usually expressed after ozone fumigation (Sharma and Davis 1994), such as glutathione S-transferases and one form of a superoxide dismutase (SOD) to increase, but others such as another form of SOD and a peroxidase gene to decrease upon aphid infestation. Moreover, Reymond et al. (2000) found that the transcriptional signature of wounding was more similar to that of dehydration than that of *Pieris rapae* feeding. Presumably, *P. rapae* has adopted a feeding strategy that minimizes the effects of water stress on gene expression. Second, a plant's response to herbivory is embedded in a large transcriptional reconfiguration of metabolism *sensu lato*. Hence herbivore attack elicits increases in the expression not only of defence genes (see PRs, alkaloids, PIs, octadecanoids), but also of genes involved in (post-) transcriptional and (post-) translational processes, protein folding and degradation, membrane transports, hormone- and second messenger synthesis, cell wall modulation and carbohydrate metabolism and photosynthesis (Arimura et al. 2000b; Reymond et al. 2000; Falco et al. 2001; Hermsmeier et al. 2001; Moran et al. 2002; Hui et al. 2003). Techniques designed to monitor changes in transcription factor abundance hold the potential of elucidating signal pathways, examining aspects of specificity and investigating large-scale transcriptional changes in the context of plant–insect interactions.

17.5 How a Molecular Understanding of Plant–Insect Interactions Can Help Elucidate Ecosystem Function

We are just beginning to appreciate the extent to which biotic interactions influence ecosystem function through the reconfiguration of plant metabolism and the resulting bottom-up ecosystem responses that arise from reconfigured plant metabolism. While ecosystem scientists have tried to measure and manipulate these biotic interactions, the techniques employed have been necessarily coarse-grained and lack the precision with which the interactions are known to occur (Hjältén, Chap. 12, this Vol.; Lehtilä and Boalt, Chap. 13, this Vol.; Schmitz, Chap. 14, this Vol.). Here, we provide examples of how molecular tools can be applied to monitor and eventually manipulate these elusive ecosystem functions.

To examine how biotic interactions, such as plant–insect interactions, affect ecosystem function requires first and foremost an understanding of the frequency and duration of these interactions. Herbivore-induced responses are a phenomenon largely studied in laboratory experiments, and while it is known to be widespread among plant taxa, it is not known whether plants growing in nature are induced most of the time. Inducible reporter systems, i.e. plants transformed with constructs, consisting of insect-inducible promoters and easy-to-measure reporter genes such as β-glucuronidase, could monitor how often particular species or guilds of insects attack plants by indicating how often their attack is 'recognized' as a transcriptional response. Promoters of genes specifically expressed in certain tissues (e.g. roots, trichomes) or catalyzing committed steps in secondary metabolite biosyntheses could report the frequency of belowground or plant-surface interactions as well as the production of certain metabolites. Promoters for genes active in the early stages in signal-transduction cascades that respond to insect attack, such as the early stages in JA biosynthesis (Fig. 17.1C), could be used to monitor the frequency of attack from herbivores that cause extensive wounding, in a manner similar to the approaches used to create reporter plants that monitor bioavailability of a specific contaminant in either soil or water.

Krizek et al. (2003) recently developed an *Arabidopsis* plant that 'reports' the bioavailability of Ni in a dose-dependent manner. The research team used the Affimetrix *Arabidopsis* GeneChip microarrays to analyze the transcriptome of seedlings exposed to Cd, Cu or Ni and identified *AHB1* (non-symbiotic haemoglobin) which was strongly upregulated by Ni. The gene is induced neither by other metals nor by other stresses including cold, dehydration, heat shock, oxidative stress or wounding. Transgenic plants expressing GUS under the *AHB1* promoter reported on the presence and concentrations of Ni in plant growth media. Although plant-based bioindicators are not as sensitive as microbial biosensors, they could serve as cheap and effective monitors of plant-available heavy metal contaminations in soils and sediments (Krizek et al. 2003).

A determination of recognition frequencies, however, does not reveal how often the perception of insect attack results in alterations of plant function, which, in turn, may lead to alterations in ecosystem function. Therefore we would need to monitor genes whose expression reflects commitments to metabolic reconfiguration, e.g. upregulation of defence and downregulation of growth processes, after insect attack. Transcription factors, such as ORCAs, that regulate genes of primary and secondary metabolism (Fig. 17.1C) are candidate genes and microarray technology will likely identify more of these major 'metabolism switch' genes. In a similar fashion, transgene-based systems have been used to monitor the consequences of exposure to environmental mutagens: transgenic *Arabidopsis* plants, which had been designed to analyze point mutations and homologous recombination events in a GUS transgene, have been specifically applied to evaluate the mutagenicity of ionizing and UV radiation and the toxicity of heavy metal ions (Kovalchuk et al. 2001).

In addition to providing the tools to monitor the frequencies and consequences of plant–insect interactions, transgenic technology will also allow for a manipulation of these interactions. To date, ecosystem consequences of plant–insect interactions have been studied experimentally by the application of insecticides (Siemann and Weisser, Chap. 18, this Vol.) and caging (Schmitz, Chap. 14, this Vol.) in order to exclude insects; by mechanical damage and leaf removal treatments (Hjältén, Chap. 12, this Vol.; Lehtilä and Boalt, Chap. 13, this Vol.) to simulate the feeding activity of insects; and by the introduction of insects to increase herbivore loads and, in some cases, create experimental insect outbreaks. While these approaches relay mainly on top-down control of insect populations, bottom-up manipulations, which are mediated by the plant, may provide a powerful manipulation that could be exploited to study ecosystem function. Three examples, in which plants have been transformed with novel genes that affect 'down-stream' resistance traits, serve to illustrate the procedure.

Genes for proteinaceous toxins from *Bacillus thuringiensis* subspecies (BT toxins), which are specifically effective against lepidopteran, dipteran and coleopteran insects, have been inserted into various crop species. These transgenic crops are widely cultivated in the USA, Australia and China and enjoy greatly reduced herbivore loads (Sharma et al. 2000). Since BT toxins are direct gene products that plants can produce without a measurable metabolic load and since BT toxins are specific to particular insect taxa, one could imagine creating ecosystems of plants transformed with different BT toxins in which herbivory from lepidopteran and coleopteran herbivores was independently manipulated. In contrast, BT-containing insecticides require repeated applications and are difficult to target to a single plant species.

At an additional level of complexity, plants have been transformed to introduce the entire pathway of a novel secondary metabolite. For example, the biosynthetic pathway of the tyrosine-derived cyanogenic glycoside dhurrin from *Sorghum bicolor* has been engineered into acyanogenic *A. thaliana*, which rendered the transgenic *Arabidopsis* plants more resistant to the flea beetle *Phyllotreta nemorum*. This beetle specifically attacks crucifers (e.g. *Arabidopsis*) and has not evolved mechanisms to detoxify or sequester cyanogenic glycosides. To transfer the dhurrin pathway, plants were transformed with the two multifunctional cytochromes P450 (CYP79A1 and CYP71E1) and a soluble UDPG-glucosyltransferase and were substantially more resistant to flea beetles as compared to plants expressing the cytochrome P450 genes only, the glucosyltransferase only or plants carrying the empty expression vectors. Thus, increased resistance could directly be attributed to the presence of dhurrin (Tattersall et al. 2001).

Herbivore resistance can also be increased by constitutively expressing the signals that elicit resistance, as illustrated in *Zea mays* plants transformed to express a wheat gene that produces hydrogen peroxide (H_2O_2). H_2O_2 contributes to cell-wall fortification, elicits pathogen-induced defensive proteins, inhibits pathogen growth and through lipid peroxidation harms insect development and reproduction. Ramputh et al. (2002) overexpressed a cell-wall-localized H_2O_2-generating enzyme (germin) and produced plants less susceptible to attack by the European corn borer *Ostinia nubilalis* (ECB). In addition to a reduction in ECB feeding and growth, stalk tunnelling damage was reduced by 50 % at plant harvest in all transgenic lines, which should decrease crop losses resulting from plant lodging caused by heavy ECB infestations.

The above three examples provide a means of excluding herbivores from particular feeding guilds (BT toxins) or with particular host-specificity (cyanogenic glycoside-intolerant insect species) or tissue specificity (stem miners) by introducing novel genes from other species. Genetic tools can also be used to increase the susceptibility of plants to particular groups of insects by antisense expression-mediated silencing of endogenous genes, which is illustrated by four examples. In 1993, Orozco-Cardenas et al. demonstrated that resistance towards insects could be modulated by genetically engineering a gene encoding a component of the inducible systemic signalling system (prosystemin) that regulates a plant defence response (proteinase inhibitors). *M. sexta* larvae grew much faster on tomato plants constitutively expressing a prosystemin antisense gene. This enhanced growth was correlated with a severe delay in prosystemin mRNA- as well as proteinase inhibitor accumulation in transgenic as opposed to wild-type plants. Similarly, antisense-mediated depletion of hydroperoxide lyase (HPL) has identified this enzyme as a major route of 13-fatty acid hydroperoxide degradation, since both of its products, hexanal and 3-hexenal, have highly reduced levels in transgenic potato plants. Although transgenic and wild-type plants did not significantly differ in the expression of wound-induced genes, *Myzus persicae* aphids feeding on the HPL-depleted plants displayed a two-fold increase in fecundity above those feeding on non-transformed plants. Thus, HPL-catalyzed production of C6 volatile aldehydes may be part of a constitutive resistance mechanism against some sucking insect pests (Vancanneyt et al. 2001). The Colorado potato beetle, a specialist leaf feeder on solanaceous plants, and the beet armyworm, a generalist feeder, have greater rates of weight gain on transgenic potato plants devoid of a specific 13-lipoxygenase isoform (LOX-H3), an enzyme involved in JA production (Royo et al. 1999). Suppressing the activity of a trichome gland-specific P450 hydroxylase, Wang et al. (2001) found a decrease in the predominant exudate component, cembatriene-diol, and an increase in its precursor, cembatriene-ol, which is particularly toxic to *Myzus nicotianae* aphids and greatly diminished aphid colonization responses. In summary, by silencing endogenous resistance genes or by expressing novel resistance genes from other species, ecologists could experimentally manipulate the species composition and the feeding behaviour of the herbivore community with a degree of specificity not possible with current techniques.

In all of the above-mentioned studies the respective transgenes have been under the control of constitutive promoters. Constitutive gene expression leads to changes in plant metabolism throughout the life cycle of the plant. For ecosystem studies conditional manipulation of biotic interactions may be advantageous. For example, an inducible BT production would allow researchers to time insect removal with a high degree of spatial and temporal precision. Several chemically inducible systems have been developed that enable a precise control over gene expression (Padidam 2003 and references therein). These systems usually contain two transcription units. The first unit encodes a transcription factor that responds to a chemical signal, while the second unit contains a response element (*cis*-element) that binds the activated transcription factor and is fused to the gene of interest. Ideally, these inducible expression systems should have a low basal but a high induced level of expression and respond rapidly to the addition as well as the removal of the inducer. The inducer itself should be non-toxic to plants, highly specific and, if intended for field use, environmentally friendly. A recent review of pros and cons of chemically inducible expression systems highlights ones with attributes conducive to the study of ecosystem processes (Padidam 2003). With the identification of regulatory promoter elements responsive to plant secondary metabolites (e.g. the diterpenoid sclareol; Grec et al. 2003), new, plant-derived inducible expression systems are imaginable. Given the recent discoveries of insect-specific gene activation (Table 17.1), the genomes of native plants are likely to harbour insect-specific promoters, which, when fused to BT genes, could allow BT expression to be triggered by attack from particular herbivores. Similar constructs for the silencing of endogenous genes would enable insect-activated gene knockouts.

Moreover, artificial transcription factors that allow the activation or suppression of endogenous genes and thereby represent an alternative to antisense-mRNA-mediated gene silencing are being developed. These artificial transcription factors are based on predefined zinc-finger modules of which each recognizes a unique 3-bp sequence of DNA (Segal et al. 2003). Six of these modules can identify an 18-bp sequence in the promoter of interest and, when fused to activation or repression domains, regulate the expression of the corresponding gene. This new technique is especially valuable for genes for which the relevant endogenous transcription factors are yet unknown.

The utility for ecosystem scientists of these potential monitoring and manipulation tools depends in large part on the discovery of candidate genes and the establishment of efficient transformation protocols for plants with natural history characteristics that are relevant for ecosystem scientists. Both requirements are within grasp, as is suggested by the increasing number of array and other transcriptional studies that are identifying insect-specific genes, as well as the development of transformation procedures for non-model plants from different functional groups (e.g. trees, herbs, grasses, nitrogen-fixing plants, etc.) in addition to plants of commercial interest, such as cereals, fruits, vegetables, ornamental, aromatic and medicinal plants (Bajaj 1999, 2000, 2001a, b). In concert with the recent developments of inducible expression systems and artificial promoters, the biotechnological preconditions for the use of molecular tools in ecosystem studies are gradually being met. Although these molecular techniques come at a substantial price, their potential to precisely monitor and manipulate plant–insect interactions may justify their costs.

Acknowledgements. We would like to thank Henrik U. Stotz, Junji Takabayashi, Jorge Zavala and Ursula Schittko who allowed us to use examples from their work to illustrate the different gene expression profiling techniques (Fig. 17.2).

References

Aerts RJ, Gisi D, Decarolis E, Deluca V, Baumann TW (1994) Methyl jasmonate vapor increases the developmentally controlled synthesis of alkaloids in *Catharanthus* and cinchona seedlings. Plant J 5:635–643

Arimura G, Ozawa R, Shimoda T, Nishioka T, Boland W, Takabyashi J (2000a) Herbivory-induced volatiles elicit defence genes in lima bean leaves. Nature 406:512–515

Arimura G, Tashiro K, Kuhara S, Nishioka T, Ozawa R, Takabayashi J (2000b) Gene responses in bean leaves induced by herbivory and by herbivore-induced volatiles. Biochem Biophys Res Commun 277:305–310

Arimura G, Ozawa R, Nishioka T, Boland W, Koch T, Kuhnemann F, Takabayashi J (2002) Herbivore-induced volatiles induce the emission of ethylene in neighboring lima bean plants. Plant J 29:87–98

Bajaj YPS (ed) (1999) Transgenic medicinal plants. In: Loerz H, Widholm JM (eds) Biotechnology in agriculture and forestry, vol 45. Springer, Berlin Heidelberg New York

Bajaj YPS (ed) (2000) Transgenic trees. In: Loerz H, Widholm JM (eds) Biotechnology in agriculture and forestry, vol 44. Springer, Berlin Heidelberg New York
Bajaj YPS (ed) (2001a) Transgenic crops II. In: Loerz H, Widholm JM (eds) Biotechnology in agriculture and forestry, vol 47. Springer, Berlin Heidelberg New York
Bajaj YPS (ed) (2001b) Transgenic crops III. In: Loerz H, Widholm JM (eds) Biotechnology in agriculture and forestry, vol 48. Springer, Berlin Heidelberg New York
Baldwin IT (1998) Jasmonate-induced responses are costly but benefit plants under attack in native populations. Proc Natl Acad Sci USA 95:8113–8118
Baldwin IT (1999) Inducible nicotine production in native *Nicotiana* as an example of adaptive phenotypic plasticity. J Chem Ecol 25:3–30
Berger S, Mitchell-Olds T, Stotz HU (2002) Local and differential control of vegetative storage protein expression in response to herbivore damage in *Arabidopsis thaliana*. Physiol Plant 114:85–91
Buchanan BB, Gruissem W, Jones RL (2000) Biochemistry and molecular biology of plants. American Society of Plant Physiologists, Rockville, Maryland
Coley PD, Bryant JP, Chapin FS (1985) Resource availability and plant antiherbivore defense. Science 230:895–899
Devoto A, Nieto-Rostro M, Xie DX, Ellis C, Harmston R, Patrick E, Davis J, Sherratt L, Coleman M, Turner JG (2002) COI1 links jasmonate signalling and fertility to the SCF ubiquitin–ligase complex in *Arabidopsis*. Plant J 32:457–466
Driebe EM, Whitham TG (2000) Cottonwood hybridization affects tannin and nitrogen content of leaf litter and alters decomposition. Oecologia 123:99–107
Falco MC, Marbach PAS, Pompermayer P, Lopes FCC, Silva MC (2001) Mechanisms of sugarcane response to herbivory. Genet Mol Biol 24:113–122
Farmer EE, Johnson RR, Ryan CA (1992) Regulation of expression of proteinase-inhibitor genes by methyl jasmonate and jasmonic acid. Plant Physiol 98:995–1002
Fidantsef AL, Stout MJ, Thaler JS, Duffey SS, Bostock RM (1999) Signal interactions in pathogen and insect attack: expression of lipoxygenase, proteinase inhibitor II, and pathogenesis-related protein P4 in the tomato, *Lycopersicon esculentum*. Physiol Mol Plant Pathol 54:97–114
Gantet P, Memelink J (2002) Transcription factors: tools to engineer the production of pharmacologically active plant metabolites. Trends Pharmacol Sci 23:563–569
Glawe GA, Zavala JA, Kessler A, van Dam NM, Baldwin IT (2003) Ecological costs and benefits correlated with trypsin protease inhibitor production in *Nicotiana attenuata*. Ecology 84:79–90
Grec S, Vanham D, De Ribaucourt JC, Purnelle B, Boutry M (2003) Identification of regulatory sequence elements within the transcription promoter region of *NpABC1*, a gene encoding a plant ABC transporter induced by diterpenes. Plant J 35:237–250
Halitschke R, Schittko U, Pohnert G, Boland W, Baldwin IT (2001) Molecular interactions between the specialist herbivore *Manduca sexta* (Lepidoptera, Sphingidae) and its natural host *Nicotiana attenuata*. III. Fatty acid–amino acid conjugates in herbivore oral secretions are necessary and sufficient for herbivore-specific plant responses. Plant Physiol 125:711–717
Hättenschwiler S, Vitousek PM (2000) The role of polyphenols in terrestrial ecosystem nutrient cycling. Trends Ecol Evol 15:238–243
Heil M, Baldwin IT (2002) Fitness costs of induced resistance: emerging experimental support for a slippery concept. Trends Plant Sci 7:61–67
Hermsmeier D, Schittko U, Baldwin IT (2001) Molecular interactions between the specialist herbivore *Manduca sexta* (Lepidoptera, Sphingidae) and its natural host *Nicotiana attenuata*. I. Large-scale changes in the accumulation of growth- and defense-related plant mRNAs. Plant Physiol 125:683–700

Hui D, Iqbal J, Lehmann K, Gase K, Saluz HP, Baldwin IT (2003) Molecular interactions between the specialist herbivore *Manduca sexta* (Lepidoptera, Sphingidae) and its natural host *Nicotiana attenuata*. IV. Microarray analysis and further characterization of large-scale changes in herbivore-induced mRNAs. Plant Physiol 131:1877–1893

Kahl J, Siemens DH, Aerts RJ, Gabler R, Kuhnemann F, Preston CA, Baldwin IT (2000) Herbivore-induced ethylene suppresses a direct defense but not a putative indirect defense against an adapted herbivore. Planta 210:336–342

Karban R, Baldwin IT (1997) Induced responses to herbivory. University of Chicago Press, Chicago

Korth KL, Dixon RA (1997) Evidence for chewing insect-specific molecular events distinct from a general wound response in leaves. Plant Physiol 115:1299–1305

Kovalchuk I, Kovalchuk O, Hohn B (2001) Biomonitoring the genotoxicity of environmental factors with transgenic plants. Trends Plant Sci 6:306–310

Krizek BA, Prost V, Joshi RM, Stoming T, Glenn TC (2003) Developing transgenic *Arabidopsis* plants to be metal-specific bioindicators. Environ Toxicol Chem 22:175–181

Li CY, Williams MM, Loh YT, Lee GI, Howe GA (2002) Resistance of cultivated tomato to cell content-feeding herbivores is regulated by the octadecanoid-signaling pathway. Plant Physiol 130:494–503

Libbert E (1993) Lehrbuch der Pflanzenphysiologie. Gustav Fischer Verlag, Jena

Martin D, Tholl D, Gershenzon J, Bohlmann J (2002) Methyl jasmonate induces traumatic resin ducts, terpenoid resin biosynthesis, and terpenoid accumulation in developing xylem of Norway spruce stems. Plant Physiol 129:1003–1018

Meinke D, Koornneef M (1997) Community standards: a new series of guidelines for plant science – community standards for *Arabidopsis* genetics. Plant J 12:247–253

Miksch M, Boland W (1996) Airborne methyl jasmonate stimulates the biosynthesis of furanocoumarins in the leaves of celery plants (*Apium graveolens*). Experientia 52:739–743

Moran PJ, Thompson GA (2001) Molecular responses to aphid feeding in *Arabidopsis* in relation to plant defense pathways. Plant Physiol 125:1074–1085

Moran PJ, Cheng YF, Cassell JL, Thompson GA (2002) Gene expression profiling of *Arabidopsis thaliana* in compatible plant–aphid interactions. Arch Insect Biochem Physiol 51:182–203

Orozco-Cardenas M, McGurl B, Ryan CA (1993) Expression of an antisense prosystemin gene in tomato plants reduces resistance toward *Manduca sexta* larvae. Proc Natl Acad Sci USA 90:8273–8276

Padidam M (2003) Chemically regulated gene expression in plants. Curr Opin Plant Biol 6:169–177

Pautot V, Holzer FM, Reisch B, Walling LL (1993) Leucine aminopeptidase – an inducible component of the defense response in *Lycopersicon esculentum* (tomato). Proc Natl Acad Sci USA 90:9906–9910

Ramputh AI, Arnason JT, Cass L, Simmonds JA (2002) Reduced herbivory of the European corn borer (*Ostrinia nubilalis*) on corn transformed with germin, a wheat oxalate oxidase gene. Plant Sci 162:431–440

Reymond P, Weber H, Damond M, Farmer EE (2000) Differential gene expression in response to mechanical wounding and insect feeding in *Arabidopsis*. Plant Cell 12:707–719

Riechmann JL, Heard J, Martin G, Reuber L, Jiang CZ, Keddie J, Adam L, Pineda O, Ratcliffe OJ, Samaha RR, Creelman R, Pilgrim M, Broun P, Zhang JZ, Ghandehari D, Sherman BK, Yu CL (2000) *Arabidopsis* transcription factors: genome-wide comparative analysis among eukaryotes. Science 290:2105–2110

Root RB (1973) Organization of a plant–arthropod association in simple and diverse habitats: the fauna of collards (*Brassica oleracea*). Ecol Monogr 43:95–124

Royo J, Leon J, Vancanneyt G, Albar JP, Rosahl S, Ortego F, Castanera P, Sanchez-Serrano JJ (1999) Antisense-mediated depletion of a potato lipoxygenase reduces wound induction of proteinase inhibitors and increases weight gain of insect pests. Proc Natl Acad Sci USA 96:1146–1151

Schittko U, Preston CA, Baldwin IT (2000) Eating the evidence? *Manduca sexta* larvae can not disrupt specific jasmonate induction in *Nicotiana attenuata* by rapid consumption. Planta 210:343–346

Schittko U, Hermsmeier D, Baldwin IT (2001) Molecular interactions between the specialist herbivore *Manduca sexta* (Lepidoptera, Sphingidae) and its natural host *Nicotiana attenuata*. II. Accumulation of plant mRNAs in response to insect-derived cues. Plant Physiol 125:701–710

Segal DJ, Stege JT, Barbas CF (2003) Zinc fingers and a green thumb: manipulating gene expression in plants. Curr Opin Plant Biol 6:163–168

Sharma YK, Davis KR (1994) Ozone-induced expression of stress-related genes in *Arabidopsis thaliana*. Plant Physiol 105:1089–1096

Sharma HC, Sharma KK, Seetharama N, Ortiz R (2000) Prospects for using transgenic resistance to insects in crop improvement. Electron J Biotechnol 3:76–95

Shen BZ, Zheng ZW, Dooner HK (2000) A maize sesquiterpene cyclase gene induced by insect herbivory and volicitin: characterization of wild-type and mutant alleles. Proc Natl Acad Sci USA 97:14807–14812

Stotz HU, Pittendrigh BR, Kroymann J, Weniger K, Fritsche J, Bauke A, Mitchell-Olds T (2000) Induced plant defense responses against chewing insects. Ethylene signaling reduces resistance of *Arabidopsis* against Egyptian cotton worm but not diamondback moth. Plant Physiol 124:1007–1017

Stout MJ, Workman J, Duffey SS (1994) Differential induction of tomato foliar proteins by arthropod herbivores. J Chem Ecol 20:2575–2594

Stout MJ, Fidantsef AL, Duffey SS, Bostock RM (1999) Signal interactions in pathogen and insect attack: systemic plant-mediated interactions between pathogens and herbivores of the tomato, *Lycopersicon esculentum*. Physiol Mol Plant Pathol 54:115–130

Strauss SY, Rudgers JA, Lau JA, Irwin RE (2002) Direct and ecological costs of resistance to herbivory. Trends Ecol Evol 17:278–285

Tamayo MC, Rufat M, Bravo JM, San Segundo B (2000) Accumulation of a maize proteinase inhibitor in response to wounding and insect feeding and characterization of its activity toward digestive proteinases of *Spodoptera littoralis* larvae. Planta 211:62–71

Tattersall DB, Bak S, Jones PR, Olsen CE, Nielsen JK, Hansen ML, Hoj PB, Moller BL (2001) Resistance to an herbivore through engineered cyanogenic glucoside synthesis. Science 293:1826–1828

Vancanneyt G, Sanz C, Farmaki T, Paneque M, Ortego F, Castanera P, Sanchez-Serrano JJ (2001) Hydroperoxide lyase depletion in transgenic potato plants leads to an increase in aphid performance. Proc Natl Acad Sci USA 98:8139–8144

van de Ven WTG, LeVesque CS, Perring TM, Walling LL (2000) Local and systemic changes in squash gene expression in response to silverleaf whitefly feeding. Plant Cell 12:1409–1423

Voelckel C, Baldwin IT (2003) Detecting herbivore-specific transcriptional responses in plants with multiple DDRT-PCR and subtractive library procedures. Physiol Plant 118:240–252

Vom Endt D, Kijne JW, Memelink J (2002) Transcription factors controlling plant secondary metabolism: what regulates the regulators? Phytochemistry 61:107–114

Walling LL (2000) The myriad plant responses to herbivores. J Plant Growth Regulation 19:195–216.

Wang EM, Wang R, DeParasis J, Loughrin JH, Gan SS, Wagner GJ (2001) Suppression of a P450 hydroxylase gene in plant trichome glands enhances natural-product-based aphid resistance. Nat Biotechnol 19:371–374

Winz RA, Baldwin IT (2001) Molecular interactions between the specialist herbivore *Manduca sexta* (Lepidoptera, Sphingidae) and its natural host *Nicotiana attenuata*. IV. Insect-induced ethylene reduces jasmonate-induced nicotine accumulation by regulating putrescine N-methyltransferase transcripts. Plant Physiol 125:2189–2202

Ziegler J, Keinanen M, Baldwin IT (2001) Herbivore-induced allene oxide synthase transcripts and jasmonic acid in *Nicotiana attenuata*. Phytochemistry 58:729–738

Section V
Synthesis

18 Testing the Role of Insects in Ecosystem Functioning

E. SIEMANN and W.W. WEISSER

18.1 Summary

Our knowledge of how herbivores impact ecosystem processes is rudimentary. More is known about how they influence plant diversity and composition than how ecosystem processes depend on herbivores and the changes they cause in plant communities. In particular, there is little theory, and few investigations, that incorporate insect herbivores, plants and ecosystem processes. We present models that make predictions about how herbivore effects on ecosystem processes vary with the diversity and functional types of plants and herbivores. In simple niche models, impacts of generalist herbivores on productivity varied little among ecosystems that differed in plant diversity. However, positive responses of non-host plant species released from competition with host plant species diminished the effects of specialist herbivores on productivity at higher levels of plant diversity. Resource models predicted that plant populations and ecosystem productivity will be more sensitive to belowground herbivory in soil resource-limited ecosystems and to aboveground herbivory in light-limited ecosystems. In the reverse situations, such as root herbivory in light-limited ecosystems, herbivory had little effect. In resource competition models, generalist root herbivory favoured plant species that are better belowground competitors. Generalist aboveground herbivory favoured better light competitors. Regardless of their mode of feeding (belowground vs. aboveground), specialist herbivores that fed on plants that were better light competitors in light-limited systems or on better soil resource competitors in soil resource-limited systems allowed competitors to invade and increased light or soil resource availability, respectively. In systems where both plant species co-existed in the absence of herbivores, specialist herbivores that fed on better light competitors favoured better soil resource competitors, allowing them to lower soil resource concentrations. Those that fed on better soil resource competitors decreased light availability by the same mechanism. Future research should place more emphasis on manipulations of insects and plants in combination and examine ecosystem responses.

Ecological Studies, Vol. 173
W.W. Weisser and E. Siemann (Eds.)
Insects and Ecosystem Function

18.2 Introduction

In the past two decades, an increasing number of ecologists have started to employ experimental approaches to investigate the ecological consequences of changes in biodiversity (Schulze and Mooney 1993; Loreau et al. 2002). A major aim of such studies is to understand the importance of species composition and species diversity for ecological processes such as energy flow and nutrient cycling, often referred to as 'ecosystem functioning'. The results obtained so far suggest that species composition and species diversity play a critical role in ecosystem productivity and other measures of ecosystem functioning (Kinzig et al. 2001; Loreau et al. 2002). A central theme is the role of plants as primary producers and the importance of a diversity of resource niches (i.e. rooting depths, phenology, form of nitrogen used; Tilman 1999). Comparatively less attention has been given to their interaction with other components of the community (Joshi et al., Chap. 6, this Vol.; Janssen and Sabelis, Chap. 9, this Vol.; Stadler et al., Chap. 11, this Vol.; Schmitz, Chap. 14, this Vol.).

Most terrestrial biodiversity–ecosystem function studies have so far concentrated on manipulating plant communities to study ecosystem processes (Loreau et al. 2002). Other important groups of organisms such as insects have been largely ignored other than as a response variable (Siemann et al. 1998; Joshi et al., Chap. 6, this Vol.). Insects are the most diverse group of organisms and therefore a major, if not the most important, component of biodiversity (Stork 1988; Siemann et al. 1999). Insects likely play a key role in mediating the relationship between plants and ecosystem processes by influencing the physiology, activity and population dynamics of plants. The scarcity of manipulative studies on the role of insects in ecosystem processes contrasts with the expert knowledge and large body of research already available, in particular, in the field of insect herbivory.

This book is the first to summarize the effects that insects have on ecosystem functioning, focusing mainly, but not exclusively, on herbivorous insects. The chapter authors with extensive experience in the field of plant–insect interactions have discussed the importance of insects for ecosystem functioning, using examples from their own work. In addition to providing an overview of known effects of insects on ecosystem functioning, it has provided a detailed discussion of the advantages and disadvantages of various techniques of manipulating insect herbivory. One of the major lessons of this book is that there is a paucity of information on how herbivore-induced changes in plant communities impact ecosystem functioning. We know something about damage levels in typical communities and the effects of herbivores on plant composition and diversity, but far less about the significance of these changes in the plant community for ecosystem processes. There are not many manipulative studies that investigate the importance of herbivores for processes other than net primary productivity. In addition, we are not aware of any manipulative studies that investigate the importance of herbivore diversity or composition for ecosystem functioning of any sort. There is also very little information on how herbivores impact changes with productivity or plant diversity. In part this may reflect a need for theory that makes clear predictions for how herbivore influences on plant populations and ecosystem functioning depend on herbivore species or functional diversity. We present here a preliminary set of models that we hope will both stimulate the development of more rigorous theory and encourage more explicit experimental investigations of insects and ecosystem functioning.

18.3 Simple Models of Niche Space

Tilman et al. (1997b) presented a model (their 'generalized niche model') that represented ecosystem productivity as the area covered by plant species in a two-dimensional niche space (Fig. 18.1A, B). In this niche space, each plant species occupied a circle with a different random location. This model is consistent with each plant species having a combination of abiotic conditions (such as temperature or soil pH) at which its performance is greatest and a decline in its performance as conditions approach its limits of tolerance. In this model, as the number of plant species increases, total ecosystem productivity increases asymptotically (Fig. 18.1C, no herbivore coverage). As the first few species are added, they are likely to have little overlap in the niche space they occupy, but the remaining unoccupied niche space is filled increasingly slowly as the number of species increases. How might the addition of herbivores change the relationship between plant diversity and productivity in this model?

18.3.1 Reduced Vigour Model

One way to incorporate herbivores in the generalized niche model is to keep the area of niche space occupied by a plant species unchanged (i.e. its range of tolerance remains the same) but to have incomplete coverage within a plant's niche space. In other words, as a herbivore is introduced into the ecosystem, the niche coverage of the plant(s) that it feeds on is discounted in the model by some amount ranging from 0 to 100% (Fig. 18.1A, shaded area). For instance, if a herbivore has a 50% impact on a species of plant, the area the plant species covers in niche space is only counted at half value when determining ecosystem functioning. The biological analogue to this model would be herbivores reducing plant vigour without restricting the conditions (types

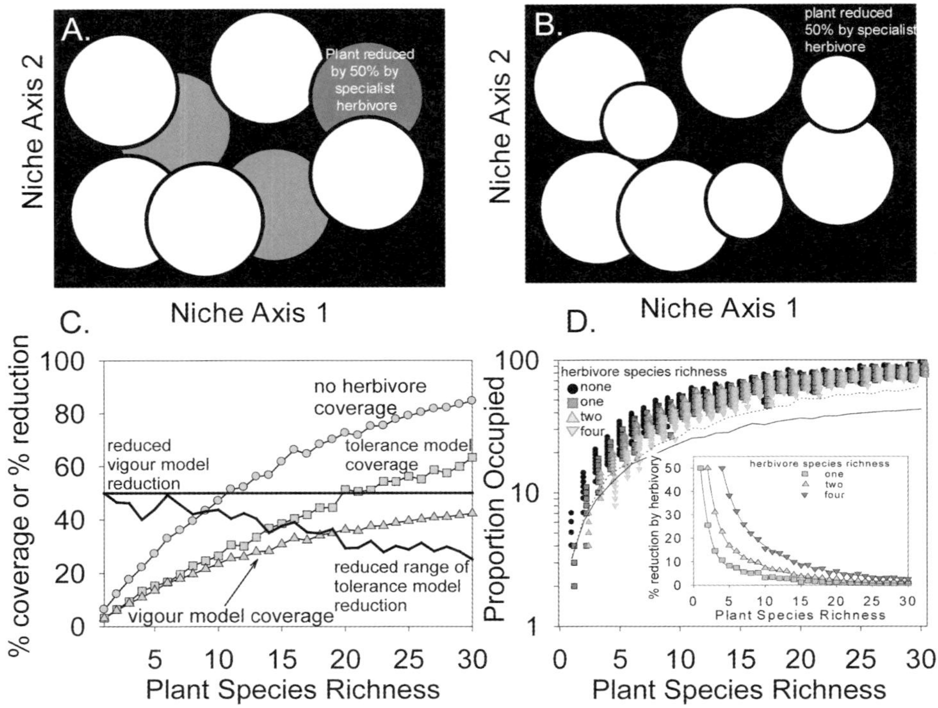

Fig. 18.1. Models of herbivore effects on the productivity of ecosystems that differ in plant diversity. **A** In the reduced vigour model, the niche coverage of plants experiencing herbivory is discounted when calculating ecosystem productivity. **B** In the reduced range of tolerance model, the niche area is reduced for plants experiencing herbivory. **C** Generalist herbivores. Without herbivores present, ecosystem productivity increases asymptotically as plant species richness increases (results from simulations with 1–30 plant species, niche radius 30, 200×200 niche array, 30 runs). In the reduced vigour model, generalist herbivores cause reductions in ecosystem productivity that are equal to the reduction of individual plant species (results from simulations identical to the no herbivore case but with 50 % reduction by herbivores). In the reduced range of tolerance model, the reductions in ecosystem productivity are relatively smaller at higher levels of plant diversity (results from simulations identical to the no herbivore case but with 50 % reduction in circle area, r=21). **D** Specialist herbivores. The zero specialist herbivore case is identical to the no herbivore case in **C**. Addition of specialist herbivore species reduces ecosystem productivity significantly in low plant diversity ecosystems but causes little change in high plant diversity ecosystems (results from simulations identical to the no herbivore case in **C** but with the addition of 1 to plant species richness number of herbivore species in the reduced vigour model with each herbivore species causing a 50 % reduction in the density of a single plant species). The reduced range of tolerance model gives almost identical results for low numbers of specialist herbivore species. As the number of herbivore species approaches the number of plant species, the specialist case is equivalent to the generalist case (no plants are free of herbivores and impacts are equivalent) and there is a greater effect of a large number of specialist herbivore species in the reduced vigour model (*solid line*) compared to the reduced range of tolerance model (*dotted line*). **D** *Inset* Reduction in productivity by the addition of 1, 2 or 4 specialist herbivore species is greatly weakened as ecosystems increase in plant diversity

of habitats, resource levels, seasons) in which they occur (e.g. Pearson et al. 2003). Perhaps the simplest case for this model is that in which herbivores are generalists that have no feeding preferences and feed on all plants such that their impact on each species of plant is equal (i.e. the discounting of niche coverage for each plant species is equivalent). This might reflect equal amounts of feeding on different plant species but might also be the result of growth vs. defence trade-offs in which smaller amounts of damage are more costly to well-defended, but less heavily fed upon, species of plants (Coley et al. 1985; Maschinski and Whitham 1989). So long as the impact of herbivores is equivalent across plant species, ecosystem productivity declines proportionally with herbivore impact. Because each plant is affected the same, there is also no effect of plant diversity on the proportional impact of herbivores on productivity. Rather, there is a larger absolute reduction in productivity at higher levels of plant diversity due only to the higher productivity in ecosystems with a greater diversity of plants. For this simple case of a generalist herbivore modelled as reducing the density of niche coverage, there is therefore little need to run simulation models. The individual species impact and community level impact of herbivory are identical at all levels of plant diversity. However, constant relative herbivore impact on productivity is not a feature of all simple models.

18.3.2 Reduced Range of Tolerance Model

If herbivores restrict the range of tolerances of plant species but have little impact on their vigour in habitats or conditions to which they are most suited (e.g. Louda and Rodman 1996; Meiners et al. 2000), reducing the size of circles for plant species in two-dimensional niche space when they experience herbivory may be an appropriate way to model herbivore impacts on plant productivity (Fig. 18.1B). To model this case we modified the Tilman et al. (1997b) generalized niche model such that the area of niche covered by each species of plant was reduced by 50 % when it was attacked by a herbivore (i.e. the radius was reduced to square root of 0.5 its original size). When generalist herbivores are modelled this way, the absolute reduction in productivity by generalist herbivores increases less rapidly with plant diversity than does productivity. As plant diversity increases, the amount of overlap between plant species is higher. Because herbivores cause the niche area covered by a particular plant species (i.e. the size of its circle) to decrease, herbivory decreases the amount of niche overlap between plant species and therefore decreases the loss in community productivity that is due to this overlap at higher plant diversities. As a consequence, the proportional reduction of plant productivity due to herbivory is smaller at higher plant diversity and the individual species level impact and community level impact are not the same (Fig. 18.1. C).

This is an interesting contrast to the pattern of plant diversity and productivity. The increasing overlap between plant species causes slow increases in productivity with plant diversity after the first several plant species are present (Tilman et al. 1997a; Hector et al. 1999), but also causes ecosystem productivity to be more resistant to impacts by generalist herbivores. This is not a resource concentration effect in the manner of Root (1973), where the mechanism of reduced herbivore impact is the result of the presence of inappropriate host plants in diverse plant assemblages which weakens specialist herbivore impacts. Rather, it is a de facto production redundancy in different abiotic conditions that reduces the impact of generalist herbivores, restricting the range of conditions that individual plant species can tolerate.

18.3.3 Specialist Herbivores

In the reduced vigour model, different scenarios are conceivable when a plant species is fed upon by a specialist herbivore. At one extreme, when one species is impacted by herbivory, the niche overlap area with other plant species could be completely utilized by the other plant species (Fig. 18.1A). In this scenario with complete compensation, herbivores mediate the outcome of competition in shared niche space such that the plant species that is not attacked by the herbivore 'takes over' the region of niche space that is shared with its competitor. When the niche overlap of the attacked plant with other plant species is larger, the impact of herbivory on community productivity is lower. This is the scenario that we modelled most intensively. The other extreme has herbivores reducing coverage in shared niche space as if there were no other plant species able to take advantage of the increase in available resources. In this scenario, community productivity decreases due to the decrease in niche coverage density of the non-overlapping area of niche space of the attacked plant, and also the niche area shared with other plants contributes less to productivity (in the analogy of Fig. 18.1A, it is also shaded to some degree). The first scenario in which compensation is strong emphasizes the buffering effect of plant diversity against herbivore impacts, whereas in the second scenario in which other plants are unable to take advantage of weakened competitors this buffering capacity is reduced.

In the reduced range of tolerance model, a specialist herbivore decreases only the niche area of the plant species on which it is specialized (Fig. 18.1B). In this model, herbivory has two effects: it reduces the niche area of the attacked plant (decreasing community productivity), and it decreases the area of overlap of this species with other plant species (increasing the contribution of these other species to productivity).

The reduced vigour model and the reduced range of tolerance models gave qualitatively similar results for all the responses we examined. Furthermore, quantitative results were almost identical when plant diversity was much greater than herbivore diversity, and the amount of area reduction and density reduction by herbivory were identical. For low numbers of specialist herbivore species we present only the results of the reduced vigour model, but the same conclusions apply for the reduced range of tolerance model. As herbivore diversity approaches plant diversity, the two models give quantitatively different results, so we present the results for both models for these extreme examples.

At low plant diversity, productivity (i.e. area covered in niche space) decreases rapidly with the addition of an increasing number of specialist herbivore species (Fig. 18.1D, inset). The simplest case with a single plant species and a single herbivore species is analogous to the case of a general herbivore and linear decreases in productivity with individual level impact. As plant diversity increases, community productivity increases (Fig. 18.1D, large graph), but for a low number of herbivore species the effect of adding a particular number of specialist herbivore species becomes increasingly weaker with increasing plant diversity (Fig. 18.1D). This reflects the greater amount of niche space that is occupied by more than one plant species at the higher levels of plant diversity. In effect, plant compensation limits herbivore impacts to changing the relative contributions of different plant species to ecosystem productivity. At higher levels of herbivore diversity relative to plant diversity, the effects of specialist and generalist herbivores converge once every plant species has a specialist herbivore species. The effect of specialist herbivores is greater in the reduced vigour model than in the reduced range of tolerance model at high herbivore diversity (Fig. 18.1D inset, lines).

Although we modelled the case of monophagous herbivores, specialists need not be restricted to a single plant species for there to be an effect of plant diversity on the ecosystem-level impacts of herbivory. Any degree of specialization whether it is polyphagous feeding or heavier feeding on different species of plants due to differences in spatial distribution, allocation above vs. belowground or phenology should allow for a positive effect of plant diversity in reducing the impact of herbivores.

18.4 Effects of Herbivores in Resource Competition Models

In a second approach, we investigated simple models of herbivore impacts in an ecosystem with two limiting resources, to analyze how plants' resource requirements may influence the impact of herbivory on plant competition and productivity. Our approach is based on the theory of resource competition (Tilman 1982, 1988). In these models, a graphical approach is used that allows predictions on the outcome of competition between plant species based on the position of the resource supply point and the zero net growth isoclines (ZNGIs) of the competing plant species (see the legend to Fig. 18.2 for a more detailed

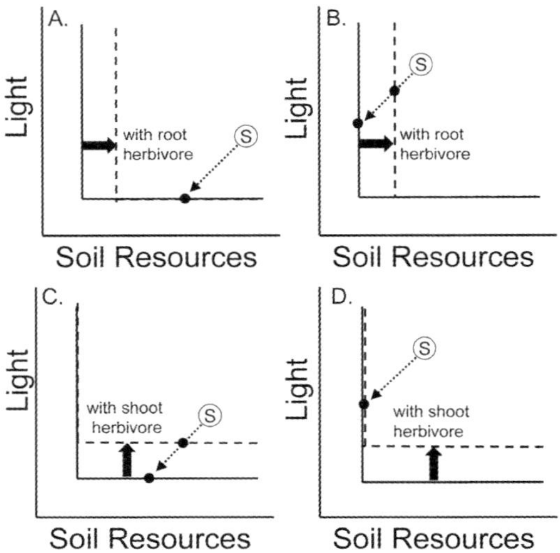

Fig. 18.2. The effect of above- and belowground herbivores in plant resource competition models. *Axes* represent available concentrations of soil resources and light. *L-shaped lines* represent zero net growth isoclines (ZNGIs) which are the concentrations of resources for which population growth is zero; population growth is negative on the origin side of the isocline and positive on the other side. The *Circle marked S* indicates the resource supply point and the circles on the lines show the resource equilibrium concentrations with plant or herbivore in the system: *dotted arraow* shows the change in resources with the plant species present, *thick arrow* shows the change in ZNGI with the herbivore present and *hatched isocline* and *dotted arrow* represent the ZNGI and equilibrium resource concentrations, respectively, when the herbivore is present. **A** Addition of a belowground (root-feeding) herbivore that feeds on a primarily light-limited plant with little ability to reallocate resources between above- and belowground structures increases the amount of soil resources the plant requires for population equilibrium (ZNGI moves to the right) but has little effect on the plant's population density or concentrations of available resources. **B** Addition of a root-feeding herbivore that feeds on a primarily soil resource limited plant reduces the plant's population density and increases both light and soil nutrient availability. **C** Addition of an aboveground (shoot-feeding) herbivore that feeds on a primarily light limited plant increases the amount of light required for population equilibrium, reduces plant population density and increases both light and soil nutrient availability. **D** Addition of an aboveground herbivore to a soil resource limited plant has no impact on plant population densities or available resource concentrations

explanation). It differs from earlier work that incorporates herbivores by modelling herbivore tolerance as a resource axis (Grover 1994, 1995; Holt et al. 1994; Leibold 1996; Chase et al. 2000) and instead models herbivore effects via changes in resource requirements. In addition, earlier attempts did not explicitly investigate how herbivore functional types or diversity affect their impacts on plant communities and ecosystem functioning.

For many terrestrial plant communities, limiting factors will be light and some soil resources such as nitrogen or water. In order to capture the effects of herbivores in ecosystems with different degrees of above- and belowground resource limitation and/or differences in plant allocation patterns, we distinguished herbivores as aboveground vs. belowground feeders in our models. So long as plants can readily adjust their allocation patterns in response to herbivore damage, the effect of herbivores on ZNGIs will be to move them towards the resource supply point but to have little effect on the relative limitation by above- vs. belowground resources or the relative available concentrations of light vs. soil resources. Of course, if the effect of herbivores is strong enough so as to move the ZNGI past the resource supply point, the plant species cannot persist. Similarly, if the main effect of herbivores is to increase plant mortality rates and/or decrease plant reproductive rates there will be a simultaneous increase in requirements for both limiting nutrients as herbivore damage increases. However, when root-feeding herbivores impact the soil nutrient gathering abilities of plants more than their light gathering abilities and aboveground herbivores (leaf or shoot feeding) primarily decrease light gathering abilities, herbivores will have differential impacts on plants in different resource conditions and plants that differ in their abilities to compete for one resource versus another (Fig. 18.2).

18.4.1 Specialist Herbivores in Resource Competition Models

The effects of specialist herbivores on plants and resources depend on the type of herbivore (above vs. belowground), the species of plant attacked (better soil resource or light competitor) and the relative supply rates of above- vs. belowground resources in resource competition models (Table 18.1). In these models, when a specialist root herbivore feeds on the plant that is a better light competitor, it reduces the ability of that plant to gather soil resources and increases the concentration of soil resources required for population equilibrium (Fig. 18.3A). In order for two plant species to co-exist with this herbivore present, there must be a higher supply of soil resources than in the absence of the herbivore (S3 vs. S2 in Fig. 18.3A). For resource supply rates that previously allowed co-existence, concentrations of soil resources will decrease when only the better soil resource competitor is now present at equilibrium. For resource supply rates that previously did not allow co-existence but that now do allow it, available concentrations of light increase.

When a specialist root herbivore feeds on the plant species that is a better soil resource competitor, it can cause the loss of that plant species at all resource supply rates in these models (Fig. 18.3B). This will happen if, under herbivory, the better soil resource competitor has soil resource requirements that exceed those of the better light competitor. Concentrations of soil resources increase for resource supply points that previously had the better soil competitor present but now have the better light competitor present (S1 in Fig. 18.3B). The concentration of light is lower for supply points that used

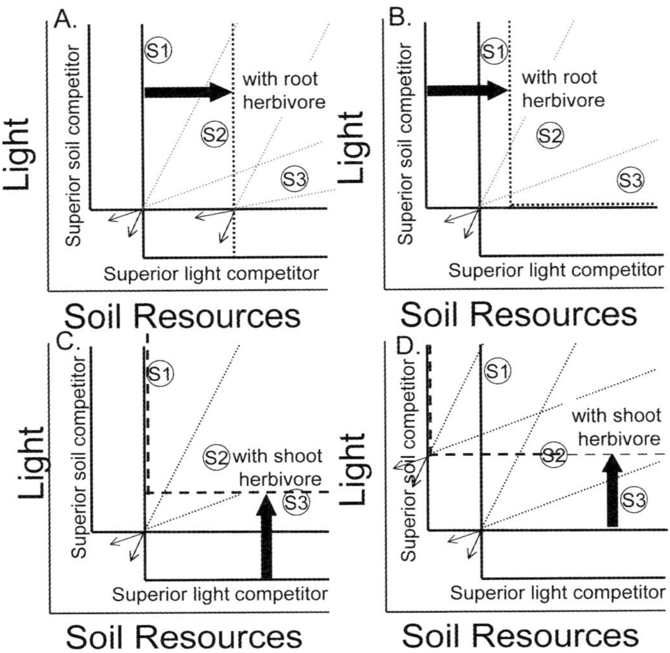

Fig. 18.3. Specialist herbivores and resource competition (cf. Table 18.1). **A** Addition of a specialist belowground herbivore, that feeds on only the plant that is the better light competitor and that has limited ability to reallocate resources between above- and belowground structures, results in competitive displacement at resource supply point *S2* (*dotted arrow*), where there was previously co-existence of both plant species (*solid arrow*) but results in co-existence at *S3* (*dotted arrow*) at which the better light competitor displaces the other species (*solid arrow*, i.e. co-existence requires higher relative rates of soil resource supply). Concentrations of limiting resources also change for some supply points – the new equilibrium point for *S3* has greater concentrations of light and *S2* has lower concentrations of soil resources, but there is little change for *S1*. **B** Addition of a specialist root herbivore that feeds on the better soil nutrient competitor results in its displacement at all resource supply points if levels of herbivory are sufficiently high (*dotted arrows* from supply points *S1–S3*). Concentrations of light decrease for *S2* and concentrations of soil resources increase for *S1*. **C** Addition of a specialist shoot herbivore that feeds on the better light competitor results in its displacement at all supply points if levels of herbivory are sufficiently high (*dotted arrows* from supply points *S1–S3*). Changes in concentrations of resources are the same as in **A**. **D** Addition of a specialist shoot herbivore that feeds on the better soil nutrient competitor is similar to the first case in terms of requirements for co-existence change (but in this case require a higher light supply) and changes in available resource concentrations are the same as in **B**. In all parts, *arrows at the intersection of isoclines* represent consumption vectors and *dotted lines* projected back from the intersections of ZNGIs represent the set of resource supply points that will lead to co-existence. For more information see Fig. 18.2 legend

to lead to co-existence but now have only the better light competitor present (S2).

Similarly, the impact of a specialist aboveground herbivore that feeds on the better light competitor may result in global displacement, but the changes in resource concentrations are similar to those in the case of a specialist root herbivore feeding on the plant (Fig. 18.3C). The dominant change in resource for points that formerly led to co-existence is to lower the concentrations of soil resources to the level of the better soil competitor. The dominant change in resource concentrations for points at which the better light competitor was the only species is to increase light concentrations to the requirements of the better soil competitor whether or not the better light competitor is present (root herbivore) or not (aboveground herbivore). For a specialist shoot herbivore that feeds on the better soil resource competitor, the changes in resource concentrations are the same as when root herbivores feed on the plant, but both species can always co-exist when the relative supply of light is high (Fig. 18.3D).

Specialist herbivores do not affect the available concentrations of soil resources when light is the main limiting resource (Fig. 18.3, S3) nor do they affect the available concentrations of light when soil resources are the main limiting resource (Fig. 18.3, S1). They may, however, *increase* the concentrations of soil resources when they are the main limiting resource. This is most likely to occur when specialist herbivores feed on the better soil resource competitor independent of the mode of feeding (belowground Fig. 18.3B, S1 vs. aboveground Fig. 18.3D, S1). Similarly, specialist herbivores have the potential to change available light concentrations when light is the main limiting resource (Fig. 18.3A, S3, C, S3).

The type of herbivory (belowground vs. aboveground) does determine the strength of the response of plant composition when one resource is strongly limiting (Table 18.1). For example, when soil resources are the main limiting resource, both belowground and aboveground specialist herbivores that feed on the better soil competitor increase the concentrations of soil resources, allowing the better light competitor to invade. This can result in either displacement (more likely with belowground herbivore, Fig. 18.3B, S1) or co-existence (more likely with aboveground herbivore, Fig. 18.3D, S1).

In cases where light and soil resources are co-limiting and there is co-existence of plant species in the absence of herbivory, the addition of a specialist herbivore always has the potential to change resource concentrations and reduces the likelihood of co-existence (Table 18.1). Regardless of the mode of feeding, specialist herbivores that feed on the better soil resource competitor will favour the plant that is the better light competitor, allowing it to *decrease* available light concentrations. Similarly, specialist herbivores that feed on the better light competitor favour plants that are better soil resource competitors and decrease available soil resource concentrations. At least in our models, the mode of feeding has little effect on the outcome of competition or resources.

Table 18.1. Summary of effects of resource competition models (Fig. 18.3). Changes are shown in *bold*. Better light competitor is indicated by *L* and better soil resource competitor is indicated by *S* in plant species composition column. *Arrows* indicate the direction of change

Resource supply point	Type of herbivore	Host plant of herbivore	Available concentrations of light	Available concentrations of soil resources	Plant species composition	Example
S1	Belowground	Light competitor	No change	No change	S	Fig. 18.3A
	Belowground	Soil competitor	No change	**Increase**	**S → L**	Fig. 18.3B
	Aboveground	Light competitor	No change	No change	S	Fig. 18.3C
	Aboveground	Soil competitor	No change	**Increase**	**S → S+L**	Fig. 18.3D
S2	Belowground	Light competitor	No change	**Decrease**	**S+L → S**	Fig. 18.3A
	Belowground	Soil competitor	**Decrease**	No change	**S+L → L**	Fig. 18.3B
	Aboveground	Light competitor	No change	**Decrease**	**S+L → S**	Fig. 18.3C
	Aboveground	Soil competitor	**Decrease**	No change	**S+L → L**	Fig. 18.3D
S3	Belowground	Light competitor	**Increase**	No change	**L → S+L**	Fig. 18.3A
	Belowground	Soil competitor	No change	No change	L	Fig. 18.3B
	Aboveground	Light competitor	**Increase**	No change	**L → S**	Fig. 18.3C
	Aboveground	Soil competitor	No change	No change	L	Fig. 18.3D

18.4.2 Generalist Herbivores in Resource Competition Models

If generalist herbivores have similar impacts on all plant species (see below for reasons why this may not be the case), there may be predictable changes in the degree of limitation by aboveground vs. belowground resources and changes in the allocation patterns of the dominant plant species depending on whether herbivores feed above vs. belowground. As a generalist belowground herbivore reduces the ability of all plant species to gather soil resources, the concentrations of available soil resources at equilibrium will increase (Fig. 18.4A). This assumes that the herbivore causes a proportional increase in the level of resources needed, not an increase up to a fixed amount independent of the previous requirements. However, so long as herbivores do not impact species differently enough to cause the order of isoclines to change, our qualitative results will not change. As the plant species in the community switch to those that favour soil resource gathering over light gathering (i.e. those with higher relative belowground allocation), the concentrations of light available at equilibrium will increase as well.

Aboveground herbivory has similar effects but in the opposite direction (Fig. 18.4B). The ability of plants to gather light is reduced, available light concentration is increased, plants with greater relative allocation aboveground

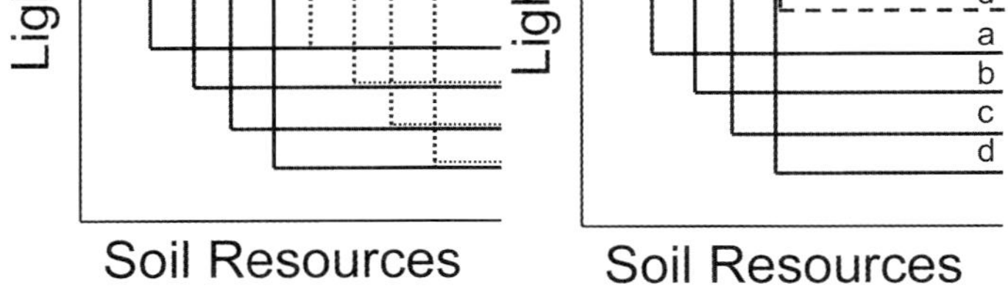

Fig. 18.4. A Generalist belowground herbivore shifts all isoclines to the right (limits the ability of all plant species to gather soil resources) in a community of plants with little ability to reallocate resources among above- and belowground structures. This results in a community dominated more by plants that are better soil resource competitors (i.e. relatively lower belowground allocation) and greater availability of both light and soil nutrients (but with the dominant plants now being relatively more efficient soil nutrient users, increases in soil resources are more pronounced). **B** Generalist aboveground herbivore shifts all isoclines upwards (limits the ability of all plant species to intercept light). This results in a community dominated by better light competitors and large increases in light availability. *Circle marked S* indicates resource supply point

are favoured and available soil resource concentrations increase. In general, aboveground herbivory makes competition for aboveground resources more important and changes in plant composition cause a corresponding change in belowground resources.

18.5 Differential Impacts on Plants with Different Traits

When herbivore impacts differ for plants that have different allocation patterns, they have the potential to change plant community composition which may in turn impact ecosystem processes when plants differ in their nutrient requirements, tissue chemistry, rooting depths or other characteristics that are relevant to production or nutrient cycling. For instance, if grazers reduce aboveground vegetation to a low level, they will have a larger impact on plants with relatively high aboveground allocation. In this case, they may shift the plant community to species with high root allocation, such as favouring grasses over forbs. With greater light availability and greater root mass they will also strengthen limitation by soil nutrients and may slow nitrogen cycling via litter feedbacks. In addition, herbivores often select host plants based on defence chemicals. Since plant growth rates and nutrient content are correlated with defence, herbivores may shift the plant community from plants with fast growth rates to those with slower growth rates. Herbivores may also select host plants based on nutrient content. When herbivores remove high nutrient foliage they may cause a shift in litter quality and nutrient cycling. These types of phenomena are poorly captured in the types of simple models we present here but should be considered in future models.

18.6 Conclusions from the Modelling Work

A theoretical framework that incorporates herbivores within traditional models of plant communities and ecosystem processes predicts different impacts of herbivores on plants and ecosystem processes depending on the types and numbers of herbivore species, types and numbers of plant species and the limiting resources in the ecosystem. One surprising result was that the changes in equilibrium resource concentrations are less sensitive to the addition of some types of herbivores than is plant species composition. Some of the predictions already have support in the literature, such as the ability of specialist insect herbivores to increase limiting resources in soil resource limited (Brown 1994) or light limited (Carson and Root 2000) ecosystems or differential impacts of mammalian grazers on composition vs. productivity for different ecosystems (Chase et al. 2000), but many remain untested. Improving on the rudimentary models presented here should give deeper insights into the relationships between insects and ecosystem processes and may motivate more explicit experimental investigations in this emerging area of research.

18.7 Suggestions for Future Studies

In our view, the main message of the chapters in this book is that herbivorous insects have a large number of effects on ecosystem functioning well beyond the simple reduction of biomass while feeding, but that we are only beginning to quantify these effects. As this was one of the main themes in the Introduction, it is necessary to spell out more specifically the areas of research where progress has been made and to identify how research should proceed in the future. In our view, significant progress has recently been made and can be made in the future using the following approaches.

18.7.1 Exploring Below- and Aboveground Interactions in More Detail

Chapters 2–5, in the second section of the book, list a large number of interactions within the belowground community and between the above- and belowground community, often mediated by effects of insects on the host plant, which affect carbon and nutrient fluxes in the community. Most of these effects have so far been quantified only at the scale of pot or mesocosm experiments, but it is already clear that, for example, the indirect effects on the soil community that herbivorous insects have by stimulating root exudation, modifying plant resource allocation and influencing the quality of litter, will have greater impact on the N cycle than the direct effect of a reduction in standing crop. Significant advances have been made by performing clever manipulative experiments (Wardle and Bardgett, Chap. 3, this Vol.; Bonkowski and Scheu, Chap. 4, this Vol.; Masters, Chap. 5, this Vol.) and by applying technology such as isotope analysis (Hartley and Jones, Chap. 2, this Vol.). The models presented here could be extended to examine linkages between above- and belowground herbivores via changes in light vs. soil resources or allocation of production above vs. belowground.

Readers may note that the chapters focused on above- and belowground interactions include not only insects, but also other soil invertebrates. As may be apparent from the chapters in Section 2, it is not always possible to disentangle the effects of insects from those of other invertebrates. In part this is because of the complex interactions among the different organisms that need to be explored further. Another reason is that, belowground, many insect species fulfil similar functions to that of other invertebrate species (e.g. feeding on fungi, feeding on other invertebrates, breaking down litter etc.) such that the specific role of a particular species or group of species can only be deduced from experiments where the presence of particular species is manipulated.

Careful experimental manipulations of the belowground fauna with and without aboveground herbivory and concomitant measurements of the nutrient fluxes in these manipulated communities are the logical next step to a fuller understanding of the role of these interactions for ecosystem functioning. Ideally, these studies should be performed at an increasingly larger scale, moving from pot experiments to mesocosms and then larger-scale field experiments. The recent surge in studies investigating these interactions (Wardle 2002) suggests that one can be optimistic about the future advancement of research in this area.

18.7.2 Measuring Herbivory Effects at Nominal Levels as Well as in Outbreak Situations

Because herbivory is most spectacular during outbreaks, much research of insect effects on ecosystem function has focused on biomass removal and the release of nutrients in outbreak situations. We believe that it is not necessarily these immediate changes in carbon and nitrogen mineralization rates where outbreaks have the greatest effects on ecosystem functioning. Rather, it is the effects of outbreaks on the plant community, i.e. a resetting of the course of succession or more generally a shift in plant community structure, that may have the longer-lasting effects on nutrient cycling (cf. Carson et al., Chap. 10, this Vol.). These effects need to be quantified in much more detail (see also below).

Quantifications of the effects of aboveground herbivores on nutrient cycling during acute herbivory are also scarce, especially for sucking insects. As Stadler et al. (Chap. 11, this Vol.) have shown for the case of forested ecosystems, insects strongly modify the throughfall to the floor, for example carbon and nitrogen concentrations. The total change in C and N fluxes with herbivory can be high even in non-outbreak situations. The plants' responses to herbivory modify the throughfall such that throughfall is a function of both current and past herbivory as well as insect densities. Further studies such as those by Stadler et al. are needed for different ecosystems, and the consequences of a changed throughfall for belowground processes and hence nutrient cycling need to be explored in more detail. This approach requires ecosystem-level measurements in situations with and without herbivory, at natural levels of herbivory as well as at experimentally augmented levels.

18.7.3 Quantifying the Effects of Plant Resource Allocation Under Herbivory for Ecosystem Functioning

The effects of insect herbivory on plants are spectacular, both in the way plants employ induced defences and how plants allocate their resources to growth or reproduction, roots or shoots, or, at a much more fundamental level, the transcription of particular genes in their genome (cf. Voelckel and Baldwin, Chap. 17, this Vol.). In the field of chemical ecology, it is almost daily that new ways in which plants respond to herbivory are discovered and it is becoming increasingly clear that not only are plants able to specifically recognize particular herbivores, but also their plasticity is much greater than previously imagined. The changes in plant resource allocation under herbivory have consequences for nutrient cycling, which may extend well beyond a decrease in plant biomass, an increase in root exudation or a change in the quality of plant litter. We know very little about these effects, but it is possible that most of the change in nutrient flux under herbivory is due to changes in plant resource allocation. Exploring these effects requires first of all the identification of the nature of plant resource allocation under herbivory (which has so far been done for only a few model species), and scaling up to the effects at the community level. This is a difficult task that may well take several years to accomplish. Nevertheless, if we are to understand the mechanisms through which herbivory affects nutrient cycling, such an approach is indispensable.

18.7.4 Combining Various Methodologies to Achieve an Understanding of Insect Effects on Ecosystem Function

In our methodology section (Chaps. 12–17), the advantages and disadvantages of different techniques to enhance, reduce or simulate insect herbivory are reviewed. All of the techniques reviewed have their disadvantages which need to be accounted for in the interpretation of the results. It is also clear from this section that there is no single best technique. While the development of molecular tools might provide ecologists with possibilities for experimental manipulations well beyond current imagination, it is not warranted to defer experimentation until this new technology is readily available. Also, insecticides, cages or simulated herbivory are, despite their negative side effects, useful for addressing particular questions. Thus, a combination of methods should be used to gain mechanistic insights. The chapters by Schmitz (Chap. 14, this Vol.) and Rogers and Siemann (Chap. 16, this Vol.) are cases in point. In addition, entomologists need to go beyond simple measurements of herbivore damage or plant biomass and should make use of isotope analysis, throughfall collectors, DOC analyzers, ion chromatography and other methods to be able to measure changes in nutrient fluxes under herbivory. Ecological entomologists use knowledge of their ecosystem to design experiments that manipulate the major components of the system in a step-wise fashion. There is a need to combine this expertise with the expertise of ecosystem ecologists which may be achieved most easily in collaborative studies. Thus, we strongly advocate the use of multiple herbivory manipulation techniques in combination with ecosystem study technologies.

To summarize, there are a number of research areas where future effort is likely to yield important results that will illuminate both the magnitude and the mechanistic basis of the way insects influence nutrient cycling. While current knowledge may be fragmentary, there is an exciting period ahead that will bring us much closer to a fuller understanding of the effects of insects on ecosystem functioning.

Acknowledgements. E.S. was supported by the US NSF (NSF DEB-9981654, DEB-0315796), EPA (R82–8903) and USDA (2003-35320-13498). W.W.W. was supported by the Deutsche Forschungsgemeinschaft, through grants FOR 456 and WE 2618/3-2618/5.

References

Brown DG (1994) Beetle folivory increases resource availability and alters plant invasion in monocultures of goldenrod. Ecology 75:1673–1683

Carson WP, Root RB (2000) Herbivory and plant species coexistence: community regulation by an outbreaking phytophagous insect. Ecol Monogr 70:73–99

Chase JM, Leibold MA, Downing AL, Shurin JB (2000a) The effects of productivity, herbivory, and plant species turnover in grassland food webs. Ecology 81:2485–2497

Chase JM, Leibold MA, Simms E (2000b) Plant tolerance and resistance in food webs: community-level predictions and evolutionary implications. Evol Ecol 14:289–314

Coley PD, Bryant JP, Chapin FS (1985) Resource availability and plant antiherbivore defense. Science 230:895–899

Grover J (1994) Assembly rules for communities of nutrient-limited plants and specialist herbivores. Am Nat 143:258–282

Grover J (1995) Competition, herbivory, and enrichment: nutrient-based models for edible and inedible plants. Am Nat 145:746–774

Hector A, Schmid B, Beierkuhnlein MC, Caldeira M, Diemer M, Dimitrakopoulou P, Finn J, Freitas H, Giller P, Good J, Harris R, Hogberg P, Huss-Danell K, Joshi J, Jumpponen A, Korner C, Leadley PW, Loreau M, Minns A, Mulder C, O'Donovan G, Otway S, Pereira J, Prinz A, Read DJ, Scherer-Lorenzen M, Schultz E-D, Siamantziouras A, Spehn E, Terry A, Troumbis A, Woodward F, Yachi S, Lawton J (1999) Plant diversity and productivity experiments in European grasslands. Science 286:1123–1127

Holt RD, Grover J, Tilman D (1994) Simple rules for interspecific dominance in systems with exploitative and apparent competition. Am Nat 144:741–771

Kinzig AP, Pacala SW, Tilman D (eds) (2001) The functional consequences of biodiversity. Princeton University Press, Princeton, NJ

Leibold MA (1996) A graphical model of keystone predators in food webs: trophic regulation of abundance, incidence, and diversity patterns in communities. Am Nat 147:784–812

Loreau M, Naeem S, Inchausti P (eds) (2002) Biodiversity and ecosystem functioning. Oxford University Press, Oxford

Louda SM, Rodman JE (1996) Insect herbivory as a major factor in the shade distribution of a native crucifer (*Cardamine cordifolia* A. Gray, bittercress). J Ecol 84:229–237

Maschinski J, Whitham T (1989) The continuum of plant responses to herbivory: the influence of plant association, nutrient availability and timing. Am Nat 134:1–19

Meiners SJ, Handel SN, Pickett STA (2000) Tree seedling establishment under insect herbivory: edge effects and interannual variation. Plant Ecol 151:161–170

Pearson TRH, Burslem D, Goeriz RE, Dalling JW (2003) Interactions of gap size and herbivory on establishment, growth and survival of three species of neotropical pioneer trees. J Ecol 91:785–796

Root R (1973) Organization of a plant–arthropod association in simple and diverse habitats: the fauna of collards (*Brassica oleracea*). Ecol Monogr 43:95–124

Schulze ED, Mooney HA (1993) Biodiversity and ecosystem function. Springer, Berlin Heidelberg New York

Siemann E, Tilman D, Haarstad J, Ritchie M (1998) Experimental tests of the dependence of arthropod diversity on plant diversity. Am Nat 152:738–750

Siemann E, Haarstad J, Tilman D (1999) Dynamics of plant and arthropod diversity during old field succession. Ecography 22:1–9

Stork NE (1988) Insect diversity: facts, fiction and speculation. Biol J Linn Soc 35:321–337

Tilman D (1982) Resource competition and community structure. Princeton University Press, Princeton, NJ

Tilman D (1988) Plant strategies and the dynamics and structure of plant communities. Princeton University Press, Princeton, NJ

Tilman D (1999) The ecological consequences of changes in biodiversity: a search for general principles. Ecology 80:1455–1474

Tilman D, Knops J, Wedin D, Reich P, Ritchie M, Siemann E (1997a) The influence of functional diversity and composition on ecosystem processes. Science 277:1300–1302

Tilman D, Lehman CL, Thomson KT (1997b) Plant diversity and ecosystem productivity: theoretical considerations. Proc Natl Acad Sci USA 94:1857–1861

Wardle DA (2002) Communities and ecosystems: linking the aboveground and belowground components. Princeton University Press, Princeton, NJ

Subject Index

Taxonomic Index

Ecological Studies

Volumes published since 1998

Volume 133
Aquatic Humic Substances: Ecology and Biogeochemistry (1998)
D.O. Hessen and L.J. Tranvik (Eds.)

Volume 134
Oxidant Air Pollution Impacts in the Montane Forests of Southern California (1999)
P.R. Miller and J.R. McBride (Eds.)

Volume 135
Predation in Vertebrate Communities: The Białowieża Primeval Forest as a Case Study (1998)
B. Jędrzejewska and W. Jędrzejewski

Volume 136
Landscape Disturbance and Biodiversity in Mediterranean-Type Ecosystems (1998)
P.W. Rundel, G. Montenegro, and F.M. Jaksic (Eds.)

Volume 137
Ecology of Mediterranean Evergreen Oak Forests (1999)
F. Rodà et al. (Eds.)

Volume 138
Fire, Climate Change and Carbon Cycling in the North American Boreal Forest (2000)
E.S. Kasischke and B. Stocks (Eds.)

Volume 139
Responses of Northern U.S. Forests to Environmental Change (2000)
R. Mickler, R.A. Birdsey, and J. Hom (Eds.)

Volume 140
Rainforest Ecosystems of East Kalimantan: El Niño, Drought, Fire and Human Impacts (2000)
E. Guhardja et al. (Eds.)

Volume 141
Activity Patterns in Small Mammals: An Ecological Approach (2000)
S. Halle and N.C. Stenseth (Eds.)

Volume 142
Carbon and Nitrogen Cycling in European Forest Ecosystems (2000)
E.-D. Schulze (Ed.)

Volume 143
Global Climate Change and Human Impacts on Forest Ecosystems: Postglacial Development, Present Situation and Future Trends in Central Europe (2001)
J. Puhe and B. Ulrich

Volume 144
Coastal Marine Ecosystems of Latin America (2001)
U. Seeliger and B. Kjerfve (Eds.)

Volume 145
Ecology and Evolution of the Freshwater Mussels Unionoida (2001)
G. Bauer and K. Wächtler (Eds.)

Volume 146
Inselbergs: Biotic Diversity of Isolated Rock Outcrops in Tropical and Temperate Regions (2000)
S. Porembski and W. Barthlott (Eds.)

Volume 147
Ecosystem Approaches to Landscape Management in Central Europe (2001)
J.D. Tenhunen, R. Lenz, and R. Hantschel (Eds.)

Volume 148
A Systems Analysis of the Baltic Sea (2001)
F.V. Wulff, L.A. Rahm, and P. Larsson (Eds.)

Volume 149
Banded Vegetation Patterning in Arid and Semiarid Environments (2001)
D. Tongway and J. Seghieri (Eds.)

Volume 150
Biological Soil Crusts: Structure, Function, and Management (2001, 2003)
J. Belnap and O.L. Lange (Eds.)

Volume 151
Ecological Comparisons of Sedimentary Shores (2001)
K. Reise (Ed.)

Volume 152
Global Biodiversity in a Changing Environment: Scenarios for the 21st Century (2001)
F.S. Chapin, O. Sala, and E. Huber-Sannwald (Eds.)

Volume 153
UV Radiation and Arctic Ecosystems (2002)
D.O. Hessen (Ed.)

Volume 154
Geoecology of Antarctic Ice-Free Coastal Landscapes (2002)
L. Beyer and M. Bölter (Eds.)

Volume 155
Conserving Biological Diversity in East African Forests: A Study of the Eastern Arc Mountains (2002)
W.D. Newmark

Volume 156
Urban Air Pollution and Forests: Resources at Risk in the Mexico City Air Basin (2002)
M.E. Fenn, L. I. de Bauer, and T. Hernández-Tejeda (Eds.)

Volume 157
Mycorrhizal Ecology (2002, 2003)
M.G.A. van der Heijden and I.R. Sanders (Eds.)

Volume 158
Diversity and Interaction in a Temperate Forest Community: Ogawa Forest Reserve of Japan (2002)
T. Nakashizuka and Y. Matsumoto (Eds.)

Volume 159
Big-Leaf Mahogany: Genetic Resources, Ecology and Management (2003)
A. E. Lugo, J. C. Figueroa Colón, and M. Alayón (Eds.)

Volume 160
Fire and Climatic Change in Temperate Ecosystems of the Western Americas (2003)
T. T. Veblen et al. (Eds.)

Volume 161
Competition and Coexistence (2002)
U. Sommer and B. Worm (Eds.)

Volume 162
How Landscapes Change: Human Disturbance and Ecosystem Fragmentation in the Americas (2003)
G.A. Bradshaw and P.A. Marquet (Eds.)

Volume 163
Fluxes of Carbon, Water and Energy of European Forests (2003)
R. Valentini (Ed.)

Volume 164
Herbivory of Leaf-Cutting Ants: A Case Study on *Atta colombica* in the Tropical Rainforest of Panama (2003)
R. Wirth, H. Herz, R.J. Ryel, W. Beyschlag, B. Hölldobler

Volume 165
Population Viability in Plants: Conservation, Management, and Modeling of Rare Plants (2003)
C.A Brigham, M.W. Schwartz (Eds.)

Volume 166
North American Temperate Deciduous Forest Responses to Changing Precipitation Regimes (2003)
P. Hanson and S.D. Wullschleger (Eds.)

Volume 167
Alpine Biodiversity in Europe (2003)
L. Nagy, G. Grabherr, Ch. Körner, D. Thompson (Eds.)

Volume 168
Root Ecology (2003)
H. de Kroon and E.J.W. Visser (Eds.)

Volume 169
Fire in Tropical Savannas: The Kapalga Experiment (2003)
A.N. Andersen, G.D. Cook, and R.J. Williams (Eds.)

Volume 170
Molecular Ecotoxicology of Plants (2004)
H. Sandermann (Ed.)

Volume 171
Coastal Dunes: Ecology and Conservation (2004)
M.L. Martínez and N. Psuty (Eds.)

Volume 172
Biogeochemistry of Forested Catchments in a Changing Environment: A German Case Study (2004)
E. Matzner (Ed.)

Volume 173
Insects and Ecosystem Function (2004)
W. W. Weisser and E. Siemann (Eds.)

Printing: Mercedes-Druck, Berlin
Binding: Stein + Lehmann, Berlin